Student Guide for

Astronomy: Observations & Theories

SECOND EDITION

for use with *Horizons: Exploring the Universe*, 10th edition

DENISE CUSANO

JOEL M. LEVINE

KENDRA SIBBERNSEN

Australia • Brazil • Canada • Mexico • Singapore • Spain • United Kingdom • United States

Coast Community College District
Kenneth D. Yglesias, Chancellor, Coast Community College District
Ding-Jo H. Currie, President, Coastline Community College
Dan C. Jones, Administrative Dean, Instructional Systems Development
Laurie R. Melby, Director of Production
Lynn M. Dahnke, Marketing Director
Judy Garvey, Electronic Media and Publishing Supervisor
Robert D. Nash, Instructional Designer
Wendy Sacket, Electronic Media and Publishing Coordinator
Linda Wojciechowski, Senior Electronic Media and Publishing Assistant
Thien Vu, Electronic Media and Publishing Assistant

The course, *Astronomy: Observations & Theories*, is produced by the Coast Community College District, in cooperation with Thomson Brooks/Cole.

Distributed by:

Coast Learning Systems
Coastline Community College
11460 Warner Avenue
Fountain Valley, CA 92708
telephone: (800) 547-4748 fax: (714) 241-6286
e-mail: CoastLearning@coastline.edu
website: www.CoastLearning.org

1 2 3 4 5 6 7 / 10 09 08 07

Printed in the United States of America.
Printer: Thomson/West

ISBN-13: 978-0-495-11365-2
ISBN-10: 0-495-11365-4

Visit Thomson Brooks/Cole online at www.thomsonedu.com

Coast

Contents

unit 1

Acknowledgments

Several of the individuals responsible for the creation of this course are listed on the copyright page of this book. In addition to these people, appreciation is expressed for the contributions of the following:

MEMBERS OF THE NATIONAL ACADEMIC ADVISORY COMMITTEE

Scott Blair, M.S., Riverside Community College, California

John Carzoli, Ph.D., Oakton Community College, Illinois

George Jacoby, Ph.D., Wisconsin-Indiana-Yale-NOAO Observatory

Kevin Murphy, M.A.E., Moraine Valley Community College, Illinois

Peter K. Schoch, M.S., Sussex County Community College, New Jersey

Michael A. Seeds, Ph.D., Professor Emeritus, Franklin & Marshall College, Pennsylvania

George R. Stanley, M.A., San Antonio College, Texas

Dennis Tabor, MPH, Cowley County Community College, Kansas

Chandra Vanajakshi, Ph.D., College of San Mateo and San Francisco State University

J. Wayne Wooten, Ed.D., Pensacola Junior College and University of West Florida

These scholars, teachers, and practitioners helped focus the approach and content of the video programs, student guide, and faculty manual to ensure accuracy, academic validity, accessibility, significance, and instructional integrity. Special thanks are extended to Mike Seeds, the author of the *Horizons* textbook, who helped focus the approach and content of the course.

Lead Academic Advisors

Stephen P. Lattanzio, M.A., Orange Coast College, California

Joel M. Levine, M.A., Orange Coast College, California

Kendra Sibbernsen, M.S., Hawkeye College, Iowa, and Metropolitan Community College, Nebraska

Many thanks go to Bob Nash for the instructional design of this course. Special thanks are also extended to Denise Cusano, who authored this guide and assisted in the instructional design, and to Kendra Sibbernsen, who reviewed the content and accuracy of this second edition of the student guide.

KEY VIDEO PRODUCTION TEAM

Peter Berkow, producer; Kris Koenig, producer-writer; Anita Berkow, production assistant; Steve Chollet, editor/assistant producer; and Bruce Kuykendall, editor. Additional thanks to Tim Doherty and Marc Foster for their graphic design work, and to David Levy and Phil Plait for their work as script editors.

Additional thanks to to Chris Hall, Sylvia Krick, and the many other supportive individuals at Thomson Brooks/Cole Publishing.

Introduction

How to Take
This Course

To the Student

Welcome to *Astronomy: Observations & Theories*. This course will provide a comprehensive survey of the fundamental observations, theories, and concepts in astronomy. Whether you are taking this course as part of your academic study or simply because you are interested in the subject, we believe that you will find this course both fascinating and useful.

Astronomy: Observations & Theories brings you the latest astronomical discoveries and astrophysical theories. In this course you will have a breathtaking view of nature, from particles smaller than the atom to giant superclusters of galaxies. As an introductory astronomy course, *Astronomy: Observations & Theories* goes beyond describing the contents of outer space. It examines the scientific method and considers both its promises and limitations. It answers many of our questions and poses new ones time after time, so that we are continually probing the innermost secrets of time and space.

Course Goals

The designers, academic advisors, and producers of this course have specified the following learning outcomes for students taking *Astronomy: Observations & Theories*. After successfully completing the course assignments, you should be able to:

1. Define astronomy, describe the process of science, and offer examples of the dynamic and ever-changing nature of this discipline.

2. Describe the appearance and motions of the sun, moon, planets, and stars that can be observed with the naked eye, and explain related phenomena such as eclipses and seasons.

3. Identify the historical contributions of Ptolemy, Copernicus, Tycho, Kepler, Galileo, and Newton, and discuss how astronomy developed from the ancient conceptions of the Greeks to a modern understanding of gravity, tides, and orbital motion.

4. Explain the interaction of light and atoms, and discuss how telescopes and associated instruments are used to gather and analyze light at different wavelengths to measure the physical characteristics of stars and galaxies.

5. Describe the interior and atmosphere of the sun, including the nuclear processes taking place in its core and surface phenomena (such as those that affect Earth).

6. Discuss how astronomers determine the basic properties of stars (such as distance from Earth, luminosity, mass, and diameter) and explain how these properties change at different stages of a star's life.

7. Explain the physical processes taking place during the birth, life, and death of stars and binary star systems.

8. Describe the structure, behavior, origin, and evolution of galaxies (including our Milky Way), explain the phenomena associated with active galactic nuclei, and discuss the distribution of galaxies throughout the universe.

9. Explain the fundamental principles of cosmology, including the shape of space-time, evidence for the big bang, and effects of inflation and acceleration of the universe.

10. Summarize the basic features of our solar system, the physical processes involved in its formation, and evidence for extrasolar planetary systems.

11. Compare and contrast the similarities and differences among the terrestrial planets, the Jovian planets, and smaller bodies of our solar system.

12. Summarize what is known about the origin and nature of life on Earth, and relate this to the search for possible life beyond our planet.

Course Components

The Course Student Guide

The course student guide is an integral part of *Astronomy: Observations & Theories*. Think of it as your "road map." This guide gives you a starting point for each lesson, as well as directions and exercises that will help you successfully navigate your way through each lesson. Reading this guide and completing the lesson exercises will provide you with information that you normally would receive in the classroom if you were taking this course on campus.

Each student guide lesson includes the following elements:

• CHECKLIST

This is an outline of all activities to be completed for the lesson. You should refer to this checklist before starting each lesson and check off each item as you complete it so you are sure you have covered everything in that lesson.

• PREVIEW

This section introduces the lesson topic, explains why it's important, and offers a quick review of previous lesson concepts that you'll need to remember.

• LEARNING OBJECTIVES

This is a list of what you will know and be able to do after you complete the lesson. After completing each lesson, you should be able to satisfy each of these learning objectives. (Note: Instructors often design test questions after learning objectives, so use them to help focus your study time.)

- Viewing Notes

This section provides a brief introduction to the lesson and offers important information that will help you understand the concepts illustrated in the video.

- Key Terms & Concepts

This is a list of the terms and concepts that are introduced in the lesson along with their definitions. You should be able to define each of the key terms and concepts for each lesson. You will also find definitions for key terms and concepts in the glossary that appears in the back of the course textbook, *Horizons: Exploring the Universe*.

- Summary

Review this section after you have read the textbook chapter and watched the lesson video. The Summary provides a final overview of the lesson by discussing key terms, offering examples, and making connections between concepts that will help you recall them at exam time.

- Review Exercises

This section includes a variety of study activities to help you learn and prepare for midterm and final examinations. They include matching and completion exercises, a multiple-choice self-test, short-answer questions, and application questions. These exercises are not typically graded by your instructor; they are designed as self-study tools to help you achieve the learning objectives in each lesson.

- Answer Key

Answers to the Review Exercises are conveniently located at the end of each lesson so that you can get immediate feedback. After completing an exercise, be sure to check the Answer Key to make sure you correctly understand the material.

- Lesson Review

This matrix at the end of each lesson identifies the textbook pages and student guide exercises that correspond to each learning objective. Use this matrix to focus your study time on those objectives you have yet to achieve.

The Textbook

The recommended textbook for this course is *Horizons: Exploring the Universe*, tenth edition, written by Michael A. Seeds (Thomson Brooks/Cole, 2008).

Horizons: Exploring the Universe invites readers on a journey of discovery. With each chapter, students learn that astronomy is much more than the study of stars and planets—it is the study of the universe in which we live, and, in turn, the study of who and what we really are. Students are encouraged to explore essential questions in the field of astronomy, including "How are stars born?," "How will the sun die?," and "Are there other worlds like Earth?"

The Video Lessons

In addition to the student guide and textbook, each lesson in this course includes a half-hour video lesson. The twenty video lessons feature leading practitioners, theoreticians, and academics in the fields of astronomy, planetary science, and astrophysics who describe and explain celestial objects and events. The video lessons will introduce you to the night sky; the life cycle of the stars; the universe of galaxies and the history of the universe; and the origin, characteristics, and evolution of the solar system. In addition, these video lessons will take you inside the data rooms of leading observatories. You will see the latest images from NASA and the Jet Propulsion Laboratory and from Earth-based telescopes, space observatories, and the Hubble Space Telescope.

Review and Study from the Video Lesson Transcripts

These transcripts, a printed copy of the audio track from each video lesson, are available as PDF files at www.CoastTranscripts.org (user name: astronomy; password: observations). The PDF files may be viewed online or printed for use as an additional study aid. You will need Adobe Acrobat Reader (which is available free of charge on the Internet) to view these files, and we have provided a link to Acrobat Reader on the transcript website.

How to Take a Distance Learning Course

If this is your first experience with distance learning, welcome. Distance learning courses are designed for busy people whose situation or schedules do not permit them to take a traditional on-campus course.

This student guide is designed to help you study effectively and learn the material presented in both the textbook and the video lessons. To complete a distance learning course successfully, you will need to schedule sufficient time to read the textbook, watch each video lesson, and study the materials in this guide.

This student guide is a complement to the textbook and the video lessons. It is not a substitute. By following the instructions in this guide, you should easily master the learning objectives for each lesson.

To complete this course successfully you will need to:

- Find and read the "syllabus" or "handbook" for this course provided by your instructor. This course syllabus will list and explain what you will be required to do to complete this course successfully. The syllabus describes assignments, explains the course grading scale, and offers a schedule for quizzes, exams, review sessions, and so on. If you have questions about what you are supposed to do in this course, look for the answer in your syllabus. If you are still unsure, contact your instructor.

- Purchase a copy of the course textbook.

- Read and study this student guide and the textbook.

- View each video lesson in its entirety. (The course syllabus or handbook should explain how you can view the lesson videos.)

- Review the Key Terms and Concepts presented in this guide and memorize their definitions.

- Focus your study time so you can achieve the Learning Objectives for each lesson.

- Complete the Review Exercises in this guide.

- Complete any additional assignments your instructor may require.

Even though you do not have scheduled classes to attend each week on campus, please keep in mind that this is a college-level course. It will require the same amount of work as a traditional, classroom version of this course and at the same level of difficulty. As a distance learner, however, it will be up to you alone to keep up with your deadlines. It's important that you schedule enough time to read, study, review, and reflect.

Do your best to keep up with the work. In a distance learning course, it's very difficult to catch up if you allow yourself to get behind schedule. We strongly recommend that you set aside specific times each week for reading, viewing the videos, completing assignments, and studying for quizzes and exams. You will be more likely to succeed if you make a formal study schedule and stick to it.

When you watch each video program, try to do so without any interruptions. If you are interrupted, you may miss an important point. If possible, view the videos in a recorded mode so you can stop the tape and take notes. Also, take some time immediately after watching the video to reflect on what you have just seen. This is an excellent time to discuss the lesson with a friend or family member. Your active thinking and involvement will promote your success.

Study Recommendations

Everyone has his or her own unique learning style. Some people learn best by studying alone the first thing in the morning, others by discussing ideas with a group of friends, still others by listening to experts and taking notes. While there is no best way to learn, psychologists and educators have identified several things you can do that will help you study and learn more effectively.

One of the advantages of distance learning is that you have many choices for how you study. You can tailor this course to fit your preferred learning style. Below are several study tips. These are proven methods that will improve your learning and retention. Please take the time to read through this list. By using one or more of these techniques, you can significantly improve your performance in this course.

Open your mind. One of the major obstacles to learning new information is that it often differs from what we already "know." To learn, you need to have an open mind. We are not suggesting that you simply believe everything you are told. In fact, we want you to think critically about everything you are told. However, try not to let old beliefs or opinions stop you from learning something new.

Reduce interference and interruptions. If possible, focus your study on one subject at a time. When you study more than one subject, you are increasing the likelihood of interference occurring. Too much information can overload your brain and make it difficult to store and recall what you want to learn. Also, do your best to avoid interruptions while you are studying. Find a quiet room with a good desk, chair, and adequate lighting. Of course, visiting with friends, listening to the radio, watching your kids, or any other distractions will also interfere with your ability to learn. Try to avoid these distractions so your mind can focus, and give yourself enough time to absorb the new information.

Don't cram. You probably already know that staying up all night cramming for an exam the following morning is not a good way to study. The opposite of cramming, is, in fact, one of the best ways to study. Spacing out your studying into smaller and more frequent study periods will improve retention. For example, instead of studying for six hours in one evening, you will learn more and retain more if you study one hour per night for six nights.

Reduce stress. In addition to being bad for your health, stress is bad for learning. Stress and anxiety interfere with learning. You will learn more and enjoy it more if you are relaxed when you study. One of the most effective ways of relaxing that does not interfere with learning is to exercise. A good brisk walk or run before you settle in to study is a good prescription for success. Ideally, you would study some, take a break, and then get some exercise while you think about what you have just learned. A little later, when you are relaxed, you can return and study some more.

Be a Smart Student. Most top students have one thing in common: excellent study habits. Students who excel have learned or were fortunate enough to have someone teach them how to study effectively. There is no magic formula for successful studying. However, there are a few universal guidelines.

- **Do** make a commitment to yourself to learn.
- **Don't** let other people interrupt you when you are studying.
- **Do** make a study schedule and stick to it.
- **Don't** study when you are doing something else, like listening to the radio.
- **Do** create a specific place to study.
- **Don't** study if you are tired, upset, or overly stressed.
- **Do** exercise and relax before you study.
- **Don't** study for extended periods of time without taking a break.
- **Do** give yourself ample time to study.
- **Don't** complain that you have to study.
- **Do** take a positive approach to learning.

Make the most of your assignments. You will master this material more effectively if you make a commitment to complete all of the assignments. The lessons will make more sense to you, and you will learn more, if you follow these instructions:

- Set aside a specific time to go read, study, and review each lesson.
- Before you read the textbook chapter for each lesson, read the Lesson Preview and Learning Objectives outlined in this guide.
- Complete all the assigned reading, both in the textbook and in this student guide, for the lesson you are studying.
- Watch the lesson video, more than once if necessary, to help you achieve the learning objectives.
- Review the Key Terms and Concepts in this guide. Check your understanding of all unfamiliar terms in the glossary.
- Complete the Review Exercises for each lesson.

Think about what you have learned. You are much more likely to remember new information if you use it. Remember that learning is not a passive activity. Learning is active. As soon as you learn something, try to repeat it to someone or discuss it with a friend. If you think about what you just learned, you will be much more likely to retain it. The reason we remember certain information has to do mostly with (1) how important that information is to us, and (2) whether or not we actively use the information. What do you do, however, when you need to learn some information that is not personally valuable or interesting to you? The best way to remember this type of information is to reinforce it—and the best reinforcer is actively using the information.

Get feedback on what you are studying. Feedback will help reinforce your learning. Also, feedback helps make sure you correctly understand the information. The Review Exercises in this guide are specifically designed to give you feedback and reinforce what you are learning. The more time and practice you devote to learning, the better you will be at recalling that information. When you take a Self-Test, make sure you immediately check your answers using the Answer Key. Don't wait to check your answers later. If you miss a question, review that part of the lesson to reinforce the correct understanding of the material. The review matrix at the end of each lesson in this guide will help guide your study.

A good gauge of how well you understand something is your ability to explain it to someone else. If you are unable to explain a term or concept to a friend, you probably need to study it further.

Contact your instructor. If you are having an especially difficult time with learning some information, contact your instructor. Your instructor is there to help you. Often a personal explanation will do wonders in helping you clear up a misunderstanding. Your instructor wants to hear from you and wants you to succeed. Don't hesitate to call, write, e-mail, or visit your instructor.

Study groups and partners. Some exercises are enhanced when done with a partner. However, some students do better studying alone. If study groups are helpful to you or you would like a partner to practice the exercises with, let your instructor know. Because study groups can sometimes turn into friendly chats, without much learning going on, you should use your group time wisely. Remember that study groups are not a substitute for individual effort.

Learn it well. One of the best methods for increasing retention is to "overlearn" the material. Just because you can answer a multiple-choice question or give a brief definition of a term doesn't mean you really understand the concept or that you'll be able to describe it in an essay or exam. Overlearning is simple. After you think you have learned a concept, spend an additional five or ten minutes actively reviewing it. Try to describe it to a friend or family member. Think up real-world examples of the concept. You will be amazed how much this will increase your long-term retention.

Enjoy learning. You do not need to suffer to learn. In fact, the opposite is true. You will learn more if you enjoy learning. If you have the attitude that "I hate to study" or "schoolwork is boring," you are doing yourself a disservice.

You will progress better and learn more if you adopt a positive attitude about learning and studying. Since you are choosing to learn, you might as well enjoy the adventure!

We are sure you will enjoy ***Astronomy: Observations & Theories***.

LESSON
1

The Study of the Universe

Checklist

For the most effective study of this lesson, complete the following activities in this sequence.

Before Viewing the Video

☑ Read the Preview, Learning Objectives, and Viewing Notes below.

☑ Read Chapter 1, "The Scale of the Cosmos," pages 2–9, in the *Horizons* textbook. Also, study Appendix A, "Units and Astronomical Data," in the textbook. Knowledge of Tables A-1 through A-6 is essential to understanding much of the content of this course. Be sure to familiarize yourself with the information in Tables A-7 through A-9.

What to Watch

☑ After reading the textbook chapter, watch the video for Lesson 1, *The Study of the Universe.*

After Viewing the Video

☑ Briefly note your answers to questions listed at the end of the Viewing Notes.
☑ ~~key terms~~
☑ Review the Summary below.

☑ Review all reading assignments for this lesson, especially the Chapter 1 summary on page 9 in *Horizons* and the Viewing Notes in this lesson.

☑ Write brief answers to the review questions at the end of Chapter 1 in *Horizons*.

☑ Complete the Review Exercises below. Check your answers with the Answer Key and review when necessary.

☑ Use the Lesson Review matrix found at the end of this lesson to review and assess your knowledge of each Learning Objective.

☒ As assigned by your instructor, complete the Applications activities and
any additional activities for this lesson.

Preview

The universe is vast and filled with a variety of objects—moons, planets, stars, and galaxies are among the most familiar to us. Astronomy is much more than the study of these familiar celestial objects—it is the study of the entire universe. Throughout this course, you will take a trip though a universe that is filled with strange and fascinating celestial objects; this lesson prepares you for this journey.

You will first take a brief tour of our solar system and the universe so you can get a sense of its scale. Astronomers today believe that the universe is infinite—it extends in all directions without limit. Because the universe is so large, we need to use special units of measurement to quantify its size. You will soon become familiar with the ways to measure such great distances.

Learning Objectives

After you complete this lesson, you should be able to:

1. Define astronomy as a science and outline its unique nature and purpose. *HORIZONS* TEXTBOOK PAGES 2–4, 8.

2. Identify the major types of objects in the universe, from planets to superclusters of galaxies, and order them according to size and scale. *HORIZONS* TEXTBOOK PAGES 4–8.

3. Describe the purpose of scientific notation and use it to write numbers. *HORIZONS* TEXTBOOK PAGE 5 AND APPENDIX A, PAGES 478–479.

4. Define the terms astronomical unit (AU) and light-year (ly), discuss their purpose, and use them to describe astronomical distances. *HORIZONS* TEXTBOOK PAGES 5–6.

At this point, read Chapter 1, "The Scale of the Cosmos," pages 2–9.

Viewing Notes

The size of the universe is so vast it's difficult to imagine. Our journey of the universe begins with Earth and expands to the edge of the known universe. The video program contains the following segments:

- ✪ Astronomy
- ✪ Our Solar System

✪ The Milky Way

✪ The Universe

The following information will help you better understand the video program:

As you tour our solar system, you will become familiar with the planets—the small, spherical, nonluminous bodies that shine by reflected light. You will also be introduced to the sun, our star—a self-luminous ball of hot gas that generates its own energy.

Your tour continues through the Milky Way Galaxy to the nearest star that is 4.2 light-years away. Recall that one light-year (ly) is the distance that light travels in one year.

As you reach the edge of the Milky Way, you will discover that galaxies are grouped in clusters. A galaxy is a group of stars, gas, and dust, held together by gravity. Galaxy clusters are large groupings of galaxies and are spread throughout the universe.

QUESTIONS TO CONSIDER

- What is the astronomical unit and when is it used? → *average distance f'm Earth to Sun*
- What is the light-year and when is it used? → *distance light travels in one yr.*
- What questions do astronomers strive to answer? → *how planets, stars, solar systems and galaxies are born, how they evolve, and how they die.*

Watch the video for **Lesson 1**, *The Study of the Universe.*

Key Terms and Concepts

Page references are keyed to the *Horizons* textbook.

1. **field of view:** The region you can see in an image. (p. 4; objective 2)

2. **scientific notation:** The system of recording very large or very small numbers by using powers of 10. (p. 5; objective 3)

3. **solar system:** The sun, and its planets, asteroids, comets, and so on. (p. 5; objective 2)

4. **planet:** A small, nonluminous body that shines by reflected light. (p. 5; objective 2)

5. **star:** A globe of gas held together by its own gravity and supported by the internal pressure of its hot gases, which generate energy by nuclear fusion. (p. 5; objective 2)

6. **astronomical unit (AU):** The average distance from Earth to the sun; 1.5×10^8 km, or 93×10^6 mi. (p. 5; objective 4)

7. **light-year (ly):** A unit of distance; the distance light travels in one year, roughly 10^{13} km or 63,000 AU. (p. 6; objective 4)

8. **galaxy:** A large system of stars, star clusters, gas, dust, and nebulae orbiting a common center of mass. (p. 7; objective 2)

9. **Milky Way:** The hazy band of light that circles our sky; it is produced by the glow of our galaxy. (p. 7; objective 2)

10. **Milky Way Galaxy:** The spiral galaxy containing our sun; it is visible in the night sky as the Milky Way. (p. 7; objective 2)

11. **spiral arm:** A long spiral pattern of bright stars, star clusters, gas, and dust. Such arms extend from the center to the edge of the disk of spiral galaxies. (p. 7; objective 2)

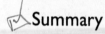Summary

The science known as astronomy began more than 2,000 years ago. The Greeks were the first to observe the sky in a systematic fashion and offer explanations for the movements in the sky. The study of astronomy looks well beyond the reaches of our solar system, into our galaxy, and into other galaxies in the universe. Astronomy attempts to answer the questions that relate to how planets, stars, solar systems, and galaxies are born, how they evolve, and how they die. Astronomy can help us understand our place in the universe and explore the possibility of life existing elsewhere in it.

Because the universe is so vast, astronomers use different units of measurement than we use in our everyday lives. When measuring distance in our solar system, astronomers typically use a unit of distance called the astronomical unit (AU). This is the average distance from the sun to Earth, or 1.5×10^{18} kilometers (about 93,000,000 miles). By using astronomical units, we can get a sense of how far away things are in our solar system by comparing them to how far away Earth is from the sun.

Since the numbers that astronomers use are large, it is more convenient to write them in scientific notation (see Appendix A in the *Horizons* textbook). Scientific notation is an exponential notation based on powers of base number 10. It is used in the sciences as a way to express large numbers more concisely and to allow for easier calculations. For example, instead of writing out the number 36,000,000,000,000, we can express this number as 3.6×10^{13}.

When measuring distances to other stars and galaxies, it's no longer convenient to measure using astronomical units. Great distances typically separate stars, so we must use an even larger measurement called the light-year (ly). As the name suggests, one light-year is the distance that light travels in one year. This is equivalent to 63,000 AU. For example, the star closest to our sun is Proxima Centauri, which is 4.2 ly from Earth—it takes 4.2 years for the light from this distant star to reach Earth.

Planets, like Earth, are small, spherical, nonluminous bodies that orbit a star and shine by reflected light. There are nine planets in our solar system, but it's likely that there are billions of other planets in the universe that orbit stars different from and similar to our sun. A star is a self-luminous body of hot gas that generates its own energy. Our sun is an average star; it and the planets that orbit it are a part of our solar system.

Our solar system is a part of the Milky Way Galaxy, which is 75,000 ly in diameter. Most of the gas and dust in the galaxy are found in the spiral arms, which are where new stars are born.

In recent times, astronomers have pushed the frontier of understanding outward, extending beyond our own galaxy. They have discovered that galaxies are commonly grouped together in clusters; clusters of galaxies are connected in a vast network called superclusters. Larger still, the superclusters are linked to form long filaments, which appear to be the largest structures in the universe.

Astronomers have discovered that the universe reveals itself not only in visible light, but also in other forms of energy. This has allowed them to probe into the vast universe and to discover the unimaginable. You will embark on this fascinating journey to explore our universe and will discover things you may never have thought possible.

Review Exercises

Matching

Match each term with the appropriate definition or description.

1. _k_ field of view		7. _d._ light-year	
2. _b_ solar system		8. _f._ Milky Way Galaxy	
3. _a_ scientific notation		9. _c._ spiral arm	
4. _g._ astronomical unit		10. _e._ Milky Way	
5. _i_ planet		11. _j_ galaxy	
6. _h._ star			

a. The system of recording very large or very small numbers by using powers of 10.

b. The sun and its planets, asteroids, comets, and other minor objects.

√ c. The area of our galaxy where stars are born in great clouds of gas and dust.

√ d. A unit of distance that light travels in one year.

√ e. The hazy band of light that circles our sky; produced by the glow of our galaxy.

√ f. The large system of billions of stars in which our sun and solar system are located.

√ g. The average distance from Earth to the sun.

√ h. A globe of gas held together by its own gravity and supported by the internal pressure of its hot gases, which generate energy by nuclear fusion.

i. A small, nonluminous body formed by accretion in a disk around a protostar.

√ j. A large system of stars, star clusters, gas, dust, and nebulae orbiting a common center of mass.

√ k. The region you can see in an image.

Completion

Fill each blank in the sentences below with the most appropriate term from the list of completion answers that follow. A term may be used once, more than once, or not at all. Check your answers with the Answer Key and review when necessary.

astronomical unit	Milky Way	sun
billions	millions	supercluster
four hours	moon	thousands
galaxies	planet	two years
light-years	star	universe

1. Distances to stars outside our solar system are commonly measured in _light years_ .

2. Light from the _Sun_ takes about _four hours_ to reach Neptune.

3. The nearest _star_ to our solar system is 4.2 light-years away.

4. Our galaxy, the _Milky Way_ , is a large group of stars, star clusters, gas, dust, and nebulae that orbit a common center of mass.

5. When astronomers refer to distances in the solar system, they do so in terms of a(n) _astronomical unit_ , which is the distance between Earth and the sun.

Self-Test

Select the best answer.

1. Of the following objects, the smallest one is
   a. a star.
   b. a galaxy.
   c. a planet.
   d. the universe.

2. Which of the following objects is the most distant from Earth?
   a. the Andromeda Galaxy
   b. Pluto
   c. the Milky Way Galaxy
   d. the sun

3. The distance between the sun and the nearest star is about
   a. 260 AU.
   b. 260,000 AU.
   c. 2.6 million AU.
   d. 2.6 billion AU.

4. Light takes approximately 8 minutes to travel from the sun to Earth. To reach Neptune, it takes light from the sun about
   a. half as long.
   b. 100 times longer.
   c. 1,000,000 times longer.
   d. 30 times longer.

5. The most convenient method by which to write extremely large or extremely small numbers is called
   a. scientific method.
   b. notation of the scientific.
   c. scientific notation.
   d. numerical analysis.

6. A shorthand way of writing 425,000,000 is
   a. 4.25×10^8.
   b. 425×10^6.
   c. 0.425×10^9.
   d. all of the above.

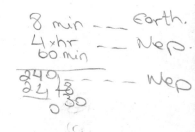

7. A light-year is
   a. the time it takes Earth to travel around the sun once.
   b. the distance Earth travels in the time it takes light from the sun to reach Earth.
   c. the distance light travels in one year.
   d. equal to 1 million astronomical units (AU).

8. The average distance between Earth and the sun is referred to as
   a. an astronomical unit.
   b. a light-year.
   c. a kilometer.
   d. a solar mile.

9. The main purpose of astronomy is to
   a. study the sun.
   b. understand planet Earth.
   c. classify objects in the Milky Way Galaxy.
   d. study the universe and understand our place it it.

10. Which of the following are topics that are the focus of the study of astronomers?
    a. stars
    b. planets
    c. the origin and evolution of galaxies
    d. all of the above

Short-Answer Questions

1. What is the astronomical unit and when is it used most conveniently?

   Average distance f/m Earth to Sun, 1.5×10^{14} m. Conviniently used to express distances within our solar system.

2. What is the study of astronomy?

   It is the study of the universe in which us humans exist, understand our place in it. Topics include stars, their birth and evolution; galaxies + clusters, origins + evolution of Universe itself; the planets - earth in particular.

3. What is the difference between a planet and star?

a star is a self-luminous ball of hot gas that generates its own energy.

a planet is a small, nonluminous body that shines by reflected light.

4. What is a light-year and when is it used most conveniently?

The distance that light travels in one year; roughly 10^{13} km or 63,000 AU.

Used most conveniently to express distances that are considered very large, like from star to star or other galaxies.

5. If the sun were to go out, how long would it be before we on Earth would know it?

8 minutes, the distance it takes light to travel one astronomical unit.

Applications

12 inches = foot
5,280 feet = mile

1. If you were to set a scale of one inch equals one astronomical unit and the nearest star is approximately 267,000 astronomical units away, how far away, using this scale, would the nearest star be? *6.64 km*

2. Write the number 2,500,000 light-years in scientific notation. Write the equivalent number of miles in scientific notation. *6×10^{12} miles in 1 ly (Appendix A)*

3. The Milky Way Galaxy is 73,000 light-years across. How long does it take light to go from one end of this galaxy to the other? *73,000 yrs.*

4. Order the following objects according to size, least to greatest: <u>Milky Way Galaxy</u>; <u>sun</u>; filaments of galaxies; <u>cluster of galaxies</u>; <u>Earth</u>; supercluster of galaxies. *Earth, sun, milky way galaxy, clusters, superclusters, filaments.*

5. Order the following objects according to distance from the sun, closest to farthest: Pluto; Proxima Centauri; Earth; Andromeda Galaxy; supercluster of galaxies; edge of the Milky Way Galaxy.

Earth, Pluto, edge of the milky way galaxy, Proxima Centauri, Andromeda Galaxy, supercluster of galaxies

Answer Key

Matching

1. k (p. 4; objective 2)
2. b (p. 5; objective 3)
3. a (p. 5; objective 2)
4. g (p. 5; objective 4)
5. i (p. 5; objective 2)
6. h (p. 5; objective 2)
7. d (p. 6; objective 4)
8. f (p. 7; objective 2)
9. c (p. 7; objective 2)
10. e (p. 7; objective 2)
11. j (p. 7; objective 2)

Completion

1. light-years (p. 6; objective 4)
2. sun; 4 hours (p. 6; objective 1)
3. star (p. 6; objective 2)
4. Milky Way (p. 7; objective 2)
5. astronomical unit (p. 5; objective 4)

Self-Test

1. c (pp. 5–8; objective 2)
2. a (video lesson; objective 2)
3. b (p. 6; objective 4)
4. d (p. 6; objective 4)
5. c (p. 5 and Appendix A, pp. 478–479; objective 3)
6. d (p. 5 and Appendix A, p. 478; objective 3)
7. c (p. 6; objective 4)
8. a (p. 5; objective 4)
9. d (p. 8; video lesson; objective 1)
10. d (pp. 2–4, 8; video lesson; objective 1)

Short-Answer Questions

1. The astronomical unit is the average distance between Earth and the sun and it's used most conveniently to represent distances in the solar system. (p. 5; objective 4)

2. Astronomy is the study of the universe. That is a very broad term. Topics in astronomy include stars, their birth and evolution; galaxies and clusters (systems of billions stars); the origin and evolution of the universe itself; and, closer to home, the planets—Earth in particular. The purpose of all this knowledge is to understand the relationship between the individual parts of this universe and our place in it. (pp. 4–8; objective 1)

3. A planet is a relatively small object that reflects light from the star about which it orbits. A star is a globe of gas of sufficient size that it is capable of producing energy by the process of thermonuclear fusion. As you will learn later in this course, we are becoming more certain about what is a planet and a star, but we have begun to discover new objects that blur the traditional distinction between a planet and a star because they have features of both. (p. 5; objective 2)

4. A light-year is the distance light travels in one year and is used most conveniently to represent distances between stars and galaxies. One light-year is equal to 63,000 astronomical units or approximately 6,000,000,000,000 miles (6×10^{12} miles). (p. 6; objective 4)

5. Eight minutes, which is the time it takes light to cover the distance of one astronomical unit (1 AU). (p. 6; objective 4)

Applications

1. The distance to Proxima Centauri on this scale is 263,000 inches. There are 12 inches to the foot, which makes the distance to the star 21,916 feet. With 5,280 feet per mile, this makes the distance to Proxima Centauri 4.15 miles. There are 1.6 km per mile, which makes the distance to this star 6.64 km. (p. 6; objective 4)

2. The answer is 2.5×10^6 light-years or 15×10^{18} miles. (Approximately 6×10^{12} miles per light-year.) (p. 5; Appendix A, p. 479; objective 4)

3. The answer is 73,000 years. (p. 6; objective 4)

4. Earth, sun, Milky Way Galaxy, cluster of galaxies, supercluster of galaxies, filaments of galaxies (pp. 2–8; objective 2)

5. Earth, Pluto, Proxima Centauri, edge of the Milky Way Galaxy, Andromeda Galaxy, supercluster of galaxies (pp. 2–8; objective 2)

Lesson Review

Lesson 1: The Study of the Universe

PLEASE NOTE: Use this matrix to guide your study and achieve the learning objectives of this lesson. It will also help you to view the video, which defines and demonstrates important concepts and principles as they relate to everyday life and actual case studies.

Learning Objective	Textbook	Student Guide
1. Define astronomy as a science and outline its unique nature and purpose.	pp. 2–4, 8	Completion: 2; Self-Test: 9, 10; Short-Answer: 2.
2. Identify the major types of objects in the universe, from planets to superclusters of galaxies, and order them according to size and scale.	pp. 4–8	Key Terms: 1, 3, 4, 5, 8, 9, 10, 11; Matching: 1, 3, 5, 6, 8, 9, 10, 11; Completion: 3, 4; Self-Test: 1, 2; Short-Answer: 3; Applications: 4, 5.
3. Describe the purpose of scientific notation and use it to write numbers.	pp. 5, 478–479	Key Terms: 2; Matching: 2; Self-Test: 5, 6.
4. Define the terms astronomical unit (AU) and light-year (ly), discuss their purpose, and use them to describe astronomical distances.	pp. 5–6	Key Terms: 6, 7; Matching: 4, 7; Completion: 1, 5; Self-Test: 3, 4, 7, 8; Short-Answer: 1, 4, 5; Applications: 1, 2, 3.

LESSON
2

Observing the Sky

Checklist

For the most effective study of this lesson, complete the following activities in this sequence.

Before Viewing the Video

☑ Read the Preview, Learning Objectives, and Viewing Notes below.

☑ Read Chapter 2, "The Sky," pages 10–21, in the *Horizons* textbook.

What to Watch

☑ After reading the textbook chapter, watch the video for Lesson 2, *Observing the Sky*.

After Viewing the Video

☑ Briefly note your answers to questions listed at the end of the Viewing Notes.

☑ Review the Summary below.

❑ Review all reading assignments for this lesson, especially the Chapter 2 summary on page 21 in *Horizons* and the Viewing Notes in this lesson.

❑ Write brief answers to the review questions at the end of Chapter 2 in *Horizons*.

❑ Complete the Review Exercises below. Check your answers with the Answer Key and review when necessary.

❑ Use the Lesson Review matrix found at the end of this lesson to review and assess your knowledge of each Learning Objective.

❑ As assigned by your instructor, complete the Applications activities and any additional activities for this lesson.

❑ ThomsonNow website → txt pg. 16+17

Preview

The journey toward understanding our universe began when ancient peoples tried to make sense of what they observed in the sky. Many ancient civilizations saw sacred meaning in the sky, and when they studied it closely, they began to detect patterns. They noticed that the sky and the stars within it appeared to turn around them every night. They also noticed that different groups of constellations appeared in different seasons. They made the connection between what they saw in the sky and what happened on the earth and they created calendars based on their observations.

Although the sky was an integral part of the lives of ancient peoples, we don't often take the time to ponder the sky today. With new technologies, we don't need the stars to aid us in navigation and bright city lights illuminate the sky so relatively few stars can be seen at night. Although we have a better understanding of the universe today, we can still find amazing things if we simply take the time to look up and appreciate the sky.

 ### Concepts to Remember

- Recall from Lesson 1 that a *light-year (ly)* is a unit of distance equal to the distance that light travels in one year. (p. 4 in this guide).

 ## Learning Objectives

After you complete this lesson, you should be able to:

1. Explain the role of constellations in ancient and modern observations of the sky. *HORIZONS* TEXTBOOK PAGES 12–13.

2. Explain how stars have been named by ancient and modern astronomers. *HORIZONS* TEXTBOOK PAGES 13–14.

3. Describe the origin of the modern stellar magnitude scale and the relative brightness between magnitudes. *HORIZONS* TEXTBOOK PAGES 14–15.

4. Describe the celestial sphere as a scientific model, and explain what astronomers mean by angular distance. *HORIZONS* TEXTBOOK PAGES 15–19.

5. Describe how the orientation of the sky depends on the observer's latitude. *HORIZONS* TEXTBOOK PAGES 16–17.

6. Describe the phenomenon of precession and its effects. *HORIZONS* TEXTBOOK PAGES 18–20.

At this point, read Chapter 2, "The Sky," pages 10–21.

Viewing Notes

The video program introduces you to the quest throughout time to understand the night sky—the largest and most mysterious laboratory known to humans. The video program contains the following segments:

- ☣ The Celestial Sphere
- ☣ The North Star
- ☣ Precession
- ☣ Constellation & Star Names
- ☣ Brightness of Stars

The following information will help you better understand the video program:

In the opening segment of the video program, you'll be introduced to common questions that people have when they think about the night sky. You'll hear some terms that may be unfamiliar to you, like black holes, variable stars, and star clusters. You'll learn about these terms and concepts in upcoming lessons.

You'll see a model of the **celestial sphere**—a crystalline sphere that surrounds the earth to which all celestial objects seem to be attached. By thinking of the sky as a giant two-dimensional sphere, it helps us locate objects and aim our telescopes. This way of thinking about the sky is a type of **scientific model**—a framework that helps scientists think about nature. Of course, the earth isn't really surrounded by a two-dimensional sphere—stars are scattered around our three-dimensional universe at vastly different distances from us.

The video provides you with an example of determining your latitude on Earth by looking at Polaris, which is near the north celestial pole. When you think about the stars being on the 360° celestial sphere, you can measure the apparent distances between the stars as angular distances. The **angular distance** is the angle formed by imaginary lines extending from the observer to each of the two objects. (By comparison, if the object appears large enough to us on Earth, we can use the **angular diameter**—the angle formed by imaginary lines extending from the observer to opposite sides of an object.) In the northern hemisphere, you can determine your latitude by measuring the angular distance from the horizon to the north celestial pole—or Polaris. Early sailors used a sextant—a measuring device—to determine their latitude on Earth.

In modern times, Polaris can be seen near the north celestial pole, but it wasn't always there. As the earth spins on its axis, it wobbles because it is not perfectly spherical. This causes the north celestial pole to change its position—

The Southern Cross I saw every night abeam. The sun every morning came up astern; every evening it went down ahead. I wished for no other compass to guide me, for these were true.

—Captain Joshua Slocum
Sailing Alone Around the World (1900)

the axis of the earth scribes a circle against the background stars in a 26,000-year cycle. This conical motion is called **precession**.

The video mentions that Hipparchus measured the positions of the "naked-eye stars." Naked-eye stars are those that can be seen without instruments.

Questions to Consider

1 • Why did ancient civilizations find it so important to study the sky?

2 • How did ancient astronomers name stars and constellations?

3 • How does the stellar magnitude scale devised by Hipparchus differ from the one astronomers use today?

4 • Does the sky look different to an observer at the North Pole than at the South Pole? If so, how?

5 • Does the sky look different to an observer standing in the same spot in winter than in summer? If so, how?

6 • What is precession? How will it change the observations that people will make 12,000 years from now?

Watch the video for Lesson 2, *Observing the Sky.*

Key Terms and Concepts

Page references are keyed to the *Horizons* textbook.

1. **constellation:** One of the stellar patterns identified by name, usually of mythological gods, people, animals, or objects. Also, the region of the sky containing that star pattern. (p. 12; video lesson; objective 1)

2. **asterism:** A named grouping of stars that is not one of the recognized constellations. Examples are the Big Dipper and the Pleiades. (p. 13; video lesson; objective 1)

3. **magnitude scale:** The astronomical brightness scale. The larger the number, the fainter the star. (p. 14; video lesson; objective 3)

4. **apparent visual magnitude (m_v):** The brightness of a star as seen by human eyes on Earth. (p. 14; video lesson; objective 3)

5. **scientific model:** A tentative description of a phenomenon for use as an aid to understanding. A mental conception of how things work, or a framework that helps scientists think about some aspect of nature. (p. 19, Window on Science 2-2; video lesson; objective 4)

6. **precession:** The slow change in the direction of Earth's axis of rotation. One cycle takes nearly 26,000 years. (p. 19; video lesson; objective 6)

7. **celestial sphere:** An imaginary sphere of very large radius surrounding Earth and to which the planets, stars, sun, and moon seem to be attached. (pp. 15–16; video lesson; objective 4)

8. **horizon:** The circular boundary between the sky and Earth. (p. 16; video lesson; objective 4)

9. **zenith:** The point on the celestial sphere directly above the observer. The opposite of the nadir. (p. 16; objective 4)

10. **nadir:** The point on the celestial sphere directly below the observer. The opposite of the zenith. (p. 16; objective 4)

11. **north celestial pole:** The point on the celestial sphere directly above Earth's North Pole. (p. 16; video lesson; objective 4)

12. **south celestial pole:** The point on the celestial sphere directly above Earth's South Pole. (p. 16; video lesson; objective 4)

13. **celestial equator:** The imaginary line around the sky directly above Earth's equator. This line divides the sky into two hemispheres. (p. 16; video lesson; objective 4)

14. **north, south, east, and west points:** The four cardinal directions. The points on the horizon directly north, south, east, and west respectively. (p. 16; objective 4)

15. **angular distance:** The angle formed by lines extending from the observer to two locations. (p. 17; video lesson; objective 4)

16. **minute of arc:** An angular measure. One-sixtieth of a degree. (p. 17; objective 4)

17. **second of arc:** An angular measure. One-sixtieth of a minute of arc. (p. 17; objective 4)

18. **angular diameter:** The angle formed by lines extending from the observer to opposite sides of an object; for example, the angular diameter of the sun, moon, or Andromeda Galaxy. (p. 17; video lesson; objective 4)

19. **circumpolar constellation (north or south):** A constellation so close to one of the celestial poles that it never sets or never rises as seen from a particular latitude. (p. 17; video lesson; objective 4)

Summary

Ancient civilizations tried to make sense of what they saw in the sky, and we continue that quest today. As we look up, we may wonder what ancient civilizations observed and why it was so important for them to understand it.

The Stars

When ancient astronomers studied the sky, they organized stars into loose groupings called **constellations**. The constellations were named for celebrated heroes, gods, and mythical beasts. But the groupings of stars look nothing like what their names represent—they simply paid homage to their namesakes. Today, constellations represent areas of the sky and have clearly defined boundaries, and any star within the constellation belongs only to that constellation. Most stars in the constellations that we see are not physically associated with one another; they simply lie in approximately the same direction from Earth.

Although modern astronomers still use many of the same Arabic names that the ancient astronomers gave to individual stars, they are more methodical in their naming conventions. We still call the brightest star in the sky Sirius, which is in the constellation called Canis Major, but modern astronomers give the star an additional name based on its brightness and the constellation in which it's located. Greek letters are used to indicate brightness, with alpha (α) being the brightest, beta (β) being the second brightest, gamma (γ) being the third, and so on. For example, since Sirius is the brightest star in Canis Major, it is named α Canis Majoris. The second brightest star in Canis Major, Murzim, is called β Canis Majoris, and so forth. (Majoris is the possessive form of Major.) To further complicate things, astronomers may also abbreviate a star's modern name: α Canis Majoris is abbreviated as α CMa A. Fainter stars often don't have names or Greek letters, but if they are located within the boundaries of the constellation, they are a part of the constellation.

While the Greek letter in a star's modern name can help you determine how bright a star is within its constellation, it cannot tell you its relative brightness compared to all other stars. To more accurately indicate the brightness of a star as seen by eye from Earth, astronomers measure the brightness of a star based on a numerical scale called **apparent visual magnitude**. This is also called just apparent magnitude or magnitude. The astronomer Hipparchus first classified stars by six magnitude classes (from 1 through 6, with 1 being the brightest, or "first class"), but modern astronomers have extended this scale. Some stars are brighter than 1.0 and are assigned magnitudes of less than one, reaching negative magnitudes. Modern telescopes can detect stars as faint as approximately magnitude 25.

It is important to note that a star's apparent visual magnitude does not take into account how far away a star is from Earth. As you think about the scale of the cosmos, realize that a faintly shining star that's nearby can *look* much brighter than a brightly shining star that's very far away. For example, Sirius is the brightest looking star in the night sky with an apparent visual magnitude of –1.47, but it's relatively close at only 8.7 ly away. By contrast, Polaris has an apparent visual magnitude of 1.97, but it's about 430 ly away. Which one is actually brighter? In Lesson 8, you'll learn how to take a star's distance into consideration when determining its actual brightness.

Because brightness as judged by eye is so subjective, astronomers most often refer to a star's intensity—the measure of the light energy that hits one square meter in one second. A simple relationship exists between magnitude and intensity: the difference in one magnitude is equal to a difference in intensity of 2.512. For example, the light coming from a star with an apparent visual magnitude (m_v) of 1 is 2.512 times more intense than the light coming from a star that has an m_v of 2. The light from a star with an m_v of 1 is two magnitudes brighter than the light coming from a star with an m_v of 3. We can take the magnitude difference between the two stars (2) and look at Table 2-1 on p. 15 in the *Horizons* textbook to determine a difference in intensity of 6.3 (or 2.512^2). A difference of 3 magnitudes would be about 16 (or 2.512^3).

The Sky and Its Motion

Although celestial objects vary in distance from Earth, it's convenient to think of the sky as a great celestial sphere that encircles Earth. Astronomers refer to this way of looking at the universe as a **scientific model**—a carefully devised mental conception of how things work. Scientists use scientific models for all types of things to help them visualize and understand things in nature. In future lessons, you'll see how scientists use many different scientific models to gain understanding.

As you envision the celestial sphere that surrounds the Earth, realize that you are able to see only half of the sphere at any given time—the half above your **horizon**. The **zenith** represents the imaginary point directly above your head. Similar to how Earth's equator divides it into northern and southern hemispheres, the **celestial equator** is a great circle that divides the celestial sphere in two. The north and south celestial poles are located directly above Earth's north and south poles, respectively. What you see in the sky depends on where you are on Earth. A person looking at the night sky in North America (the top half of the sphere) will see different stars and constellations than someone in Australia (the bottom half of the sphere).

In the scientific model, the celestial sphere rotates on the axis from the north celestial pole to the south celestial pole. In reality, the eastward rotation of the earth makes the sun, moon, and the stars appear to move westward in the sky. If you stood at the earth's North Pole and looked into the night sky, you would see Polaris nearly overhead. If you looked at the sky for hours, you'd see the other stars seemingly rotate around Polaris—around the north celestial pole.

When thinking about the stars being on an imaginary 360° sphere, we can measure the apparent distances between them as angular distances or angular separation. If an observer's latitude is 50°, we can say that Polaris—known as the North Star—is 50° above the horizon. Of course, this is not the actual distance between Polaris and the horizon—it's simply a way of providing a relative distance between points on the sphere. The distance measured between two points on the sphere is known as the **angular distance**. Angular distances are measured in degrees, minutes, and seconds of arc. The **angular diameter** of a single object is the angular distance from one edge of the object to the other. For example, the angular diameter of the sun is about half a degree meaning that it appears to be 0.5° in size.

Thousands of years ago, as people looked north in the night sky, the star closest to the celestial pole was not Polaris—it was Thuban (α Draconis). And, about 12,000 years from now, the star nearest the celestial pole will be Vega (α Lyrae). You may wonder how this can happen. Earth's axis of rotation traces a cone in a period of 26,000 years. The angle the cone makes with respect to the vertical of Earth's orbit is 23.5°. As a result, Earth's rotation axis slowly changes the direction in which it is pointing, and hence, the stars toward which the axis points change. Today, the axis points to the star Polaris in the constellation Ursa Minor. Thousands of years ago, it pointed to the star Thuban in the constellation Draco. As Earth rotates on its axis, the gravitational forces of the moon and the sun cause Earth to wobble—or **precess**—much like a spinning top. Earth wobbles because it is slightly bulged in the center; if it were perfectly spherical, it would not precess.

In future lessons, your understanding of the celestial sphere and the scientific model will come into play. As you study the cycles of the sky in Lesson 3, you can use the model of the celestial sphere to help you understand it.

Review Exercises

Matching

Match each term with the appropriate definition or description.

E. → D

1. _a._ constellation	9. _c._ angular diameter
2. _e._ asterism	10. _m._ zenith
3. _j._ magnitude scale	11. _f._ precession
4. _d._ apparent visual magnitude	12. _q._ minute of arc
	13. _p._ celestial equator
5. _h._ celestial sphere	14. _n._ second of arc
6. _i._ north celestial pole	15. _k._ scientific model
7. _b._ south celestial pole	16. _l._ angular distance ← L.
8. _o._ north point	17. _g._ circumpolar constellation

(1) a. A stellar pattern identified by name, usually of mythological gods, people, or animals.

(7) b. Point on the celestial sphere directly above Earth's South Pole.

(9) c. The angle formed by lines extending from the observer to opposite sides of an object.

(4) d. The brightness of a star as seen by human eyes on Earth.

(2) e. A named group of stars that is not one of the traditional constellations.

(11) f. The slow change in direction of Earth's axis of rotation.

(17) g. A constellation so close to one of the celestial poles that it never sets or never rises as seen from a particular latitude.

(5) h. An imaginary sphere of very large radius surrounding Earth and to which the planets, stars, sun, and moon seem to be attached.

(6) i. The point on the celestial sphere directly above Earth's North Pole.

(3) j. System for describing brightness of a star.

(15) k. A mental conception, an idea that helps us think about and understand nature.

(16) l. The angle formed by lines extending from the observer to two locations.

(10) m. The point in the sky directly overhead.

(14) n. An angular unit of measure—each minute has 60 of them.

(8) o. One of the four cardinal directions; the point on the horizon directly north.

(13) p. Imaginary line around the sky directly above Earth's equator.

(12) q. An angular unit of measure—each degree has 60 of them.

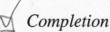 *Completion*

Fill each blank in the sentences below with the most appropriate term from the list of completion answers that follow. A term may be used once, more than once, or not at all. Check your answers with the Answer Key and review when necessary.

~~celestial equator~~	horizon	north
celestial sphere	latitude	~~precession~~
circumpolar	~~magnitude~~	south
constellations	~~minutes of arc~~	west
east	nadir	~~zenith~~

1. The _zenith_ is the point on the celestial sphere directly above an observer, regardless of where the observer is located on Earth.

2. The Greek astronomer Hipparchus sorted the stars in the sky into six classes according to their _magnitude_.

3. _Precession_ describes the slow change in the direction of Earth's axis of rotation.

4. For an observer located on the surface of Earth, the full moon has an angular diameter of approximately 30 _minutes of arc_.

5. The _celestial equator_ always crosses the horizon at the east point and west point.

 Self-Test

Select the best answer.

1. Astronomers today use the names of constellations to
   a. refer to groups of stars first identified by the ancient Greeks.
   b. identify asterisms.
   c. identify a specific area of the sky.
   d. refer to only the brightest star in a constellation.

2. The principal reason ancient civilizations gave names to groups of stars in the sky was probably to
   a. predict the seasons.
   b. provide a common reference among different cultures.
   c. develop a calendar.
   d. honor gods, heroes, and animals.

3. When the cosmos is described in terms of the celestial sphere, Earth is located
   a. at the center of the sphere.
   b. one-third from the edge of the sphere.
   c. beyond the limits of the sphere and in orbit around it.
   d. equidistant from the edge of the sphere as the sun.

4. Many of the brightest stars in the sky have names that were derived from ancient
   a. Latin.
   b. Greek.
   c. Arabic.
   d. the Romance Languages.

5. Some star names are based on letters of the Greek alphabet and the constellation in which they are located. The brightest star is usually referred to by the letter
   a. a.
   b. α.
   c. β.
   d. A.

6. The great circle that divides the celestial sphere into two equal hemispheres is called the
   a. horizon.
   b. celestial meridian.
   c. lines of magnitude.
   d. celestial equator.

7. According to the magnitude scale established by Hipparchus, a fifth-magnitude star is
   a. fainter than a fourth-magnitude star.
   b. brighter than a fourth-magnitude star.
   c. the brightest type of star.
   d. the faintest type of star.

8. Stars brighter than a first-magnitude star are identified by
   a. high positive numbers.
   b. fractions less than one or negative numbers.
   c. letters of the Arabic alphabet.
   d. letters of the Greek alphabet.

9. The celestial sphere with Earth at its center is
   a. a correct description of the cosmos.
   b. a scientific model.
   c. incorrect and this picture offers no help in visualizing nature.
   d. the model used by ancient Greek astronomers to explain how the Earth orbits the sun.

10. The moon and sun have an angular diameter of approximately
    a. 2,500 miles.
    b. 0.5 degree.
    c. 2 degrees.
    d. 4,000 kilometers.

11. As you, an observer, travel northward, the celestial north pole is an angular distance above the horizon equal to
    a. your latitude.
    b. your longitude.
    c. the position of the zenith.
    d. the position of nadir.

12. At far northern or southern latitudes some constellations do not rise above or set below the horizon. These constellations are said to be
    a. asterisms.
    b. the brightest in the sky.
    c. circumpolar.
    d. at rest and are not observed to move at all.

13. As Earth rotates on its axis and revolves around the sun, its axis slowly wobbles and traces a cone in space. This motion is called
    a. oscillation of the axis.
    b. precession.
    c. angular motion.
    d. revolution of the axis.

14. One of the most noticeable consequences of precession (given enough time) is the
    a. change of the pole star.
    b. change of the period of Earth's orbit.
    c. disappearance of certain constellations during the year.
    d. appearance of different stars on the celestial equator during the year.

Short-Answer Questions

1. What role did the stars Sirius and Procyon play in the culture of ancient Egypt?

→ The appearance of Sirius on the horizon in the morning sky coincided w/ the flooding of the Nile River. Procyon appeared in the morning sky a month earlier.

2. What do the star names α Orionis and β Orionis signify?

→ Brightest (β) and second (α) brightest stars in the Orion constellation.
α Orionis = Betelgeuse   β Orionis = Rigel.
Usually, α is brightest and β 2nd brightest, but this is an exception.

3. Describe the effect precession of Earth has on the appearance of the night sky.

→ the pole star changes. One cycle is about 26,000 years
Reference points in the sky (celestial poles, celestial equator) shift their position.
Today, the star nearest to the N cel. pole is Polaris. 5,000 yrs ago was Thuban (α Draco).
In 12,000 years, will be Vega (α Lyra)

4. Use the star charts in Appendix B of *Horizons* to identify five most likely north circumpolar constellations observed from northern middle latitudes.

TODO

5. Use the star charts in Appendix B of *Horizons* to identify five most likely south circumpolar constellations observed from southern middle latitudes.

6. What is the difference between the horizon and celestial equator?

→ the horizon is det. by observer's location; separates sky and earth; only objects above it are visible. Celestial equator divides celestial sphere in 2 equal hemispheres. Located directly above Earth equator. Is the reference circle f/m which positions of stars are determined; observer only needs to adjust latitude.

ToDo

Applications

1. Using Table A-9 on page 481 of *Horizons*, determine the intensity ratio of the stars Sirius and Beta Crucis. Use the apparent visual magnitude for the two stars.

2. Two of the nearest stars to the sun, ε Eridani and τ Ceti, have a magnitude difference of 0.2. What is the intensity ratio of these two stars?

3. How many diameters of the full moon can fit within the bowl of the Big Dipper?

4. How far above the horizon (angular distance) is the star Polaris if the observers latitude is 34° north?

5. You are standing on Earth's equator. What is the angular distance between your zenith and the celestial equator?

Answer Key

Matching

1. a (p. 12; video lesson; objective 1)

2. e (p. 13; video lesson; objective 1)

3. j (p. 14; video lesson; objective 3)

4. d (p. 14; video lesson; objective 3)

5. h (p. 16; video lesson; objective 4)

6. i (p. 16; video lesson; objective 4)

7. b (p. 16; video lesson; objective 4)

8. o (p. 16; objective 4)

9. c (p. 16; video lesson; objective 4)

10. m (p. 16; objective 4)

11. f (pp. 18–19; video lesson; objective 6)

12. q (p. 17; objective 4)

13. p (p. 16; video lesson; objective 4)

14. n (p. 17; objective 4)

15. k (p. 18, Window on Science 2-2; video lesson; objective 4)

16. l (p. 17; video lesson; objective 4)

17. g (p. 17; video lesson; objectives 4 & 5)

Completion

1. zenith (p. 16; objective 4)

2. magnitude (p. 14; objective 3)

3. precession (pp. 18–19; video lesson; objective 6)

4. minutes of arc (p. 17; objective 4)

5. celestial equator (p. 16; video lesson; objective 4)

Self-Test

1. c (pp. 12–13; video lesson; objective 1)

2. d (p. 12; objective 1)

3. a (pp. 15–17; video lesson; objective 4)

4. c (p. 13; objective 2)

5. b (p. 13; objective 2)

6. d (p. 16; video lesson; objective 4)

7. a (p. 14; video lesson; objective 3)

8. b (p. 14; video lesson; objective 3)

9. b (p. 19; video lesson; objective 4)

10. b (p. 17; objective 4)

11. a (p. 17; objective 4)

12. c (p. 17; objective 5)

13. b (pp. 18–19; video lesson; objective 6)

14. a (pp. 18–19; video lesson; objective 6)

Short-Answer Questions

✓1. The appearance of the star Sirius on the horizon in the morning sky signaled the flooding of the Nile River in ancient Egypt. The appearance of Procyon in the morning sky warned of the arrival of Sirius one month later. It needs to be understood that the star Sirius does not cause the flooding of the river; however, the timing of the flood was linked to the appearance of Sirius in ancient Egypt. (video lesson; objective 1)

2. α Orionis is the name of the second brightest star in the constellation of Orion. This star is also known as Betelgeuse. β Orionis is the name of the brightest star in the constellation of Orion also known as Rigel. It's important to understand that the star designated by the Greek letter α usually refers to the brightest star and β usually refers to the second brightest star, but these two stars are exceptions. (p. 13; objective 2)

3. Precession causes reference points in the sky, such as the celestial poles and celestial equator, to shift their positions. This effect is not immediately noticeable over the course of a few nights; over longer periods of time, however, precession is very noticeable. Today, the star nearest the north celestial pole is Polaris. It is the brightest star in the constellation of Ursa Minor. About 5,000 years ago, the star nearest the north celestial pole was Thuban, the brightest star in the constellation of Draco. In about 12,000 years, the star nearest the pole will be Vega, the brightest star in the constellation of Lyra. (pp. 18–19; objective 6)

4. Ursa Minor, Ursa Major, Cepheus, Cassiopeia, Draco (answers may vary depending on how far or how close the constellations are to the north celestial pole). (*Horizons* Appendix B; objective 5)

5. Chamaeleon, Hydrus, Octans, Mensa, Apus (answers may vary depending on how far or how close the constellations are to the south celestial pole). (*Horizons* Appendix B; objective 5)

6. The horizon is the circle that separates the sky from the earth. It is determined by the observer's location. As a result, the observer only sees half of the sky—the sky above the observer's head. Only objects above the horizon are visible. The celestial equator divides the celestial sphere into to equal hemispheres. The celestial equator is located directly above Earth's equator. The celestial equator is the reference circle from which positions of stars are determined. Using this reference, "all" observers have a standard method of locating objects in the sky. The individual needs only to adjust for his or her latitude to determine if the desired object is above the horizon. (p. 16; video lesson; objective 4)

Applications

1. The answer is 12.6 (p. 15, Reasoning with Numbers 2-1; video lesson; objective 3).

   Sirius appears 12.6 times brighter in the sky than Beta Crucis. From Table A-8 on page 480 in *Horizons*, the magnitude of Sirius is 1.47 and the magnitude of Beta Crucis is 1.28. Using the equation on page 15, $IA/IB = (2.512)^{(1.28-(-1.47))} = 12.6$.

2. The answer is 1.20 (p. 15, Reasoning with Numbers 2-1; objective 3).

   The intensity ratio is 1.20. If you look at the list of the nearest stars in Table A-9 of Appendix A, page 481 in *Horizons*, you see that ε Eri (Eridanus) has an apparent visual magnitude of 3.7 and τ Ceti (Cetus) has an apparent visual magnitude of 3.5. Since τ Ceti has the smaller magnitude, it is the brighter of the two stars. You can also calculate this as $(2.512)^{0.2} = 1.20$

3. The answer is 20. The bowl of the Big Dipper has an angular diameter of 100 and the angular diameter of the full moon is 0.5°. Divide 10 by 0.5 and the result is 20. (p. 17; objective 4)

4. 34° (p. 17; objective 5)

5. 0°. If the observer is on the equator of Earth, the celestial equator passes directly overhead and through the observer's zenith. (p. 17; objective 5)

Notes:

MTCN.060-558-4509.

Q. GF's name.
A. Deanna.

Flm marylin Sanders.
To Cory Hansen.

$75.00

Agenda

Friday Jan. 28 ...... check e-mail / Lesson 2 Applications +finish Lesson 2 checklist / finish Lesson 3,4,5,6 / nails / wash hair / walk dogs / pay Mom Quest 4,5 Short Answer

Sal. Jan. 29 ....

Sun. Jan. 30 ....

Mon Jan. 31 ....

End of Jan. plan & Finish Astro / Fix shoes / Finish last sem. papers
End of Feb. plan & Finish class + exams / Get some money.

Lesson Review

Lesson 2: Observing the Sky

PLEASE NOTE: Use this matrix to guide your study and achieve the learning objectives of this lesson. It will also help you to view the video, which defines and demonstrates important concepts and principles as they relate to everyday life and actual case studies.

Learning Objective	Textbook	Student Guide
1. Explain the role of constellations in ancient and modern observations of the sky.	pp. 12–13	Key Terms: 1, 2; Matching: 1, 2; Self-Test: 1, 2; Short-Answer: 1.
2. Explain how stars have been named by ancient and modern astronomers.	pp. 13–14	Self-Test: 4, 5; Short-Answer: 2.
3. Describe the origin of the modern stellar magnitude scale and the relative brightness between magnitudes.	pp. 14–15	Key Terms: 3, 4; Matching: 3, 4; Completion: 2; Self-Test: 7, 8; Applications: 1, 2.
4. Describe the celestial sphere as a scientific model, and explain what astronomers mean by angular distance.	pp. 15–19	Key Terms: 5, 7, 8, 9, 10, 11, 12, 13, 14, 15, 16, 17, 18, 19; Matching: 5, 6, 7, 8, 9, 10, 12, 13, 14, 15, 16, 17; Completion:1, 4, 5; Self-Test: 3, 6, 9, 10, 11; Short-Answer: 6; Applications: 3.
5. Describe how the orientation of the sky depends on the observer's latitude.	pp. 16–17	Matching: 17; Self-Test: 12; Short-Answer: 4, 5; Applications: 4, 5.
6. Describe the phenomenon of precession and its effects.	pp. 18–20	Key Terms: 6; Matching: 11; Completion: 3; Self-Test: 13, 14; Short-Answer: 3.

LESSON

3

Celestial Cycles

Checklist

For the most effective study of this lesson, complete the following activities in this sequence.

Before Viewing the Video

☑ Read the Preview, Learning Objectives, and Viewing Notes below.

☑ Read Chapter 3, "Cycles of the Sky," pages 22–43, in the *Horizons* textbook.

What to Watch

To Do → ❑ After reading the textbook chapter, watch the video for Lesson 3, *Celestial Cycles*.

After Viewing the Video

❑ Briefly note your answers to questions listed at the end of the Viewing Notes.

❑ Review the Summary below.

❑ Review all reading assignments for this lesson, especially the Chapter 3 summary on page 42 in *Horizons* and the Viewing Notes in this lesson.

❑ Write brief answers to the review questions at the end of Chapter 3 in *Horizons*.

❑ Complete the Review Exercises below. Check your answers with the Answer Key and review when necessary.

❑ Use the Lesson Review matrix found at the end of this lesson to review and assess your knowledge of each Learning Objective.

❑ As assigned by your instructor, complete the Applications activities and any additional activities for this lesson.

❑ Review Assignments (marks) for this class

❑ Book exams

FINISH THIS CLASS SOON !!!

Preview

The universe is in constant motion. As Earth spins on its axis, the moon travels around it. Earth orbits the sun and the sun revolves around the center of the galaxy. Our galaxy is moving through space and the universe is expanding. Some of these motions are impossible to detect in a human's lifetime; but others manifest themselves to a patient observer. We can detect the daily motions of Earth and the moon, and the effect of Earth's yearly trip around the sun.

Throughout history, these cycles have had a great influence on society—our calendars and clocks are based on them and the ancients wondered about and sometimes feared them. Today, we understand their origin, predict their reoccurrence, and remain in awe of their beauty. These are known as the "celestial cycles."

We may take the daily and yearly cycles for granted, but humans are fascinated with lunar and solar eclipses—some of the most remarkable celestial events. This alignment of the moon, Earth, and the sun has been observed for thousands of years and such alignments, too, occur in cycles. Many ancient civilizations believed that an eclipse served as a warning for fateful events.

In Lesson 2, you studied how using a model of the sky—the celestial sphere—can help us measure daily motions reflected in the sky as a result of Earth's rotation. In this lesson, you'll continue your study of celestial motion.

CONCEPTS TO REMEMBER

- Recall from Lesson 2 that *angular diameter* is the angle formed by lines extending from the observer to opposite sides of an object. Astronomers use angular diameter to refer to the apparent size of an object as seen from Earth. Both the sun and the moon have an angular diameter of about 0.5 degrees (p. 17 in this guide).

- In Lesson 2, you learned that the *celestial equator* is the imaginary line around the sky directly above Earth's equator. The position of the celestial equator relative to the ecliptic is related to the change of seasons and the length of daylight (p. 17 in this guide).

- In Lesson 2, you also learned that the *celestial sphere* is an imaginary sphere of very large radius surrounding Earth to which the planets, stars, sun, and the moon seem to be attached. Astronomers use this model of the universe to better understand it (p. 17 in this guide).

☑ Learning Objectives

After you complete this lesson, you should be able to:

1. Explain how the daily and annual motion of Earth causes what we see in the day and nighttime sky. *HORIZONS* TEXTBOOK PAGES 22–28; VIDEO PROGRAM.

2. Describe the tilt of Earth's axis with respect to the ecliptic and explain why this tilt causes the seasons. *HORIZONS* TEXTBOOK PAGES 25–27.

3. Describe the motion of the moon relative to Earth and the sun, explain how this causes us to see phases of the moon, and identify these phases by name. *HORIZONS* TEXTBOOK PAGES 29–33.

4. Explain the necessary conditions for a lunar eclipse, describe what causes total and partial eclipses, and describe their appearance. *HORIZONS* TEXTBOOK PAGES 29, 32–33.

5. Explain the necessary conditions for a solar eclipse; describe what causes total, partial, and annular eclipses; and describe their appearance. *HORIZONS* TEXTBOOK PAGES 33–39.

6. Discuss what the Milankovitch hypothesis tells us about astronomical influences on Earth's climate. *HORIZONS* TEXTBOOK PAGES 39–41.

☑ **At this point, read Chapter 3, "Cycles of the Sky," pages 22–43.**

☑ Viewing Notes

The video program introduces you to the celestial cycles. You will see how the rotation of Earth and its revolution around the sun affect what we see in the sky.

The video program contains the following segments:

- ☢ Daily & Yearly Cycles
- ☢ The Cycle of the Moon
- ☢ Eclipses
- ☢ Lunar Eclipse
- ☢ Solar Eclipse

The following information will help you better understand the video program:

The video program discusses the daily motions of the sky that are caused by Earth's **rotation** on its axis; the yearly cycle is caused by Earth's **revolution** around the sun. If Earth did not have a yearly cycle, the sun would rise and set at the same time every day.

The **ecliptic** is the apparent path of the sun as it moves across the sky. The sun follows an apparent path that is lower in the sky in the winter and more

Two things inspire me to awe—the starry heavens above and the moral universe within.
—Albert Einstein
(1879–1955)

directly overhead in the summer. This changes the amount of solar energy that reaches each hemisphere during different times of the year.

The video program mentions that the moon raises tides on Earth—the effect that is manifested in the extreme maximum and minimum positions of the water on our beaches known as high and low tides. Earth also has a gravitational effect on the moon. The gravitational force exerted by Earth on the moon does produce a tug on the side of the moon closest to Earth that is greater than the tug on the far side of the moon. This force had the effect of slowing the moon's rotation to the point that only one side of the moon now points toward Earth at all times.

As the moon revolves around Earth, it rotates once on its axis in the same number of days, causing the same side of the moon to always face Earth. The 29.53-day **synodic** month is the period in which the moon goes through its phases; the 27.32-day **sidereal** month is the period in which the moon goes from one point in its orbit back to the same point.

An eclipse occurs when the shadow of any astronomical object falls on to another. A **lunar eclipse** occurs at full moon when the moon moves through the shadow of the earth. A **solar eclipse** occurs at new moon when the shadow of the moon falls on the earth.

QUESTIONS TO CONSIDER

(handwritten note: already wrote down & answer them)

1. • What causes the apparent daily motions of the sun and stars across the sky?
2. • What causes the seasons?
3. • Why would a city in northern Alaska have nearly 24 hours of daylight in the summer and nearly 24 hours of darkness in the winter?
4. • What is the difference between rotation and revolution?
5. • What causes the phases of the moon?
6. • What conditions are necessary for a lunar eclipse to occur?
7. • What conditions are necessary for a solar eclipse to occur?
8. • Why don't lunar and solar eclipses happen twice a month?

Watch the video for Lesson 3, *Celestial Cycles*.

Know these:

Key Terms and Concepts

Page references are keyed to the *Horizons* textbook.

1. **rotation:** Motion around an axis passing through the rotating body. (p. 22; video lesson; objective 1)

2. **revolution:** Orbital motion about a point located outside the orbiting body. (p. 22; video lesson; objective 1)

3. **ecliptic:** The apparent path of the sun around the sky. (p. 24; video lesson; objective 2)

4. **zodiac:** A band centered on the ecliptic and encircling the sky. (p. 25; video lesson; objective 2)

5. **vernal equinox:** The place on the celestial sphere where the sun crosses the celestial equator moving northward. Also, the time of year when the sun crosses this point, about March 20, and spring begins in the northern hemisphere. (p. 26; video lesson; objective 2)

6. **summer solstice:** The point on the celestial sphere where the sun is at its most northerly point. Also, the time when the sun passes this point, about June 22, and summer begins in the northern hemisphere. (p. 26; video lesson; objective 2)

7. **autumnal equinox:** The point on the celestial sphere where the sun crosses the celestial equator going southward. Also, the time when the sun reaches this point and autumn begins in the northern hemisphere—about September 22. (p. 26; objective 2)

8. **winter solstice:** The point on the celestial sphere where the sun is farthest south. Also the time of year when the sun passes this point, about December 22, and winter begins in the northern hemisphere. (p. 26; video lesson; objective 2)

9. **perihelion:** The orbital point of closest approach to the sun. (p. 27; objective 2)

10. **aphelion:** The orbital point of greatest distance from the sun. (p. 27; objective 2)

11. **evening star:** Any planet visible in the sky just after sunset. (p. 28; objective 1)

12. **morning star:** Any planet visible in the sky just before sunrise. (p. 28; objective 1)

13. **lunar eclipse:** The darkening of the moon when it moves through Earth's shadow. (p. 29; video lesson; objective 4)

14. **umbra:** The region of a shadow that is totally shaded. (p. 29; video lesson; objective 4)

15. **penumbra:** The portion of a shadow that is only partially shaded. (p. 32; video lesson; objective 4)

16. **sidereal period:** The time a celestial body takes to turn once on its axis or revolve once around its orbit relative to the stars. (p. 31; objective 5)

17. **synodic period:** The time a solar system body takes to orbit the sun once and return to the same orbital relationship with Earth. That is, orbital period referenced to Earth. For the moon, this is the time for the moon to orbit Earth and return to the same position relative to the sun. This is the period associated with the phases of the moon—new moon to new moon. (p. 31; objective 5)

18. **solar eclipse:** The event that occurs when the moon passes directly between Earth and the sun, blocking our view of the sun. (p. 33; video lesson; objective 5)

19. **photosphere:** The bright visible surface of the sun. (p. 34; objective 5)

20. **chromosphere:** Bright gases just above the photosphere of the sun. (p. 34; objective 5)

21. **corona:** On the sun, the faint outer atmosphere composed of low-density, high-temperature gas. (p. 34; objective 5)

22. **prominence:** Eruption on the solar surface; most visible during total solar eclipses. (p. 34; objective 5)

23. **diamond-ring effect:** During a total solar eclipse, the momentary appearance of a spot of photosphere at the edge of the moon, producing a brilliant glare set in the silvery ring of the corona. (p. 35; objective 5)

24. **annular eclipse:** A solar eclipse in which the solar photosphere appears around the edge of the moon in a bright ring, or annulus. The corona, chromosphere, and prominences cannot be seen. (p. 35; video lesson; objective 5)

25. **perigee:** The point closest to Earth in the orbit of a body circling Earth. (p. 35; objective 5)

26. **apogee:** The point farthest from Earth in the orbit of a body circling Earth. (p. 35; video lesson; objective 5)

27. **node:** The points where an object's orbit passes through the plane of Earth's orbit. (p. 37; video lesson; objective 5)

28. **Saros cycle:** An 18-year, 11.33-day period after which the pattern of lunar and solar eclipses repeats. (p. 38; video lesson; objective 5)

29. **Milankovitch hypothesis:** Suggestions that Earth's climate is determined by slow periodic changes in the shape of its orbit, the angle of its axis, and precession. (p. 39; objective 6)

Summary

The constant motion in our solar system creates the periodic celestial events such as the seasons, phases of the moon, and eclipses. Earth moves in nearly a circular orbit around the sun in a 365.25-day period. This **revolution** of the earth around the sun is what we define as the year. As it revolves around the sun, it rotates—turns on its axis—once every 24 hours.

The Cycle of the Sun

Although the sun does not revolve around Earth, it appears to move along the **ecliptic**—the apparent path of the sun in the sky. If the sun weren't so bright and you could see the stars behind it, you would detect the change in position of the sun against the backdrop of the stars. Because Earth travels in a complete 360° circle around the sun every 365.25 days, the sun appears to change its position in the sky about 1° per day in the eastward direction.

If you imagine the path of Earth around the sun, you can understand the seasons. Earth holds its axis fixed—at a 23.5° tilt with respect to the plane of its orbit—as it revolves around the sun. Therefore, the sunlight reaching different parts of its surface varies in intensity. When Earth is on one side of the sun, the northern hemisphere is tipped toward the sun and it receives more direct sunlight—it is summer in the northern hemisphere and winter in the southern hemisphere. Six months later, Earth is on the opposite side of the sun, the northern hemisphere is tipped away from the sun and receives less direct sunlight—it is winter in the northern hemisphere and summer in the southern hemisphere.

In Lesson 2, you learned about the *celestial sphere*; astronomers use this model to help describe the seasons. The *celestial equator* is directly above Earth's equator. Because Earth is tipped in its orbit, the celestial equator and the ecliptic are inclined to each other 23.5° causing them to intersect at the autumnal equinox and vernal equinox. When the sun—in its path along the ecliptic—crosses the celestial equator moving northward, the point is called the **vernal equinox**. This occurs on or around March 20 and marks the beginning of spring. About three months later, around June 22, the sun is at its farthest point north along the ecliptic called the **summer solstice**. This is the longest day of the year in the northern hemisphere and marks the beginning of summer. Another three months pass and the sun crosses the celestial equator moving southward—around September 22—the **autumnal equinox** marks the beginning of autumn in the northern hemisphere. Around December 22, the sun reaches its southernmost point along the ecliptic called the **winter solstice**. This marks the beginning of winter in the northern hemisphere and is the shortest day of the year. In the southern hemisphere, the names are reversed— the equinox that occurs in March is the autumnal and the one in September is the vernal; the solstice occurring in June is called winter and the one in December is called summer.

As Earth makes its nearly circular trip around the sun, so do the other planets; the planets farthest from the sun move the slowest. The planets lie on nearly the same plane as Earth's orbit and therefore appear to move eastward along the ecliptic, or counterclockwise if you were looking down at our solar system from space and above Earth's north pole. Venus and Mercury are always seen close to the sun because their orbits lie inside of Earth's orbit—they are closer to the sun. Depending on the time of year, these planets can sometimes be seen just before sunrise or just after sunset. Any planet visible in the sky shortly before sunrise is called a **morning star**, and any planet visible in the evening sky just after sunset is called an **evening star**, even though they are planets—not stars.

The Cycles of the Moon

If you spend a few hours studying the night sky, you can detect the moon's eastward motion against the backdrop of the stars. The moon moves fairly quickly in the sky—nearly one lunar diameter per hour. The moon orbits Earth every 27.32 days—this is known as its **sidereal period**. The moon also spins once as it orbits Earth and therefore keeps the same side facing Earth—we can never see the far side of the moon when we observe it from Earth.

Perhaps the most obvious cycle of the sky is the phases of the moon—from new moon to full moon and back again. A common misconception is that a portion of Earth's shadow falling upon the moon causes them. Rather, the phases of the moon are caused by the relative positions of the moon, Earth, and sun in the sky (see p. 30 in the *Horizons* textbook).

If you think about the moon's orbit around Earth, it will help you understand the phases. Sunlight always illuminates half of the moon, but as the moon orbits Earth we get to see more or less of the illuminated half. The new moon marks the beginning of the lunar cycle. The moon is between Earth and the sun and the sunlight falls on the side of moon that we cannot see—the side that we can see is dark. During the new moon, the moon rises at dawn and sets at sunset. A few days later, we see a waxing crescent as the moon grows fatter and the portion of the illuminated surface that we see grows from 0 to 50 percent over the period of about a week. About one week into the cycle, the first quarter moon occurs—the sun and moon are about 90 degrees apart in the sky relative to Earth and we see half of the illuminated side of the moon—it looks to us like it is illuminated on one half of its face. But it's called first quarter anyway because it's a quarter of the way through its cycle. A few days later, we see a waxing gibbous moon as the moon grows fatter and the illuminated portion of the surface that we see grows from 50 to 100 percent.

Two weeks into the four-week cycle the full moon occurs—Earth is between the sun and the moon and we see all of the side that is illuminated. The full moon rises about the time of sunset and sets about the time of sunrise. As the moon moves through the second week of its cycle, we see a waning gibbous— the moon gets thinner as it gets closer to its third quarter phase; the portion of the surface that we see as illuminated "wanes" from 100 to 50 percent. Three weeks through the cycle, the moon is in the third quarter—we again see half of the illuminated side of the moon—it looks to us like it's illuminated on the opposite side of its face compared to what we saw when it was at first quarter. As the moon moves through the third week of its cycle, we see the waning crescent as the illuminated portion that we can see gets thinner still—it wanes from 50 to 0 percent as observed from Earth. The cycle is completed when we once again see a new moon—the side that we see is dark.

The complete cycle of lunar phases takes 29.53 days—this is known as the moon's **synodic period**. You might wonder how the cycles of the moon can differ from the moon's orbital period of 27.32 days. This is because Earth keeps orbiting the sun while the moon is going through its phases and orbiting Earth. To go through one lunar cycle (new moon to new moon), the moon must travel *more than* 360° along its orbit. Therefore, the synodic month—or lunar cycle— is approximately two days longer than the sidereal month—or orbital period.

At every new moon and full moon, Earth, the sun, and the moon are in nearly a straight line. When they are in a perfectly straight line, a solar or lunar eclipse can occur. You might wonder why an eclipse doesn't occur every month. Eclipses occur infrequently because the moon's orbit is tilted a little more than 5° compared to Earth's orbit; therefore the moon doesn't follow the ecliptic exactly. Because of this tilt, the new moon and full moon usually occur when the moon is either above or below Earth's orbital plane. The points where the moon passes through the plane of Earth's orbit are called the **nodes** of the moon's orbit; the line connecting these nodes is called the line of nodes. When the moon, Earth, and the sun are all lined up on this line of nodes, we have the conditions for an eclipse.

A **lunar eclipse** occurs when the moon is full and passes through Earth's shadow. Earth's shadow consists of two parts—the umbra and the penumbra. The

umbra is the region of total shadow and the **penumbra** is the region of partial shadow. If the entire moon passes through the umbra, we would see a total lunar eclipse. If the moon passes a bit too far north or south, it may partially enter the umbra and we'll see a partial lunar eclipse. If the moon passes only through the penumbra and misses the umbra entirely, a penumbral eclipse occurs; this type of eclipse is barely noticeable since the penumbra is not very dark.

During a total lunar eclipse, the moon takes several hours to pass through Earth's shadow. The moon first moves through the penumbra dimming more as it moves deeper into the penumbra. After about an hour, the moon reaches the umbra and the umbral shadow darkens the moon. Depending on where the moon crosses the shadow, the period of total eclipse—called totality—can last as long as 1 hour, 45 minutes. It's easy to see a lunar eclipse because you just have to be standing on the side of Earth that's facing the moon.

When the moon is between the sun and Earth, the moon can cast its shadow on Earth and a **solar eclipse** can be observed. As you discovered in Lesson 2, both the sun and the moon have an *angular diameter* of 0.5°. Because the angular diameters are the same at most times of the year, the moon is just the right size to cover the disk of the sun. Even though the sun is 400 times larger than the moon, it's about 400 times farther away from Earth, on average, and therefore can be totally eclipsed by the moon.

Although there are more solar eclipses than lunar eclipses during the course of one year, the shadow produced by the moon falls on a very small area of Earth—most of the time on the oceans. There are different types of solar eclipses: total, partial, and a special type of partial eclipse called an annular eclipse. In a total eclipse, the sun is completely covered by the moon. Total eclipses occur yearly but are rare for any individual to see because the umbra of the moon's shadow barely reaches Earth; it casts a small circular shadow that is never larger than 270 km and moves at a rate of 1600 km/hr. If you were standing in that small umbral area—in the path of totality—you would see a total eclipse. If you were standing just outside the umbral area, in the penumbra, you would see a partial eclipse up to 3,000 miles from the path of totality. If you were standing outside of the penumbra, you wouldn't see the eclipse at all.

A total solar eclipse is a spectacular event lasting an average of 2 or 3 minutes with a maximum of 7.5 minutes. When the moon covers the brilliant surface of the sun called the **photosphere**, you are able to see the faint outer layers of the sun's atmosphere—the **chromosphere** and the **corona**. The chromosphere is a bright pink ring that surrounds the disk of the sun; the corona envelops the sun and its pale white glow extends far past the surface of the sun. Because the photosphere is so bright, the chromosphere and the corona are visible only during a solar eclipse, when the photosphere is obstructed.

Earth, the moon, and the sun may line up to cause a solar eclipse, but at certain times the moon may appear slightly smaller in the sky and may not completely cover the sun. This is called an **annular eclipse**, which occurs because both Earth's orbit and the moon's are slightly elliptical. The moon looks significantly larger when it's at **perigee**—the point in its orbit closest to Earth than it does when it's at **apogee**—the most distant point in its orbit. Because Earth's orbit is also slightly elliptical, the distance between the sun and Earth can vary as well. If the moon is at the farthest part of its orbit during totality, the angular diameter of the moon is less than that of the sun. Therefore, the moon doesn't completely cover the sun and a ring (or annulus) of the photosphere is visible (see Figure 3-10 on p. 36 in the *Horizons* textbook).

Depending on the positions of the moon as it covers the sun during an eclipse, its imperfect surface may allow a small amount of light from the photosphere to peek out from behind it. A small part of the photosphere forms a brilliant spot as it slips out through a lunar valley or crag at the moon's edge while the inner corona forms a silvery ring of light around the sun. This is called the **diamond-ring effect** (see Figure 3-9 on p. 35 in the *Horizons* textbook).

Astronomical Influences on Earth's Climate

You've already discovered how the movement of Earth causes seasons to occur. This movement may also cause changes in climate over decades and centuries.

The **Milankovitch hypothesis** proposes that small changes in Earth's orbit—in precessions and in inclination—can have drastic effects on Earth's climate. This hypothesis was proposed in the 1920s and is still being tested.

Earth's orbit is slightly elliptical, but over the course of about 100,000 years, Earth's orbit changes slightly and this can affect climate. If Earth's orbit became more elliptical over time, summers in the northern hemisphere may become too cool to melt the ice from the previous winter and glaciers would grow larger.

In Lesson 2, you learned about *precession*—the slow change in the direction of Earth's axis of rotation over a period of about 26,000 years. At present, Earth reaches **perihelion** during winter in the northern hemisphere—when it is 1.7 percent closer than average to the sun, its axis is tipped away. When Earth is **aphelion** and is 1.7 percent farther away than average, its axis is tipped toward the sun and it is summer in the northern hemisphere. About 13,000 years from now, summers in the northern hemisphere will occur at perihelion and winters at aphelion—the opposite of the present day positions. In this scenario, summers may become warmer and not only melt the previous winter's snow but also may prevent new glaciers from forming. Remember that there is more land area relative to ocean in the northern hemisphere, so the precessional changes in the two hemispheres do not cancel each other out.

The inclination of Earth's axis relative to its orbital plane is another factor that may affect climate changes according to the Milankovitch hypothesis. Currently, the inclination of Earth's equator to the plane of its orbit is 23.5°. Over the course of about 41,000 years, this angle can vary from 22° to 24°. When the inclination is greater, the seasons become more severe. This, in combination with the effects of precession, can cause climate changes on Earth.

So far, you've learned about observing the sky with the unaided eye—what we call *naked-eye astronomy*. In the next two lessons you'll learn about the history of astronomy and how astronomers developed tools and methods to better observe the sky.

Review Exercises

Matching

Match each term with the appropriate definition or description.

1. _____ sidereal period		12. _____ autumnal equinox
2. _____ synodic period		13. _____ diamond-ring effect
3. _____ aphelion		14. _____ apogee
4. _____ evening star		15. _____ annular eclipse
5. _____ lunar eclipse		16. _____ nodes
6. _____ penumbras		17. _____ ecliptic
7. _____ umbra		18. _____ Saros cycle
8. _____ solar eclipse		19. _____ perigee
9. _____ photosphere		20. _____ Milankovich hypothesis
10. _____ summer solstice		21. _____ vernal equinox
11. _____ corona		

a. Suggestions that Earth's climate is determined by slow periodic changes in the shape of its orbit, the angle of its axis, and precession.

b. The event that occurs when the moon passes directly between Earth and the sun, blocking our view of the sun.

c. The points where an object's orbit passes through the plane of Earth's orbit.

d. The point on the celestial sphere where the sun crosses the celestial equator going southward. Also, the time when the sun reaches this point and autumn begins in the northern hemisphere.

e. The point closest to Earth in the orbit of a body circling Earth.

f. The point on the celestial sphere where the sun is at its most northerly point; when the sun passes this point, summer begins in the northern hemisphere.

g. The portion of a shadow that is only partially shaded.

h. The time a celestial body takes to turn once on its axis or revolve once around its orbit relative to the stars.

i. The apparent path of the sun around the sky.

j. The point farthest from Earth in the orbit of a body circling Earth.

k. The bright visible surface of the sun.

l. An 18-year, 11-day period after which the pattern of lunar and solar eclipses repeats.

m. An object's orbital period with respect to the sun.

n. The region of total shadow, i.e. completely in the shade.

o. A solar eclipse in which the photosphere appears around the edge of the moon in a brilliant ring, or annulus.

p. During a total solar eclipse, the momentary appearance of a spot of photosphere at the edge of the moon, producing a brilliant glare set in the silvery ring of the corona.

q. The darkening of the moon when it moves through Earth's shadow.

r. The place on the celestial sphere where the sun crosses the celestial equator moving northward; the time of year when spring begins in the northern hemisphere.

s. Any planet visible in the sky just after sunset.

t. The orbital point of greatest distance from the sun.

u. On the sun, the faint outer atmosphere composed of low-density, high temperature gas.

Completion

Fill each blank in the sentences below with the most appropriate term from the list of completion answers that follow. A term may be used once, more than once, or not at all. Check your answers with the Answer Key and review when necessary.

annular	morning	sidereal
aphelion	new	synodic
ecliptic	night	vernal equinox
evening	perihelion	winter

1. Seasons occur on Earth because Earth's axis of rotation is tilted with respect to the _____.

2. The period of the phases of the moon is referred to as the _____ month.

3. For a solar eclipse to occur, the moon must be at _____ phase.

4. _____ is the point in Earth's orbit when it is closest to the sun.

5. A(n) _____ star is the name used to identify a planet visible in the sky just after sunset.

Self-Test

Select the best answer.

1. When the moon is located in a direct line and between Earth and the sun, the phase is called
   a. full moon.
   b. last quarter.
   c. new moon.
   d. first quarter.

2. The angle formed by the sun, Earth, and the moon with Earth at the vertex during the third-quarter phase of the moon is
   a. 90°.
   b. 180°.
   c. 45°.
   d. 0°.

3. During the month of January, the constellation most readily visible around midnight in the northern hemisphere is
   a. Cancer.
   b. Gemini.
   c. Taurus.
   d. all of the above.

4. The rising of the stars in the east and the setting of these stars in the west is the result of
   a. Earth's motion around the sun.
   b. the moon's motion around Earth.
   c. Earth's rotation on its axis.
   d. individual stellar motions unrelated to the motions of Earth.

5. The vernal equinox, the first day of spring, occurs when the sun
   a. reaches its highest daytime position in the northern hemisphere.
   b. reaches its lowest daytime positioning the southern hemisphere.
   c. crosses the celestial meridian.
   d. moving northward crosses the celestial equator.

6. The winter solstice occurs when the sun is
   a. at its lowest position at noon when viewed by northern hemisphere observers.
   b. near the full moon as viewed by southern hemisphere observers.
   c. crossing the celestial equator and moving southward.
   d. rising due east and setting due west.

7. In order for a total lunar eclipse to occur,
    a. Earth must pass through the penumbra of the moon's shadow.
    b. the moon must pass through the penumbra of Earth's shadow.
    c. the moon must pass through the umbra of Earth's shadow.
    d. Earth must pass through the umbra of the moon's shadow.

8. A lunar eclipse can occur only if the phase of the moon is
    a. new.
    b. last quarter.
    c. waxing crescent.
    d. full.

9. In order to observe a total solar eclipse, an individual must be located completely within the
    a. penumbra of the moon's shadow.
    b. umbra of the moon's shadow.
    c. umbra of Earth's shadow.
    d. penumbra of Earth's shadow.

10. Twice a year, the line of nodes points toward the sun. When a full or new moon approaches the line of nodes at this time,
    a. nothing happens.
    b. a lunar eclipse is followed by a solar eclipse within two weeks.
    c. a solar eclipse is followed by a lunar eclipse within four weeks.
    d. an eclipse of one kind or another is possible.

11. The Milankovitch hypothesis, stating that changes in Earth's climate have an astronomical origin,
    a. has been universally accepted.
    b. has been universally rejected.
    c. has never been taken seriously.
    d. probably contains some truth and remains the subject of debate and testing.

12. Which of the following are astronomical influences at work in the Milankovitch hypothesis?
    a. small changes in Earth's orbit
    b. precession
    c. inclination of Earth's axis
    d. all of the above

Short-Answer Questions

1. What would an observer on the moon see looking back at Earth during the new moon phase? First quarter? Waxing crescent? Full moon?

2. Why isn't there a lunar or solar eclipse every month?

3. Identify six of the twelve constellations that lie along the ecliptic.

4. During what time of the evening would one be able to observe the waxing crescent phase of the moon?

5. Identify the three main factors that make up the Milankovitch hypothesis to explain variations in Earth's climate.

6. What is the angular distance of the sun from the celestial equator on the summer solstice? What is the explanation of this angular separation?

Applications

1. The sun is 149.6×10^6 km from Earth and its angular diameter is 0.50°. What is the actual linear diameter of the sun?

2. The equatorial radius of Jupiter is 71,494 km. At a distance of 629.3×10^6 km from Earth, Jupiter appears as a small banded disk in a telescope. What is the angular diameter of Jupiter as observed from Earth?

Answer Key

Matching

1. h (p. 31; objective 5)
2. m (p. 31; objective 5)
3. t (p. 27; objective 2)
4. s (p. 28; objective 1)
5. q (p. 29; objective 4)
6. g (p. 32; objective 4)
7. n (p. 29; objective 4)
8. b (p. 33; objective 5)
9. k (p. 34; objective 5)
10. f (p. 26; objective 2)
11. u (p. 34; objective 5)
12. d (p. 26; objective 2)
13. p (p. 35; objective 5)
14. j (p. 35; objective 5)
15. o (p. 35; objective 5)
16. c (p. 37; objective 5)
17. i (p. 24; objective 2)
18. l (p. 38; objective 5)
19. e (p. 35; objective 5)
20. a (p. 39; objective 6)
21. r (p. 26; objective 2)

Completion

1. ecliptic (p. 24; objective 2)
2. synodic (p. 31; objective 5)

3. new (p. 37; objective 5)

4. perihelion (p. 27; objective 2)

5. evening (p. 28; objective 1)

Self-Test

1. c (pp. 30–31; objective 3)

2. a (pp. 30–31; objective 3)

3. d (p. 24; video lesson; objective 1)

4. c (p. 24; video lesson; objective 1)

5. d (pp. 26–27; video lesson; objective 2)

6. a (pp. 26–27; objective 2)

7. c (pp. 29, 37; video lesson; objective 4)

8. d (pp. 29, 37; objective 4)

9. b (p. 33; objective 5)

10. d (p. 37; objective 5)

11. d (pp. 39–40; objective 6)

12. d (pp. 39–40; objective 6)

Short-Answer Questions

1. During the new moon phase, the moon is between Earth and the sun. Looking back toward Earth, assuming the shadow of the moon does not fall on Earth, Earth would be totally illuminated. Earth would be "full phase." During first quarter, Earth would be half illuminated or "first quarter" as well. During the waxing crescent phase, three-quarters of Earth's illuminated side would be visible from the moon. Earth would be in "gibbous phase." During full moon, assuming the moon is not within Earth's shadow, Earth's night time side would be visible from the moon. Earth would be "new." (pp. 30–31; objective 3)

2. The moon's orbit is tipped approximately 5° with respect to the orbit of Earth about the sun, the ecliptic plane. The moon's orbit intersects the ecliptic plane in two places called nodes. Only if the moon is at or near a node is an eclipse (lunar or solar) possible. An additional condition is that the phase of the moon must either be new (solar eclipse) or full (lunar eclipse). (pp. 36–38; objectives 4 & 5)

3. The ecliptic is the path the sun appears to follow across the sky. The ecliptic intersects twelve constellations known as the zodiac. These constellations are Pisces, Aries, Taurus, Gemini, Cancer, Leo, Virgo, Libra, Scorpius, Sagittarius, Capricornus, and Aquarius. (p. 24; objective 1)

4. The waxing crescent phase of the moon occurs a few days after the new moon phase. This phase is visible just after sundown. The observers would direct their view toward the western horizon. (pp. 30–31; video lesson; objective 3)

5. There are three factors that contribute to the Milankovitch hypothesis suggesting that climatic changes and ice ages are the result of variations in Earth's orbit, the tilt of its axis, and the wobble of its axis of rotation. Earth's orbit is elliptical. It is known that this elliptical shape changes slightly over the millennia, causing Earth to vary its distance from the sun during the summer and winter. The axis of rotation of Earth is presently tilted 23.5° with respect to its orbit or the ecliptic plane. This tilt varies from 22° to 24°. Finally, the third factor is precession. The wobble of Earth's axis causing the northern hemisphere to point towards the sun during summer months and away from the sun during winter months. The hypothesis suggests that changes in distance from the sun as well as the direction and tilt of the rotation axis may be sufficient to cause climate changes. Farther from the sun during winter months along with a greater tilt in the axis may be enough to create colder winters and the onset of ice ages. (pp. 39–40; objective 6)

6. The angular distance between the sun and the celestial equator is 23.5°. This angular distance is the result of Earth's axis tipped with respect to the plane formed by Earth's orbit. If you imagine the ellipse of Earth's orbit as being a plane and drawing a line perpendicular to that plane, Earth's axis is tilted 23.5° with respect to that perpendicular line. (pp. 26–27; objective 2)

Applications

1. Using the small-angle formula given on page 34 of the *Horizons* textbook (Reasoning with Numbers 3-1):

$$\frac{angular\ diameter = 1,800}{206,265} = \frac{linear\ diameter}{149.6 \times 10^6\ km}$$

linear diameter = 1,305,505 km [Remember: the angular diameter must be expressed in seconds of arc. 3,600 seconds = 1 degree] (p. 34; objective 1)

2. Using the small-angle formula given on page 34 of the *Horizons* textbook (Reasoning with Numbers 3-1):

$$\frac{angular\ diameter}{206,265} = \frac{linear\ diameter = 71,494\ km}{629 \times 10^6\ km}$$

angular diameter = 23.4 seconds of arc. Divide this by 3,600 seconds/ degree, the angular diameter of Jupiter is approximately 0.007 degrees or about 130 times smaller than the angular diameter of the sun as seen without a telescope. (p. 34; objective 1)

Lesson Review

Lesson 3: Celestial Cycles

PLEASE NOTE: Use this matrix to guide your study and achieve the learning objectives of this lesson. It will also help you to view the video, which defines and demonstrates important concepts and principles as they relate to everyday life and actual case studies.

Learning Objective	Textbook	Student Guide
1. Explain how the daily and annual motion of Earth causes what we see in the day and nighttime sky.	pp. 22–28	Key Terms: 1, 2, 11, 12; Matching: 4; Completion: 5; Short-Answer: 3; Self-Test: 3, 4; Applications: 1, 2.
2. Describe the tilt of Earth's axis with respect to the ecliptic and explain why this tilt causes the seasons.	pp. 25–27	Key Terms: 3,4,5,6,7,8,9,10; Matching: 3, 10, 12, 17, 21; Completion: 1, 4; Self-Test: 5, 6; Short-Answer: 6.
3. Describe the motion of the moon relative to Earth and the sun, explain how this causes us to see phases of the moon, and identify these phases by name.	pp. 29–33	Self-Test: 1, 2; Short-Answer: 1, 4.
4. Explain the necessary conditions for a lunar eclipse, describe what causes total and partial eclipses, and describe their appearance.	pp. 29, 32–33	Key Terms: 13, 14, 15; Matching: 5, 6, 7; Self-Test: 7, 8; Short-Answer: 2.
5. Explain the necessary conditions for a solar eclipse; describe what causes total, partial, and annular eclipses; and describe their appearance.	pp. 33–39	Key Terms: 16, 17, 18, 19, 20, 21, 22, 23, 24, 25, 26, 27, 28; Matching: 1, 2, 8, 9, 11, 13, 14, 15, 16, 18, 19; Completion: 2, 3; Self-Test: 9, 10; Short-Answer: 2.
6. Discuss what the Milankovitch hypothesis tells us about astronomical influences on Earth's climate.	pp. 39–41	Key Terms: 29; Matching: 20; Self-Test: 11, 12; Short-Answer: 5.

LESSON
4

The Birth of Astronomy

Checklist

For the most effective study of this lesson, complete the following activities in this sequence.

Before Viewing the Video

❑ Read the Preview, Learning Objectives, and Viewing Notes below.

❑ Read Chapter 4, "The Origin of Modern Astronomy," pages 44–71, in the *Horizons* textbook.

What to Watch

❑ After reading the textbook chapter, watch the video for Lesson 4, *The Birth of Astronomy.*

After Viewing the Video

❑ Briefly note your answers to questions listed at the end of the Viewing Notes.

❑ Review the Summary below.

❑ Review all reading assignments for this lesson, especially the Chapter 4 summary on page 70 in *Horizons* and the Viewing Notes in this lesson.

❑ Write brief answers to the review questions at the end of Chapter 4 in *Horizons.*

❑ Complete the Review Exercises below. Check your answers with the Answer Key and review when necessary.

❑ Use the Lesson Review matrix found at the end of this lesson to review and assess your knowledge of each Learning Objective.

❑ As assigned by your instructor, complete the Applications activities and any additional activities for this lesson.

☐ Book exams

Preview

This lesson introduces you to several key astronomers who studied the universe and developed models of planetary motion. Very early astronomers believed that the gods controlled the stars and the planets. China and Ancient Babylonia have written records of astronomical observations dating back to 3500 B.C. But, it wasn't until the ancient Greeks began to use observations, reasoning, and mathematics that scientific models of the universe were developed.

The ancient Greek astronomers believed that Earth was the center of the universe. Most also believed that the heavens were perfect and all celestial objects orbited Earth in perfect circles and with uniform speeds. Astronomers tried to use their observations to substantiate these principles, which made for some very interesting and complicated models of the universe.

Some later astronomers developed models of the universe that went against popular beliefs and the teachings of the church. These forward-thinkers proposed that Earth revolved around the sun; some were persecuted for their ideas. It wasn't until the invention of the telescope that astronomers were able to offer substantial evidence that Earth, and other known planets, orbited the sun.

Concepts to Remember

- Recall from Lesson 2 that the *celestial sphere* is an imaginary sphere of very large radius surrounding Earth to which the planets, stars, sun, and the moon seem to be attached. Astronomers use this model of the universe to better understand it (p. 17 in this guide).

- In Lesson 2, you also learned that a *scientific model* is a carefully devised mental conception of how something works. Such models help scientists think about how some aspect of nature, like how the sun creates its energy, or why solar activity occurs (p. 16 in this guide).

Learning Objectives

After you complete this lesson, you should be able to:

1. Describe the motions of the planets as seen from Earth and define the concept of retrograde motion. *HORIZONS* TEXTBOOK PAGE 48.

2. Describe the geocentric models of the solar system developed by Aristotle and Ptolemy and identify the assumptions on which these models were based. *HORIZONS* TEXTBOOK PAGES 46–49.

3. Explain the Copernican model and hypothesis, and compare the Copernican model with the Ptolemaic model. *HORIZONS* TEXTBOOK PAGES 47–52.

4. Discuss the importance of Tycho Brahe's observations. *HORIZONS* TEXTBOOK PAGES 52–54.

5. List and explain Kepler's three laws of planetary motion. *HORIZONS* TEXTBOOK PAGES 55–56.

6. Describe how the telescopic observations of the solar system made by Galileo provided evidence against the Ptolemaic model of the universe. *HORIZONS* TEXTBOOK PAGES 57–61.

7. Distinguish between a scientific paradigm, hypothesis, model, theory, and law. *HORIZONS* TEXTBOOK PAGES 52, 56.

8. List Newton's laws of motion and universal gravitation, and describe how they explain orbital motion. *HORIZONS* TEXTBOOK PAGES 61–65.

9. Describe how the moon and the sun produce tides on Earth and explain the effect of tidal forces on Earth's rotation and the orbital motion of the moon. *HORIZONS* TEXTBOOK PAGES 66–68.

At this point, read Chapter 4, "The Origin of Modern Astronomy," pages 44–71.

The passions of astronomy are no less profound because they are not noisy.
—John Steinbeck
The Short Reign
of Pippin IV
(1957)

Viewing Notes

The video program introduces you to some of the most influential thinkers in the history of astronomy. Although the early astronomers developed an inaccurate model of the universe, their observations allowed later astronomers to build a model based on observations rather than religious beliefs. The video program contains the following segments:

- ❂ Ancient Astronomy
- ❂ Aristotle & Ptolemy
- ❂ The Copernican Model
- ❂ Tycho Brahe
- ❂ Kepler
- ❂ The Telescope
- ❂ Newton

The following information will help you better understand the video program:

Plato's model of the universe included a *celestial sphere*. You learned about the celestial sphere in Lesson 2 (pp. 15–17 in the *Horizons* textbook). Most astronomers of this time believed that the universe contained a number of celestial spheres to which the planets and stars were attached. These spheres rotated around Earth in uniform motion. The model of the geocentric—or Earth-centered—universe caused astronomers to come up with elaborate models to explain what they observed.

It wasn't until Copernicus that a **heliocentric**—or sun-centered—model of the universe was developed.

Tycho Brahe is perhaps the best *naked-eye* astronomer in recorded history. A naked-eye astronomer is one who observes the planets and stars without special equipment like a telescope.

In 1572, when Brahe was 26 years old, a *supernova* occurred—an explosion of a star that increased its brightness tremendously. Brahe could not detect parallax in the position of the supernova and thus concluded that the event occurred far beyond the moon. This provided evidence that contradicted Aristotle's assertion that the heavens were perfect. This observation, along with Brahe's documented positions of the planets and stars, helped Kepler develop his three laws of planetary motion.

The video program continues to present and evaluate the contributions of later astronomers—Kepler, Galileo, and Newton—each of whom built upon the work of those that came before.

QUESTIONS TO CONSIDER

- How did the Ptolemaic model of the universe account for retrograde motion?

- How did the Copernican model of the universe account for retrograde motion?

- How did Tycho Brahe's observations discredit the first principle that stated the heavens were perfect?

- How did Galileo's observation of the moons that orbit Jupiter support the heliocentric model of the universe?

- How did Galileo's observation of the phases of Venus support the heliocentric model of the universe?

- How did Newton's calculations support the heliocentric model of the universe?

Watch the video for Lesson 4, *The Birth of Astronomy*.

Key Terms and Concepts

Page references are keyed to the *Horizons* textbook.

1. **heliocentric universe:** A model of the universe with the sun at the center, such as the Copernican universe. (p. 47; video lesson; objective 3)

2. **uniform circular motion:** The classical belief that the perfect heavens could only move by the combination of uniform motion along circular orbits. (p. 48; video lesson; objective 2)

3. **geocentric universe:** A model universe with Earth at the center, such as the Ptolemaic universe. (p. 48; video lesson; objective 2)

4. **parallax:** The apparent change in position of an object due to a change in the location of the observer. (p. 48; objective 2)

5. **retrograde motion:** The apparent backward (westward) motion of planets as seen against the background of stars. (p. 48; objective 1)

6. **epicycle:** The small circle followed by a planet in the Ptolemaic theory. The center of the epicycle follows a larger circle (the deferent) around earth. (p. 49; video lesson; objective 2)

7. **deferent:** In the Ptolemaic theory, the large circle around Earth along which the center of the epicycle was thought to move. (p. 49; objective 2)

8. **equant:** In the Ptolemaic theory, the point off center in the deferent from which the center of the epicycle appears to move uniformly. (p. 49; objective 2)

9. **paradigm:** A commonly accepted set of scientific ideas and assumptions. (p. 52; video lesson; objective 7)

10. **ellipse:** A closed curve around two points called the foci such that the total distance from one focus to the curve and back to the other focus remains constant. Shape of the orbit of a planet around the sun according to Kepler's first law. (p. 55; video lesson; objective 5)

11. **semimajor axis:** Half of the longest diameter of an ellipse. (p. 55; objective 5)

12. **eccentricity:** A number between 1 and 0 that describes the shape of an ellipse. The distance from one focus to the center of the ellipse divided by the semimajor axis. The eccentricity of a circle (a perfect ellipse) equals 0; as the ellipse becomes "more oval," the eccentricity approaches 1. (p. 55; objective 5)

13. **hypothesis:** A conjecture, subject to further tests, that accounts for a set of facts or observations. (p. 56; video lesson; objective 7)

14. **theory:** A system of assumptions and principles applicable to a wide range of phenomena that have been repeatedly verified. (p. 56; objective 7)

15. **natural law:** A theory that is almost universally accepted as true. (p. 56; objective 7)

16. **mass:** A measure of the amount of matter making up an object, as opposed to the weight of an object. Weight is a measure of the gravitational force acting on an object. (p. 62; objective 8)

17. **inverse square relation:** A rule that the strength of an effect, such as gravity, decreases in proportion as the distance squared increases. (p. 62; objective 8)

18. **circular velocity:** The velocity an object needs to stay in orbit around another object. (p. 64; objective 8)

19. **geosynchronous satellite:** A satellite that orbits eastward around Earth with a period of 24 hours and remains above the same spot on Earth's surface. (p. 64; objective 8)

20. **center of mass:** The balance point of a body or system of masses. The point about which a body or system of masses rotates in the absence of external forces. (p. 65; objective 8)

21. **closed orbit:** An orbit that returns to the same starting point over and over. Either a circular orbit or an elliptical orbit. (p. 65; objective 8)

22. **escape velocity:** The initial velocity an object needs to escape from the surface of a celestial body. (p. 65; objective 8)

23. **open orbit:** An orbit that carries an object away, never to return to its starting point. (p. 65; objective 8)

24. **spring tide:** Ocean tide of high amplitude that occurs at full and new moon. At this position, the moon is aligned with Earth and the sun so that the sun's gravity works with the moon's and amplifies the tides. (p. 68; objective 9)

25. **neap tide:** Ocean tide of low amplitude occurring at first- and third-quarter moon. At this position, the moon's and the sun's gravity work against each other, decreasing the size of the tides. (p. 68; objective 9)

Summary

Because classic astronomers were bound to the idea that Earth was the center of the universe and the heavens were perfect, they came up with very elaborate ideas to explain planetary motion.

One thing that ancient astronomers had difficulty explaining was **retrograde motion**—the apparent backward motion of planets as seen against the backdrop of the stars. The planets appeared to move eastward then stop in their orbits and move westward before continuing their eastward motion. Instead of changing their belief that Earth was the center of the universe, well-known astronomers developed intricate scientific models to describe what they observed.

Pre-Copernican Astronomy

Although the chapter in the *Horizons* textbook begins with Aristotle's study of the universe, both China and Ancient Babylonia have written records of astronomical observations dating back to 3500 B.C. The Babylonians were careful observers of the sky: they used complex tables to predict the motion of the moon and Venus. The ancient Greeks were able to use these tables as a foundation for their own study of the universe.

Many ancient astronomers reasoned from *first principles*—things that were obviously true. Plato (427?–347 B.C.) developed a first principle that stated that the heavens were perfect. Because the only perfect geometrical shape is the circle and the only perfect motion is uniform motion, Plato concluded that all motion in the heavens was made of a combination of circles turning at uniform rates.

Plato's student Aristotle (384–322 B.C.) reasoned from another first principle that stated Earth was the center of the universe. Because he observed no **parallax**—the apparent motion of an object because of the motion of the observer—he concluded that Earth did not move. He devised a model of the universe in which Earth was imperfect and lay at the center. His model contained 55 spheres to which the seven known planets were attached. At the time, the known "planets" included Mercury, Venus, Mars, Jupiter, and Saturn, as well as the sun and Earth's moon.

Ptolemy—a follower of Aristotle's ideas—constructed an elaborate model based on mathematics to describe how other planets and the sun revolved around Earth. In his model, each planet followed small circles—called **epicycles**—which slid around a **deferent**, a larger circle that surrounded Earth. By adjusting the size and rate of rotation of the circles, he could approximate the retrograde motion of a planet.

Nicolaus Copernicus (1473–1543)

For nearly 2,000 years, astronomers were firm in their beliefs that Earth was the center of the universe and the heavens were perfect. But, in 1530, Copernicus developed a model of the universe in which the sun was the center. Although his **heliocentric universe** was highly controversial, it explained the retrograde motion of the planets in a simple and straightforward way. In his model, Earth moves faster along its orbit and passes the planets that orbit farther from the sun. Although Copernicus' model was able to explain retrograde motion, it still could not accurately predict the positions of the planets any better than Ptolemy's model. Therefore, it wasn't widely accepted.

Tycho Brahe (1546–1601)

Tycho Brahe rejected both the Copernican and the Ptolemaic models because of their inaccurate predictions relating to planetary motion. He devised his own model in which the sun and moon orbited the stationary Earth and the other planets circled the sun.

Brahe's most significant contribution to astronomy was the development of instruments to aid in observation. The telescope wasn't invented yet, but he developed a quadrant that was used to measure the position of the stars with precision. His contribution would allow Kepler to understand the orbits of the planets.

Johannes Kepler (1571–1630)

Kepler believed in Copernican's heliocentric universe. He used Brahe's data to analyze the motion of the planets and developed three fundamental laws of planetary motion. His first law states that each planet orbits the sun in an **ellipse** with the sun at one of the foci. An ellipse is a closed curve, like an oval, around two points called the foci. This was revolutionary because for thousands of years, it was accepted that the planets had perfectly circular orbits.

Kepler's second law states that a line from the planet to the sun sweeps over equal areas in equal intervals of time. In other words, when a planet is closer to the sun, the imaginary line from the planet to the sun is shorter than when it's farther from the sun. Therefore, the planet has to speed up to sweep over the same amount of area of the ellipse.

Kepler's third law states that a planet's orbital period squared is proportional to its average distance from the sun cubed. In other words, the farther away a planet is from the sun, the longer it takes to make its orbit around the sun. It not only has a greater distance to travel but also moves slower than if it were closer.

Galileo Galilei (1564–1642)

The telescope was invented during Kepler's lifetime, and his contemporary, Galileo, was the first person to use it to systematically observe the sky. Galileo's observations helped finally to discredit the Ptolemaic **geocentric**, or Earth-centered, model of the universe.

By far, Galileo's most important discovery was that Jupiter had moons. This helped support the Copernican model in which the planets orbited the sun and moons orbited the planets. He was able to measure the orbital periods of Jupiter's moons and found that the innermost moon had the shortest period and the moons farther from Jupiter had proportionally longer periods. He saw similarities between the orbital motion of Jupiter's moons and the orbital motion in the heliocentric model. Critics of the heliocentric model said that if Earth moved, it would lose its moon because the people of the time didn't yet have an understanding of gravity. But, since Galileo observed that Jupiter moved and didn't lose its moons, he reasoned that Earth could also move and keeps its moon in orbit.

Galileo also discovered that the moon had mountains and valleys, which contradicted the assumption that the heavens were perfect. He also discovered there were stars in the universe too faint to see with the naked eye. When he observed the sun, he noticed sunspots, which indicated that the sun also wasn't perfect. The movement of the spots indicated that the sun was a sphere that rotated on its axis.

When he observed Venus, he noticed that it went through a complete set of phases and compared the phases of Venus to those of Earth's moon. If both Venus and the moon orbited Earth, then their phases would be similar. But they were vastly different. This observation also supported the heliocentric model of the universe.

Galileo's observations and published works created a storm of controversy. Galileo's contribution to science was important because it asserted that modern science—rather than religious faith—could provide rational ways to understand the universe. Other scientists used their own observations to try to understand the universe and when their observations contradicted religious Scripture, the scientists assumed that their own observations of reality were correct.

Isaac Newton (1642–1727)

Newton provided the final pieces to solve the puzzle of planetary motion. His laws of motion and gravity contributed to the study of astronomy and helped explain how planets and moons stayed in their orbits.

Newton's first law states that a body at rest remains at rest or a body in motion continues in a straight line with uniform speed unless acted upon by some unbalanced force. This, in conjunction with his law of universal gravitation, helped explain why the moon stayed in its orbit around Earth: The force of gravity between the moon and Earth holds the moon in orbit. If the force of gravity didn't exist, the moon would keep moving in a straight line and would leave Earth.

His second law states that a body's change of motion is proportional to the force acting on it and the change of motion is in the direction of the force. This was the first clear statement in the scientific community of the principle of cause and effect.

Newton's third law of motion states when one body exerts a force on a second body, the second body exerts an equal and opposite force back on the first body.

Many people are familiar with the tale of Newton sitting beneath an apple tree. When considering why apples fell from a tree toward Earth, he discovered gravity. He realized that gravitation is a force of mutual attraction and that the distance between objects was important to the amount of force exerted between them. As two objects come closer together, the force of gravity increases. For example, if the distance between two objects is halved, the force between them is four times greater.

His laws of gravity help us understand how the moon affects the tides on Earth. The gravitational forces between Earth, the sun, and the moon cause tides. According to Newton's laws, the gravitational force between two bodies increases when the distance between the two bodies decreases. Therefore, the gravitational force between Earth and the moon is a bit stronger on the side of Earth that is closest to the moon and the force is less on the side of Earth that is farthest away from the moon. The force pulls on the oceans and causes the water to flow into a bulge on the side of Earth closest to the moon. There is also a bulge on the side of Earth farthest away from the moon since the moon pulls more strongly on the center of Earth than on the far side of it. As Earth rotates on its axis, the tidal bulges remain fixed with respect to the moon and the landmass of Earth rotates into the tidal bulges. Since there are two tidal bulges on opposite sides of Earth, the tides rise and fall twice a day on most coasts.

The sun also creates tides on Earth, but since the sun is farther away, the tides caused by the sun are less than half those caused by the moon. During the new moon and the full moon, the sun and the moon create increased tidal bulges that produce dramatic tidal changes, called spring tides. **Spring tides** actually occur with every new and full moon—not only in the spring. At these times, high tides are higher and low tides lower. Conversely, during the first- and third-quarter moons, the sun and moon's gravitational pulls are working at right angles to one another on Earth and make their effects less dramatic. At these times, the tides caused by the sun reduce the tides caused by the moon and they are less extreme than usual; these are called **neap tides**.

As you have learned in this lesson, Copernicus, Kepler, Newton, and others revolutionized science and the understanding of our universe. For thousands of years, astronomers believed that the sun and all celestial objects revolved around Earth. Instead of reasoning from "first principles" as the ancient astronomers had, these scientists made careful observations and constructed models of the universe that supported their observations. This method of scientific thinking made huge contributions not only to astronomy but to all of the sciences.

Review Exercises

Matching I

Match each term with the appropriate definition or description.

1. _____ parallax	9. _____ hypothesis
2. _____ geocentric universe	10. _____ theory
3. _____ uniform circular motion	11. _____ natural law
4. _____ retrograde motion	12. _____ ellipse
5. _____ epicycle	13. _____ paradigm
6. _____ deferent	14. _____ spring tide
7. _____ equant	15. _____ neap tide
8. _____ heliocentric universe	

a. A model of the universe with the sun at the center, such as the Copernican universe.

b. In the Ptolemaic theory, a small circle in which a planet moves.

c. A model of the universe with Earth at the center.

d. A closed curve around two points called foci, such that the total distance from one focus to the curve and back to the other focus remains constant.

e. The apparent change in the observed position of an object due to a change in the location of the observer.

f. A conjecture, subject to testing, to explain certain facts and observations.

g. The apparent backward (westward) motion of planets as seen against the background stars.

h. The belief that objects in the sky move uniformly along perfectly circular orbits.

i. A theory that is almost universally accepted as true.

j. In the Ptolemaic theory, the point off center in the deferent from which the center of the epicycle appears to move uniformly.

k. A network of assumptions and principles providing a framework for formulating theories.

l. A system of assumptions and principles applicable to a wide range of phenomena that have been repeatedly verified.

m. In the Ptolemaic theory, the large circle around Earth along which the center of the epicycle moved.

n. Ocean tide of high amplitude that occurs at full and new moon.

o. Ocean tide of low amplitude occurring at first- and third-quarter moon.

Matching II

Match each term with the appropriate definition or description.

1. _____ center of mass	5. _____ geosynchronous satellite
2. _____ circular velocity	6. _____ inverse square relation
3. _____ closed orbit	7. _____ mass
4. _____ escape velocity	8. _____ open orbit

a. A measure of the amount of matter making up an object, as opposed to the weight of an object.

b. A rule that the strength of an effect, such as gravity, decreases in proportion as the distance squared increases.

c. The velocity an object needs to stay in orbit around another object.

d. A satellite that orbits eastward around Earth with a period of 24 hours and remains above the same spot on Earth's surface.

e. The balance point of a body or system of masses.

f. An orbit that returns to the same starting point over and over.

g. An orbit that carries an object away, never to return to its starting point.

h. The initial velocity an object needs to escape from the surface of a celestial body.

Completion

Fill each blank in the sentences below with the most appropriate term from the list of completion answers that follow. A term may be used once, more than once, or not at all. Check your answers with the Answer Key and review when necessary.

Tycho Brahe	geocentric	retrograde
Earth	geosynchronous	rip tides
epicycles	heliocentric	sun
Galileo	neap tides	uniform circular

1. Ptolemy developed what is commonly called the
_____ or _____-
centered system.

2. Johannes Kepler used the observational data of _____ to devise three laws that described the motions of the planets.

3. The first observations of objects in the solar system that did not orbit the sun or Earth were made by _____.

4. Low ocean tides that occur at first- and third-quarter moons are called _____.

5. A _____ satellite orbits eastward around Earth with a period of 24 hours and remains above the same location on the surface of Earth.

Self-Test

Select the best answer.

1. The usual motion of the planets against the background stars is
   a. east to west.
   b. north to south.
   c. west to east.
   d. south to west.

2. When the planet Mars is observed to undergo retrograde motion, Earth
   a. remains fixed in its orbit.
   b. moves faster in its orbit and passes Mars.
   c. is on the far side of the sun.
   d. slows down in its orbit around the sun.

3. Ptolemy assumed that the planets moved through the perfect heavens in perfect motion. This perfect motion is called
   a. uniform accelerated motion.
   b. motion without parallax.
   c. retrograde motion.
   d. uniform circular motion.

4. In order to account for some of the observed peculiar motions of the planets, Ptolemy's model required that the planets move in
   a. small circles called deferents.
   b. straight lines at constant velocity.
   c. small circles called epicycles.
   d. the equant.

5. In *De Revolutionibus*, Copernicus proposed that the sun was located at
   a. the center of the universe.
   b. one of the foci of an ellipse.
   c. directly above Earth.
   d. the equant.

6. The movements of the planets in the heliocentric system of Copernicus required that the planets move
   a. in elliptical orbits.
   b. on epicycles.
   c. on deferents.
   d. in circular orbits.

7. The success of Tycho Brahe is attributed to his ability to
   a. make accurate observations with large and well-constructed instruments.
   b. construct theoretical models of physical phenomena.
   c. receive financial support from the king for his astronomical endeavors.
   d. make excellent observations with a telescope.

8. According to the Tychonic model of the solar system,
   a. Earth was the center of the universe and the planets orbited Earth.
   b. Earth was stationary, the sun and moon orbited Earth, and the planets orbited the sun.
   c. the sun was the center of the universe and the planets orbited the sun in elliptical paths.
   d. the sun was the center of the universe and the planets orbited the sun in circular orbits.

9. A planet moves fastest in its orbit when it is
   a. closest to Earth.
   b. farthest from the sun.
   c. at one of the foci.
   d. closest to the sun.

10. According to Kepler's first law of planetary motion, the orbits of the planets are
    a. circles, with Earth located at the center.
    b. ellipses, with the sun located at one of the foci.
    c. ellipses, with Earth located at one of the foci.
    d. closest to the sun.

11. Galileo's most convincing observation that proved Venus orbited the sun was the discovery of
    a. moons in orbit around Jupiter.
    b. the highly reflective atmosphere of Venus.
    c. the phases of Venus.
    d. evidence for the rotation of the planet.

12. Galileo observed and recorded four moons orbiting Jupiter. This evidence showed that
    a. Jupiter was not a planet.
    b. Earth was not the center of all orbital motion in the solar system.
    c. Jupiter's moons were like Earth's moon.
    d. Jupiter orbited the sun.

13. A single conjecture or assertion expressed in a way such that it can be tested best describes a
    a. law.
    b. theory.
    c. hypothesis.
    d. paradigm.

14. A natural law can best be described as a
    a. conjecture subject to further tests.
    b. representation of what actually exists in nature.
    c. correct physical representation of a conceptual idea.
    d. theory that is almost universally accepted as true.

15. If an object is moving and it remains undisturbed by outside forces, it will
    a. stop.
    b. continue to move in a circular path.
    c. continue to move in a straight line with a uniform velocity.
    d. stop and begin to move in the opposite direction.

16. According to Newton's law of universal gravitation, if the distance between two objects is doubled, the force of gravity between them will
    a. double.
    b. be cut in half.
    c. be one fourth as great.
    d. remain unchanged.

17. The bulge in Earth's oceans that occurs on the side of Earth nearest the moon results from
    a. the moon's gravitational influence being greater on the side of Earth nearest the moon than it is on Earth's center.
    b. the moon's gravitational influence being weaker on the side of Earth nearest the moon than it is on Earth's center.
    c. Earth's gravitational influence on the oceans being weaker than the moon's.
    d. Earth's rotation, which creates centrifugal forces.

18. The extreme tides of the month that occur when the sun's and moon's gravitational influence operate together and increase the net force of gravity on Earth's oceans are referred to as
   a. neap tides.
   b. spring tides.
   c. lunar tides.
   d. solar tides.

Short-Answer Questions

1. Retrograde motion is usually described in terms of the motion of the planet Mars. Describe retrograde motion for the planet Venus or Mercury.

2. What were the limitations of the Copernican model of the solar system?

3. Why did the phases of Venus help to contribute to the verification of the heliocentric model of the solar system?

4. What does Newton's first law of motion tell us about moving objects and how that motion can be changed?

5. The tides are produced by the gravitational influence of both the sun and the moon. Why then does the moon exert the greater influence?

6. Briefly describe the retrograde motion of Mars as it is observed in the sky, as explained by the geocentric model of Ptolemy, and as explained by the heliocentric model of Copernicus.

Applications

1. The semimajor axis of the orbit of Mars is 1.88 AU. Find its period of revolution about the sun.

2. The space shuttle orbits Earth at an altitude of 600 km or 6.00×10^5 m above Earth's surface. This makes the radius of the orbit 6.90×10^6 m. Use the equation for orbital velocity to calculate the space shuttle's velocity around Earth.

3. Use the answer for Question 2 above and determine the period of revolution of the space shuttle around Earth.

4. What is the difference between a model and a hypothesis? Using this difference, read the paragraph on page 52 of the *Horizons* textbook (under Building Scientific Arguments) and explain why the model failed but the hypothesis was correct.

5. How did the craters on the moon and the phases of Venus argue against the Ptolemaic model of the solar system? (See Building Scientific Arguments on pages 60–61 of the *Horizons* textbook.)

Answer Key

Matching I

1. e (p. 48; objective 2)

2. c (p. 48; video lesson; objective 2)

3. h (p. 48; video lesson; objective 2)

4. g (p. 48; objective 1)

5. b (p. 49; video lesson; objective 2)

6. m (p. 49; objective 2)

7. j (p. 49; objective 2)

8. a (p. 47; video lesson; objective 3)

9. f (p. 56; video lesson; objective 7)

10. l (p. 56; objective 7)

11. i (p. 56; objective 7)

12. d (p. 55; video lesson; objective 5)

13. k (p. 52; video lesson; objective 7)

14. n (p. 68; objective 9)

15. o (p. 68; objective 9)

Matching II

1. e (p. 65; objective 8)

2. c (p. 64; objective 8)

3. f (p. 65; objective 8)

4. h (p. 65; objective 8)

5. d (p. 64; objective 8)

6. b (p. 62; objective 8)

7. a (p. 62; objective 8)

8. g (p. 65; objective 8)

Completion

1. geocentric, Earth (pp. 46–48; objective 2)

2. Tycho Brahe (pp. 53–55; objective 4)

3. Galileo (pp. 58–59; objective 6)

4. neap tides (p. 68; objective 9)

5. geosynchronous (p. 64; objective 8)

Self-Test

1. c (p. 48; objective 1)

2. b (pp. 48, 50; objective 1). Mars appears to drift backward.

3. d (p. 48; video lesson; objective 2). The circle was the perfect geometrical figure.

4. c (p. 49; video lesson; objective 2). The velocity around the epicycles was constant.

5. a (pp. 47, 50–51; video lesson; objective 3).

6. d (pp. 50–51; video lesson; objective 3) Copernicus still believed in the classics. The perfect geometrical figure was the circle.

7. a (pp. 52–53; video lesson; objective 4). Although Tycho made excellent observations, he never used a telescope.

8. b (p. 53; video lesson; objective 4). Tycho was hoping to create a compromise between the geocentric and heliocentric models.

9. d (pp. 55–56; video lesson; objective 5). Kepler's second law.

10. b (p. 55; video lesson; objective 5). Kepler's first law and first to diverge from the perfect geometrical figure of the circle.

11. c (p. 59; video lesson; objective 6). Not only were the phases observed, but their location relative to the sun and the size of the image demonstrated that Venus orbited the sun and not Earth.

12. b (pp. 58–59; video lesson; objective 6). Jupiter did not lose its moons as it moved through space. This led to the belief that Earth wouldn't lose its moon.

13. c (p. 56; objective 7). Theory and law are far more general and can be applied to many more phenomena

14. d (p. 56; objective 7). A paradigm is a set of ideas.

15. c (p. 62; video lesson; objective 8). Newton's first law of motion.

16. c (p. 62; objective 8). The force of gravity is inversely proportional to the square of the distance of separation of the two objects.

17. a (p. 66; objective 9). Because of the inverse square law of gravity, the moon's gravitational influence on the side of Earth nearest the moon is greater and therefore creates the bulge.

18. b (p. 68; objective 9)

Short-Answer Questions

1. Venus and Mercury orbiting closer to the sun than Earth move faster in their orbits. Therefore they complete a revolution around the sun in a shorter period of time. As a result, Venus and Mercury can be observed to "pass" Earth by. Unfortunately, the two planets remain close to the sun in the sky and are only observed in the early evening or early morning. As they emerge from behind the sun, they appear to move easterly in the sky. Their smaller orbits cause them to begin to move westerly as they eventually move closer to the sun in the sky. This is the apparent change in direction as seen from Earth. (pp. 48–50; objective 2)

2. The Copernican model required that the orbits of the planets were circular and that the planets moved with constant velocities. In other words, uniform circular motion. Ptolemy's model also required this constraint. As a result, the ability to predict the locations of the planets was limited. (pp. 48–50; objective 2)

3. The phases of Venus provide the evidence that the planet orbits the sun. Even though the moon goes through phases orbiting Earth, the size of

the moon as it appears in the sky remains constant. Galileo not only saw that Venus went through all of the phases like those of the moon, he also saw that the crescent phase of Venus was larger (closer means larger) than the gibbous phase. This indicated that the planet had to be closer to Earth during this phase than during the gibbous phase. In the Ptolemaic system, Venus orbited around an imaginary point located between Earth and the sun, which would result in an incomplete phase cycle with the phase never appearing more than half full. Since Galileo's observations showed that Venus had a complete phase cycle with the smallest angular diameter at full phase, Venus had to orbit the sun. (p. 59; objective 6)

4. Newton's first law of motion tells us that motion is a natural state. In other words, objects do not need extra forces to keep them moving. This is somewhat contradictory to what we experience. Give an object a push and it eventually slows down. However, that is exactly what Newton's first law predicts. If there are unbalanced forces acting on an object, such as friction from a surface, the object's motion will change. Think about what happens on an icy road. Objects cannot stop unless there is sufficient friction or an unbalanced force to stop them. Furthermore, objects cannot get started (they remain at rest) if the road is too icy. Wheels spin but no forward motion. To change the state of motion of an object, there must exist an unbalanced force acting on it. (p. 62; objective 8)

5. The tides are the result of the difference in gravitational force exerted by the sun and moon on the near and far sides of Earth. Since the moon is closer to Earth than the sun, the difference in gravitational force produced by the moon is larger and therefore has a greater effect on Earth's oceans than the sun. (pp. 66–68; objective 9)

6. The motion of Mars against the background stars is mostly from west to east. However, when retrograde motion begins, Mars appears to stop relative to the background stars and move east to west. According to the geocentric model of Ptolemy, Mars is moving at a constant speed on an epicycle as the epicycle moves with constant speed on the deferent around Earth. As Mars moves on the portion of the epicycle nearest Earth, it appears to move westerly against the background stars. This is retrograde motion. (See figure on page 48 of *Horizons*.) According to the heliocentric model of Copernicus, Earth is moving at a constant speed and faster than Mars. As they orbit the sun in circular paths, Earth catches up to and passes the slower moving Mars. This causes Mars to appear to drift backward (westerly) against the "fixed" stars. (See Figure 4-2 on p. 50 of *Horizons*.) Again, the westerly motion is called retrograde. The same thing occurs if you pass a slower moving vehicle on the highway. As you catch up to and pass this slower moving vehicle, it appears to move backward relative to your

vehicle and the road, even though you know both vehicles continue to travel in the same direction. (pp. 48–50; objective 1)

Applications

1. The answer is 2.58 years. (p. 56; objective 5)

2. The answer is 7,603 meters per second. (p. 63; objective 8)

3. The answer is 5,702 seconds or 1.58 hours.

   The period of the orbit = the circumference of the orbit ($2\pi r$) divided by the velocity. (p. 63; objective 8)

4. A hypothesis is a statement or conjecture regarding a phenomenon. An example would be the statement that the sun is the center of the solar system. This hypothesis, if it is to be considered scientific, must be tested. A model is a description and mental construct that can be used to test the hypothesis. The model of Copernicus was to put the sun at the "center" of the solar system and the planets in "circular" orbits moving with "constant" or "uniform" velocity. The idea or hypothesis, the planets orbiting the sun with the sun at the center is correct; however, the observations used to test the model showed that the model did not predict the positions of the planets any better than the geocentric model. The model of Copernicus was inadequate. A better model was needed. (pp. 52, 56; objectives 3 & 7)

5. The Ptolemaic model of the solar system described the planets orbiting Earth and held that the heavenly bodies were perfect. Galileo's observation of Venus showed that the planet orbited the sun. The craters on the moon showed that the heavenly bodies were not perfect. (pp. 49, 58–59; objectives 2 & 6)

Lesson Review

Lesson 4: The Birth of Astronomy

PLEASE NOTE: Use this matrix to guide your study and achieve the learning objectives of this lesson. It will also help you to view the video, which defines and demonstrates important concepts and principles as they relate to everyday life and actual case studies.

Learning Objective	Textbook	Student Guide
1. Describe the motions of the planets as seen from Earth and define the concept of retrograde motion.	p. 48	Key Terms: 5; Self-Test: 1, 2; Short-Answer: 6.
2. Describe the geocentric models of the solar system developed by Aristotle and Ptolemy and identify the assumptions on which these models were based.	pp. 46–49	Key Terms: 2, 3, 4, 6, 7, 8; Matching I: 1, 2, 3, 5, 6, 7; Completion: 1; Self-Test: 3, 4; Short-Answer: 1, 2; Applications: 5.
3. Explain the Copernican model and hypothesis, and compare the Copernican model with the Ptolemaic model.	pp. 47–52	Key Terms: 1; Matching I: 4, 8; Self-Test: 5, 6; Applications: 4.
4. Discuss the importance of Tycho Brahe's observations.	pp. 52–54	Completion: 2; Self-Test: 7, 8.
5. List and explain Kepler's three laws of planetary motion.	pp. 55–56	Key Terms: 10, 11, 12; Matching I: 13; Self-Test: 9, 10; Applications: 1.
6. Describe how the telescopic observations of the solar system made by Galileo provided evidence against the Ptolemaic model of the universe.	pp. 57–61	Completion: 3; Self-Test: 12; Short-Answer: 3; Applications: 5.

Learning Objective	Textbook	Student Guide
7. Distinguish between a scientific paradigm, hypothesis, model, theory, and law.	pp. 52, 56	Key Terms: 9, 13, 14, 15, 16; Matching I: 9, 10, 11, 13; Self-Test: 13, 14; Applications: 4.
8. List Newton's laws of motion and universal gravitation, and describe how they explain orbital motion.	pp. 61–65	Key Terms: 17, 18, 19, 20, 21, 22, 23, 24; Matching II: 1, 2, 3, 4, 5, 6, 7, 8; Completion: 5; Self-Test: 15, 16; Short-Answer: 4; Applications: 2, 3.
9. Describe how the moon and the sun produce tides on Earth and explain the effect of tidal forces on Earth's rotation and the orbital motion of the moon.	pp. 66–68	Key Terms: 25, 26; Matching I: 14, 15; Completion: 4; Self-Test: 17, 18; Short-Answer: 5.

Notes:

LESSON
5

Astronomical Tools

Checklist

For the most effective study of this lesson, complete the following activities in this sequence.

Before Viewing the Video

- ❑ Read the Preview, Learning Objectives, and Viewing Notes below.
- ❑ Read Chapter 5, "Astronomical Telescopes," pages 72–97, in the *Horizons* textbook.

What to Watch

- ❑ After reading the textbook chapter, watch the video for Lesson 5, *Astronomical Tools.*

After Viewing the Video

- ❑ Briefly note your answers to questions listed at the end of the Viewing Notes.
- ❑ Review the Summary below.
- ❑ Review all reading assignments for this lesson, especially the Chapter 5 summary on page 96 in *Horizons* and the Viewing Notes in this lesson.
- ❑ Write brief answers to the review questions at the end of Chapter 5 in *Horizons.*
- ❑ Complete the Review Exercises below. Check your answers with the Answer Key and review when necessary.
- ❑ Use the Lesson Review matrix found at the end of this lesson to review and assess your knowledge of each Learning Objective.
- ❑ As assigned by your instructor, complete the Applications activities and any additional activities for this lesson.

❑ Book exams

Preview

In previous lessons you discovered things about the stars and planets—such as the phases of the moon, eclipses, and seasons—that can be observed without sophisticated equipment. You also learned how ancient astronomers created models of the universe that helped explain what they saw. Ancient astronomers observed the universe using *naked-eye astronomy*—studying the sky without the aid of special equipment. As you learned in Lesson 4, the turning point in astronomy was Galileo's use of the telescope. Astronomy has come a long way since Galileo turned his telescope toward the heavens. Today, astronomers use much more sophisticated equipment to observe the universe.

Modern astronomy's quest is to collect more light and more data about the universe. Every technological advance in astronomy has been made to further this quest and has provided astronomers with more information to analyze. Astronomers can study visible light coming from stars using optical telescopes far more powerful than Galileo's. But visible light is only one kind of energy that is emitted by celestial objects. In addition to collecting and analyzing visible light, astronomers can use specialized telescopes and instruments to detect infrared and ultraviolet light, radio waves, X rays, and gamma rays. You will learn about these types of energy in this lesson.

In this lesson, you will examine the tools that modern astronomers use to gather not only visible light that is emitted by celestial objects, but other forms of energy as well. In this and future lessons you'll discover how these tools can help us understand much more about the universe than ancient astronomers ever dreamed of.

Concepts to Remember

- Recall from Lesson 4 that *naked-eye astronomy* is the observation of the planets and stars without the use of special equipment, such as telescopes (p. 54 in this guide).

Learning Objectives

After you complete this lesson, you should be able to:

1. Describe the nature of electromagnetic radiation, explain its characteristics at different wavelengths, and discuss how this influences the way astronomers make observations. *HORIZONS* TEXTBOOK PAGES 74–76.

2. Compare and contrast the different optical telescope designs, including their relative advantages and disadvantages. *HORIZONS* TEXTBOOK PAGES 76–78 AND 82–83.

3. Discuss how astronomers use technology to enhance the performance of optical telescopes. *HORIZONS* TEXTBOOK PAGES 81–86.

4. Describe the factors that determine the light-gathering, magnifying, and resolving powers of a telescope, noting the effects of Earth's atmosphere. *HORIZONS* TEXTBOOK PAGES 78–80.

5. Describe the instrumentation used for photography and spectroscopy with a telescope. *HORIZONS* TEXTBOOK PAGES 86–88.

6. Explain the operation, advantages, and limitations of radio telescopes, including the method of radio interferometry. *HORIZONS* TEXTBOOK PAGES 88–91.

7. Describe the advantages of placing telescopes in space and the types of observations they are designed to make. *HORIZONS* TEXTBOOK PAGES 91–95.

At this point, read Chapter 5, "Astronomical Telescopes," pages 72–97.

The first telescope was not so good because I didn't know that much about it, but I learned rapidly. You learn from mistakes.
—Clyde Tombaugh
U.S. astronomer
(1906–1997)

Viewing Notes

In the video program for this lesson, you will have the opportunity to visit some of the greatest observatories in the world. You will see how modern astronomers use powerful instruments to better understand the universe.

The video program contains the following segments:

- Radiation
- Optical Telescopes
- Detectors & Spectrographs
- Infrared
- Space-Based Observatories
- Radio Telescopes

The following information will help you better understand the video program:

In the opening segment of the video program, you'll learn that visible light is a form of **electromagnetic radiation** and that it can behave both as a particle and a wave. The discipline to study this phenomenon is called quantum mechanics—the study of matter and radiation at an atomic level. You will learn more about quantum mechanics in Lesson 6.

Electromagnetic radiation is able to travel though a *vacuum*—a space devoid of matter—and can reach the uppermost limit of Earth's atmosphere. Energy within certain ranges of the electromagnetic spectrum is able to reach Earth's surface passing through regions called **atmospheric windows**. It's important to note that these "windows" don't refer to geographic regions—they refer to wavelength regions in the spectrum to which Earth's atmosphere is transparent. Only electromagnetic radiation of certain wavelengths can penetrate Earth's atmosphere. Ozone absorbs far-ultraviolet waves and water vapor absorbs far-infrared waves, while gamma rays and X rays are absorbed in the upper layers of the atmosphere.

In the video program, you'll see several different kinds of telescopes. You'll recall from the textbook that a **reflecting telescope**—or reflector—uses a concave mirror to reflect the incoming light. A refracting telescope—or refractor—uses a lens that bends the incoming light. The **refracting telescope** suffers from a variety of limitations such as chromatic aberration (the edges of the lens act as prisms producing a rainbow around the image) and size (the lens can only be so large before it begins to sag under its own weight).

The **spectrograph** is a tool that astronomers can use to study a star or planet's spectrum to learn more about it. In the video program, you will learn about single-slit and multi-slit spectrographs. A single-slit spectrograph allows astronomers to view the spectra of only a few stars, planets, or galaxies at a time. A multi-slit spectrograph allows astronomers to view as many as 50 or 60 galaxies in a single observation.

QUESTIONS TO CONSIDER

- How do scientists explain how light acts both like a particle and a wave?
- What are the advantages and disadvantages of making optical telescopes with large lenses?
- What types of technological advances have enhanced the performance of optical telescopes?
- Why is a spectrograph an important tool for astronomers?
- What is the purpose of placing telescopes on mountaintops and remote locations?
- What are the advantages of radio telescopes over optical telescopes?

Watch the video for Lesson 5, *Astronomical Tools.*

Key Terms and Concepts

Page references are keyed to the *Horizons* textbook.

1. **electromagnetic radiation:** Changing electric and magnetic fields that travel through space at the speed of light and transfer energy from

one place to another; examples are light and radio waves. (p. 74; video lesson; objective 1)

2. **wavelength:** The distance between successive peaks or troughs of a wave; usually represented by λ. (p. 74; video lesson; objective 1)

3. **photon:** A quantum of electromagnetic energy; carries an amount of energy that depends inversely on its wavelength. (p. 74; video lesson; objective 1)

4. **nanometer (nm):** A unit of distance equaling one-billionth of a meter (10^{-9}); commonly used to measure the wavelength of light. (p. 75; objective 1)

5. **angstrom (Å):** A unit of distance ($1 \text{ Å} = 10^{-10}$ m); commonly used to measure the wavelength of light. (p. 75; objective 1)

6. **atmospheric window:** Wavelength region in which our atmosphere is transparent—at visual, infrared, and radio wavelengths. (p. 76; video lesson; objective 4)

7. **focal length:** The focal length of a lens is the distance from the lens to the point where it focuses parallel rays (light coming from very great distances) of light. (p. 76; objective 2)

8. **primary lens or mirror:** The largest lens or mirror in a telescope. (p. 76; video lesson; objective 2)

9. **objective lens or mirror:** In a refracting telescope, the long-focal-length lens that forms an image of the object viewed; the lens closest to the object. In a reflecting telescope, the principal mirror (reflecting surface) that forms an image of the object viewed. (p. 76; objective 2)

10. **eyepiece:** A short-focal-length lens used to enlarge the image in a telescope; the lens nearest the eye. (p. 77; objective 2)

11. **refracting telescope:** A telescope that forms images by bending (refracting) light with a lens. (p. 76; objective 2)

12. **reflecting telescope:** A telescope that uses a concave mirror to focus light into an image. (p. 76; objective 2)

13. **chromatic aberration:** A distortion found in refracting telescopes because lenses focus different colors at slightly different distances. Consequently, images are surrounded by color fringes. (p. 77; objectives 2 & 4)

14. **achromatic lens:** A telescope lens composed of two lenses ground from different kinds of glass and designed to bring two selected colors to the

same focus and correct for chromatic aberration. (p. 77; objectives 2 & 4)

15. **light-gathering power:** The ability of a telescope to collect light. Proportional to the area of the telescope's objective lens or mirror. (p. 78; objective 4)

16. **resolving power:** The ability of a telescope to reveal fine detail. Depends on the diameter of the telescope objective. (p. 78; objectives 2 & 4)

17. **diffraction fringe:** Blurred fringe surrounding any image, caused by the wave properties of light. Because of this, no image detail smaller than the fringe can be seen. (p. 78; objective 4)

18. **seeing:** Atmospheric conditions on a given night. When the atmosphere is unsteady and produces blurred images, the seeing is said to be poor. (p. 78; video lesson; objective 4)

19. **magnifying power:** The ability of a telescope to make an image larger. (pp. 79–80; objective 4)

20. **light pollution:** The illumination of the night sky by wasted light from cities and outdoor lighting, which prevents the observation of faint objects. (p. 80; objective 4)

21. **prime focus:** The point at which the objective mirror forms an image in a reflecting telescope. (p. 82; objective 2)

22. **secondary mirror:** In a reflecting telescope, a mirror that directs the light from the primary mirror to a focal position. (p. 82; video lesson; objective 2)

23. **Cassegrain focus:** The optical design in which the secondary mirror reflects light back down the tube through a hole in the center of the objective mirror. (p. 82; objective 2)

24. **Newtonian focus:** The optical design in which a diagonal mirror reflects light out the side of the telescope tube for easier access. (p. 82; objective 2)

25. **Schmidt-Cassegrain focus:** The optical design that uses a thin corrector plate at the entrance to the telescope tube. A popular design for small telescopes that allows for wide-angle viewing. (p. 82; objective 2)

26. **sidereal drive:** The motor and gears on a telescope that turn it westward to keep it pointed at a star. (p. 83; objective 2)

27. **equatorial mounting:** A telescope mounting that allows motion parallel to and perpendicular to the celestial equator. (p. 83; objective 2)

28. **polar axis:** In an equatorial telescope mounting, the axis that is parallel to Earth's axis. (p. 83; objective 2)

29. **alt-azimuth mounting:** A telescope mounting that allows the telescope to move in altitude (perpendicular to the horizon) and in azimuth (parallel to the horizon). (p. 83; objective 2)

30. **active optics:** Thin telescope mirrors that are controlled by computers to maintain proper shape as the telescope moves. (p. 83; video lesson; objective 3)

31. **adaptive optics:** A computer-controlled optical system used to partially correct for seeing in an astronomical telescope. (p. 83; video lesson; objective 3)

32. **interferometry:** The observing technique in which separated telescopes are combined to produce a virtual telescope with the resolution of a telescope much larger in diameter. (p. 85; video lesson; objective 6)

33. **charge-coupled device (CCD):** An electronic device consisting of a large array of light-sensitive elements used to record very faint images. (p. 86; video lesson; objectives 3 & 5)

34. **false-color image:** A representation of graphical data with added or enhanced color to reveal detail. (p. 86; video lesson; objective 3)

35. **spectrograph:** A device that separates light by wavelengths to produce a spectrum—a fingerprint of the source producing the light. (p. 86; video lesson; objective 5)

36. **grating:** A piece of material in which numerous microscopic parallel lines are scribed. Light encountering a grating is dispersed to form a spectrum. (p. 87; video lesson; objective 5)

37. **comparison spectrum:** A spectrum of known spectral lines used to identify unknown wavelengths in an object's spectrum. This helps to identify the fingerprint of the source. (p. 87; objective 5)

38. **radio interferometer:** Two or more radio telescopes that combine their signals to achieve the resolving power of a larger telescope. (p. 89; video lesson; objective 6)

Summary

In order to understand how astronomers use their tools, you must first understand the nature of light and other forms of electromagnetic radiation. **Electromagnetic radiation**—energy in the form of interacting electric and magnetic waves—travels through space at the speed of light and transfers energy from one place to another. It is measured by **wavelength**, or the distance between the successive peaks of the electromagnetic wave.

Visible light is one form of electromagnetic radiation. Gamma rays and X rays are types of electromagnetic radiation that have very short wavelengths and contain a great deal of energy. Radio waves are another form, but they have very long wavelengths and carry considerably less energy. There is an inverse relationship between energy and wavelength. As energy increases, wavelength decreases.

The Electromagnetic Spectrum

A **spectrum** is the result of electromagnetic radiation that is spread out in order of wavelength. You are familiar with one type of spectrum—a rainbow, which is a spectrum of only visible light ranging in wavelength from the violet to the red. In the full electromagnetic spectrum, light is only one small part.

Only a very small portion of electromagnetic radiation from space actually reaches Earth's surface. Water vapor and ozone in Earth's atmosphere absorb most of the infrared and ultraviolet radiation respectively. The balance is absorbed by layers of the upper atmosphere. Only visible light, some shorter-wavelength infrared and some radio waves reach Earth's surface through two wavelength regions called **atmospheric windows**. From Earth's surface, astronomers can study the visible light that comes from space using optical telescopes and they can study the radio waves that come from space using radio telescopes.

The wavelength of visible light is so small that it is measured using the **nanometer (nm)**, or one billionth of a meter (10^{-9} m). It can also be measured using the **angstrom (Å)**, which equals (10^{-10} m). Other types of electromagnetic radiation such as radio waves can be measured using millimeters, centimeters, or meters.

Optical Telescopes

Optical telescopes allow astronomers to observe the visible portion of the electromagnetic spectrum (some sophisticated optical telescopes can also detect infrared). There are two types of optical telescopes: reflecting and refracting. A **reflecting telescope** uses a concave mirror to reflect the incoming light. A **refracting telescope** uses a convex lens to bend the incoming light. Both types of telescopes form a small inverted image that can be viewed through an **eyepiece** that magnifies the image.

Refracting telescopes distort the light as it is refracted through the lens. Short-wavelength light bends more than long-wavelength light; therefore, blue light and red light don't focus at exactly the same location. As a result, the image being viewed appears to have color "fringes" or a slight fuzziness or blurring around it. This **chromatic aberration**—or color separation—in the refracting telescope makes it less preferable than the reflecting telescope to astronomers.

The reflecting telescope can produce a much clearer image than the refracting telescope because it does not produce chromatic aberration. Reflecting telescopes are also less expensive because only the highly reflective surface of the mirror has to be flawless—not the entire piece of glass. Unlike the lens in a refracting telescope, a mirror can be supported from behind. This allows telescope designers to build much bigger telescopes that gather much more light.

The three powers of a telescope that determine its quality are light-gathering power, resolving power, and magnifying power. **Light-gathering power** refers to the telescope's ability to collect light. This is important because most celestial objects are very faint and the telescope has to gather what little light reaches it. The primary purpose of a telescope is to gather light and to allow us to see things that are too faint to see with the naked eye. Light-gathering power is proportional to the area of the telescope's objective: a large telescope can catch more light than a smaller one (see Reasoning with Numbers 5-1 on page 84 in the *Horizons* textbook).

The second power that affects the quality of a telescope is its **resolving power**—the ability for the telescope to show fine detail. A larger telescope has more resolving power because it reduces the amount of **diffraction fringe**—the blurred fringe surrounding an image that's caused by the wave properties of light. Resolving power is affected by the quality of the mirror or lens and also by **seeing**—the atmospheric conditions that cause an image to become blurred.

The magnifying power of a telescope is less important than its light-gathering power or resolving power, but still affects the quality of the images that we can see. **Magnifying power** refers to the ability to make an image larger. No matter how great the magnifying power of a telescope, it cannot make up for a lack of light-gathering or resolving power.

Advances in technology have improved the performance of optical telescopes. Computers can move the telescope to follow the apparent westward motion of the stars as Earth rotates. Computers can also control the shape of a thin, lightweight mirror or segmented mirrors by using **active optics**. Astronomers also depend on computers to reduce seeing distortion caused by Earth's atmosphere by using **adaptive optics**.

The method known as **optical interferometry** allows astronomers to connect multiple telescopes together to increase resolving power. In an optical interferometer, several small telescopes combine their light through a network of mirrors. As with every telescope on Earth, turbulence in the atmosphere distorts the light and high-speed computers must continuously make adjustments to correct for the distortion. Building an interferometer is very challenging, but it can allow for increased resolving power while maintaining the optical quality that smaller mirrors can provide.

Special Instruments

In the early twentieth century, astronomers captured images of the universe on a photographic plate—a flat sheet of metal or glass on which a photographic image can be recorded by allowing light to collect on a light-sensitive emulsion. By increasing how long light is exposed to the plate, astronomers could capture images from very faint light sources.

Today, astronomers use special instruments such as electronic imaging systems and spectrographs to collect and analyze the light the telescope gathers. Astronomers can record images with a **charge-coupled device (CCD)**—a specialized computer chip that contains about one million microscopic light detectors arranged in rows and columns in an array about the size of a postage stamp. They are superior to the photographic plate in detecting both bright and faint objects and are much more sensitive. Most digital cameras also use a CCD to record images to be transferred to the memory chip inside the camera.

One of the most important of all astronomical tools is the **spectrograph**—an imaging device used to analyze light from a celestial object. Modern spectrographs use a **grating**—a piece of glass with thousands of microscopic parallel lines etched into the surface. The grating reflects and diffracts the light so that it spreads out and the spectrum can be more easily recorded and analyzed. As you saw in the video program, fiber optics can be used to create a multi-slit spectrograph. The fiber optics carry the light from the image of the star to the diffraction grating, which creates the spectrum. As you will learn in Lesson 6, a celestial object's spectrum can reveal its chemical composition and allow us to learn so much more about it.

Radio Telescopes

Whereas optical telescopes gather visible light from celestial objects, radio telescopes detect radiation in the radio wavelengths. A radio telescope looks like a giant dish that you might have on your house to receive satellite television. The dish reflector of the radio telescope collects and focuses electromagnetic radiation like a reflecting optical telescope does. The radio telescope's antenna transmits the radiation collected by the dish and sends it to an amplifier and on to a recording instrument.

Because humans can't detect radio waves like we can light waves, astronomers have to convert the radio waves into something that we can perceive. Astronomers can measure the strength of the radio waves in various places in the sky and create a map that outlines the areas with uniform radio intensity. This allows astronomers to detect patterns in the radio waves. They use **false-color images** to help illustrate the variation of different intensities and/or energies emitted that are not in the visible portion of the electromagnetic spectrum.

Radio waves have very low energy compared to visible light and especially compared to gamma rays and X-ray radiation, which makes it especially challenging to receive signals. To compensate for this, the collecting dishes on the radio telescopes must be incredibly large; since the wavelength of radio waves is very large, a very large collector is needed to resolve more details. Just as optical telescopes have to overcome interference in the form of light pollution, radio telescopes must overcome human-made radio interference. To overcome this, radio telescopes are built in very remote locations far away from civilization.

The radio telescope is limited by its poor resolution, by low intensity of the received energy, and by interference from human-made objects on Earth. The ability to collect and analyze electromagnetic radiation in a radio telescope depends on its resolving power. Radio waves are very long and the images become fuzzy because of large diffraction fringes. Radio telescopes must be very large to increase their resolution power, but they are still quite limited. Several telescopes can be linked together to form a **radio interferometer** to increase the resolution.

With all the disadvantages that radio telescopes have to overcome, you might wonder why astronomers use them. Unlike an optical telescope, a radio telescope can detect clouds of cool hydrogen in the universe, which don't produce visible light and would otherwise be undetectable. Cool hydrogen emits a radio signal at a specific wavelength and the ability to detect it allows astronomers to determine where stars are born. Some of the most interesting objects in the universe produce no visible light and are only detectable at radio wavelengths.

Another advantage to using a radio telescope is that it is able to collect electromagnetic radiation that can travel through space unhindered by dust particles. Since visible light has relatively short wavelengths, it can be scattered as it passes thorough dust and never reach Earth. But, since radio waves are so long, they can pass through dust without being scattered. They can also be detected both in the day and at night.

Astronomy from Space

Much of what we can learn about the universe is lost to ground-based observers because we can detect primarily visible and radio wavelengths from Earth's surface. For this reason, astronomers have developed specialized telescopes that can operate at very high altitudes or orbit beyond Earth's atmosphere.

Astronomers can detect near-infrared, some far-infrared and near-ultraviolet radiation by placing telescopes at very high altitudes where the effects of Earth's atmosphere are diminished. Most infrared radiation coming from space is absorbed by water vapor and carbon dioxide in Earth's atmosphere. Near-infrared radiation is just beyond the red end of the visible spectrum and has a wavelength between 1.2 and 40 micrometers or microns. Near-infrared telescopes can be used in high altitudes where the air is thin and dry, such as on a mountaintop. Far-infrared radiation includes wavelengths of more than 40 micrometers and can provide astronomers information about cool planets, forming stars, and other cool objects. Far-infrared can be detected by airborne observatories that travel above most of Earth's water vapor. Near ultra-violet radiation can be recorded by photographic plates or CCDs.

To detect electromagnetic radiation that doesn't come to Earth through the visible and radio windows, astronomers can place telescopes in space. A telescope in space can observe a wide range of wavelengths, and the images are not blurred or distorted by Earth's atmosphere. These telescopes can detect higher-energy wavelengths, such as gamma rays and X rays that cannot reach the ground. These telescopes are critical to our understanding of the universe: X rays are emitted by black holes that would otherwise be undetectable. Enhanced far-infrared detection from space reveals how stars are born.

The astronomer's tools are highly specialized pieces of equipment and are very expensive to build and operate. Yet, they can provide us with information about the universe that we couldn't acquire without them. In the next lesson, you'll learn how astronomers use this specialized equipment—including the spectrograph—to learn more about the universe

Review Exercises

Matching I

Match each term with the appropriate definition or description.

1. _____ electromagnetic radiation	13. _____ Newtonian focus	
2. _____ wavelength	14. _____ active optics	
3. _____ photon	15. _____ equatorial mounting	
4. _____ atmospheric window	16. _____ alt-azimuth mounting	
5. _____ refracting telescope	17. _____ light-gathering power	
6. _____ focal length	18. _____ resolving power	
7. _____ objective	19. _____ seeing	
8. _____ eyepiece	20. _____ magnifying power	
9. _____ chromatic aberration	21. _____ charge-coupled device (CCD)	
10. _____ reflecting telescope	22. _____ spectrograph	
11. _____ prime focus	23. _____ grating	
12. _____ Cassegrain focus	24. _____ radio interferometer	

    a. Mounting in which one axis points to a celestial pole.

    b. Mounting in which one axis points vertically upward.

    c. A quantum of electromagnetic energy.

    d. Fields that transfer energy, such as light waves, radio waves, and gamma rays.

    e. Inversely related to the energy of a photon.

    f. Wavelength region in which a particular photon can reach the ground from space.

    g. Computerized control of the shape of a mirror.

    h. The point at which the objective mirror forms an image.

    i. A focus location at the upper side of a telescope.

    j. A focus location directly beneath the objective mirror.

    k. The lens or mirror that first produces a focused image, as opposed to modifying, enlarging, or redirecting the image.

    l. A telescope that has a lens for an objective.

    m. A telescope that focuses light with a mirror.

    n. A lens for visually enlarging a focused image.

    o. A color distortion specific to simple lenses but not to mirrors.

    p. The distance between an objective and its focused image of a very distant object.

    q. Combining of signals from two of more radio telescopes to increase resolving power.

    r. An electronic device for recording faint images.

    s. An alternative to a prism as a way of dispersing light into its component colors.

t. The instrument consisting of lenses and a grating that allows for light to be analyzed and dispersed into its component colors.
u. A term for the quality of atmospheric conditions on a given night.
v. The ability of a telescope to make starlight brighter.
w. The ability of a telescope to make an image larger.
x. The ability of a telescope to reveal fine detail.

Matching II
Match each term with the appropriate definition or description.

1. _____ achromatic lens	9. _____ nanometer
2. _____ adaptive optics	10. _____ polar axis
3. _____ angstrom	11. _____ primary lens
4. _____ comparison spectrum	12. _____ refracting telescope
5. _____ diffraction fringe	13. _____ Schmidt-Cassegrain focus
6. _____ false-color image	14. _____ secondary mirror
7. _____ interferometry	15. _____ sidereal drive
8. _____ light pollution	

a. A unit of distance equaling one-billionth of a meter (10^{-9}); commonly used to measure the wavelengths of light.
b. A unit of distance equaling 10^{-10} m.
c. The largest lens or mirror in a telescope.
d. A telescope that forms images by bending (refracting) light with a lens.
e. A telescope lens composed of two lenses ground from different kinds of glass and designed to bring two selected colors to the same focus.
f. Blurred fringe surrounding any image, caused by the wave properties of light.
g. The illumination of the night sky by wasted light, which prevents the observation of faint objects.
h. Part of a reflecting telescope that directs the light from the primary mirror to a focal position.
i. The optical design that uses a thin corrector plate at the entrance to the telescope tube.
j. The motor and gears on a telescope that turn it westward to keep it pointed at a star.
k. In an equatorial telescope mounting, the axis that is parallel to Earth's axis.
l. A computer-controlled optical system used to partially correct for seeing in an astronomical telescope.
m. The observing technique in which separated telescopes are combined to produce a virtual telescope with the resolution of a telescope much larger in diameter.

n. A representation of graphical data with added or enhanced color to reveal detail.

o. A spectrum of known spectral lines used to identify unknown wavelengths in an object's spectrum. This helps to identify the fingerprint of the source.

Completion

Fill each blank in the sentences below with the most appropriate term from the list of completion answers that follow. A term may be used once, more than once, or not at all. Check your answers with the Answer Key and review when necessary.

chromatic	interferometer	refracting telescopes
electromagnetic	radio	resolving
radiation	reflecting telescopes	ultraviolet radiation

1. One problem faced by astronomers is that only a small portion of _____ reaches Earth.

2. One of the main problems of refracting telescopes is the inability to focus all colors of light simultaneously, which is known as _____ aberration.

3. One of the principal advantages of _____ is that light does not pass through a significantly thick piece of glass.

4. Astronomers overcome the relatively poor _____ power of a radio telescope by combining two or more such telescopes into a radio _____.

Self-Test

Select the best answer.

1. Of the types of electromagnetic radiation listed below, the one with the shortest wavelength is
   a. X ray.
   b. radio waves.
   c. visible light.
   d. ultraviolet light.

2. The energy associated with an electromagnetic wave is inversely proportional to the wave's
   a. frequency.
   b. speed.
   c. value for Planck's constant.
   d. wavelength.

3. In addition to visible light, Earth's atmosphere is transparent to
   a. X rays.
   b. radio waves.
   c. gamma rays.
   d. long-wavelength infrared radiation.

4. The 10-meter Keck telescope in Hawaii is the world's largest
   a. segmented refracting telescope.
   b. segmented reflecting telescope.
   c. interferometer.
   d. single dish mirror.

5. The newest, largest ground-based optical telescopes are supported by
   a. equatorial mountings.
   b. alt-azimuth mountings.
   c. stationary mountings with moving coelostat mirrors.
   d. none of the above.

6. For an optical telescope used on Earth under normal seeing conditions, the main value of making larger telescopes is greater
   a. magnifying power.
   b. resolving power.
   c. light-gathering power.
   d. maneuverability.

7. Large, thin mirrors, sometimes called "floppy mirrors," sag under their own weight. To correct for this,
   a. a rigid support system is attached to the back of the mirror.
   b. computer-controlled thrusters are placed under the mirror to change the shape of the mirror several times per second.
   c. a secondary mirror is used to focus the light.
   d. a correcting lens is placed in front of the objective mirror.

8. Earth's atmosphere causes the images seen in telescopes to be blurred. This can be corrected by a process known as
   a. adaptive optics.
   b. Schmidt lens corrector.
   c. active optics.
   d. interferometry.

9. If the eyepiece of a telescope is replaced with an eyepiece with a focal length twice as long, the magnifying power is
   a. four times greater.
   b. two times greater.
   c. four times less.
   d. two times less.

10. A device that analyzes light by spreading it out according to wavelength is called a(n)
    a. interferometer.
    b. camera.
    c. photometer.
    d. spectrograph.

11. A grating is a device that
    a. bends light and allows it to focus at a point.
    b. reflects light into a charge-coupled device (CCD) to enhance the details of an image.
    c. disperses light into its component colors using thousands of microscopic parallel grooves.
    d. connects optical telescopes to an interferometer.

12. One of the disadvantages of a radio telescope compared to a reflecting optical telescope is its inability to
    a. reveal finer detail.
    b. detect cool hydrogen.
    c. see through interstellar dust.
    d. see some of the most distant objects in the universe.

13. The main reason for creating an interferometer with two separate radio dishes is to
    a. monitor two different radio wavelengths simultaneously.
    b. increase the intensity of the detected signal.
    c. compare objects in two different directions simultaneously.
    d. increase the resolution in the data.

14. A high-altitude airplane is used to make observations
    a. at gamma-ray wavelengths.
    b. at X-ray wavelengths.
    c. in the ultraviolet.
    d. in the infrared.

15. The Hubble Space Telescope, orbiting above Earth's atmosphere, eliminates the blurring effect of Earth's atmosphere and permits the telescope to
    a. observe gamma rays.
    b. observe wavelengths from the near infrared to the near ultraviolet.
    c. receive radio wavelengths.
    d. transmit radar pulses that reflect off the surface of the moon.

Short-Answer Questions

1. How is the method used by the Chandra X-ray telescope to focus electromagnetic waves different from Earth-based optical telescopes?

2. Identify the types of electromagnetic energy that can only be observed by space-based telescopes. What properties of Earth's atmosphere prevent this radiation from reaching the surface?

3. Identify all of the various forms of electromagnetic radiation and relate these forms to their wavelengths and energy.

4. What features of a telescope determine magnifying, light-gathering, and resolving power respectively?

5. What information does a spectrograph reveal about an object? What is the purpose of the grating and comparison spectrum?

Applications

1. Compare the light gathering power of the 10-meter Keck telescope with the 8-meter Gemini telescope.

2. The resolving power of a 10-centimeter (4-inch) telescope is 1.16 seconds of arc. What is the resolving power of the 40-inch (100-cm) Yerkes refractor?

3. What is the magnification (M) of a telescope with a focal length of 1 meter combined with an eyepiece of 25 mm?

4. Calculate the energy of a red photon of wavelength 700 nm and compare it to the energy of a violet photon of wavelength 400 nm. [The unit associated with energy is the joule.]

5. Cool hydrogen emits a radio signal at the specific wavelength of 21 cm. What is the energy associated with this wavelength and how does it compare to the energy emitted by a photon of red light of wavelength 700 nm?

Answer Key

Matching I

1. d (p. 74; video lesson; objective 1)

2. e (p. 74; video lesson; objective 1)

3. c (p. 74; video lesson; objective 1)

4. f (p. 76; video lesson; objective 4)

5. l (p. 76; objective 2)

6. p (p. 76; objective 2)

7. k (p. 76; objective 2)

8. n (p. 77; objective 2)

9. o (p. 77; objectives 2 & 4)

10. m (p. 76; objective 2)

11. h (p. 82; objective 2)

12. j (p. 82; objective 2)

13. i (p. 82; objective 2)

14. g (p. 83; video lesson; objective 3)

15. a (p. 83; objective 2)

16. b (p. 83; objective 2)

17. v (p. 78; objective 4)

18. x (p. 78; objectives 2 & 4)

19. u (p. 78; video lesson; objective 4)

20. w (pp. 79–80; objective 4)

21. r (p. 86; video lesson; objectives 3 & 5)

22. t (p. 86; video lesson; objective 5)

23. s (p. 87; video lesson; objective 5)

24. q (p. 89; objective 6)

Matching II

1. e (p. 77; objectives 2 & 4

2. l (p. 83; video lesson; objective 3)

3. b (p. 75; objective 1)

4. o (p. 87; objective 5)

5. f (p. 78; objective 4

6. n (p. 86; video lesson; objective 3)

7. m (p. 85; video lesson; objective 6)

8. g (p. 80; objective 4)

9. a (p. 75; objective 1)

10. k (p. 83; objective 2)

11. c (p. 76; video lesson; objective 2)

12. d (p. 76; video lesson; objective 2)

13. i (p. 82; objective 2)

14. h (p. 82; video lesson; objective 2)

15. j (p. 83; objective 2)

Completion

1. electromagnetic radiation (p. 74; video lesson; objective 1)

2. chromatic (p. 77; objectives 2 & 4)

3. reflecting telescopes (p. 76; objective 2)

4. resolving, interferometer (p. 89; objective 6)

Self-Test

1. a (pp. 75–76; objective 1)

2. d (p. 75; objective 1). Use the formula at the top of page 75—the wavelength is in the denominator and therefore inversely proportional to the energy.

3. b (pp. 75–76; objective 1)

4. b (p. 83; objective 2)

5. b (p. 83; objective 2)

6. c (p. 78; objective 4)

7. b (p. 83; objective 3)

8. a (p. 83; objective 3)

9. d (p. 84, Reasoning with Numbers 5-1; objective 4)

10. d (pp. 86–87; objective 5)

11. c (pp. 86–87; objective 5)

12. a (p. 89; objective 6). A single radio dish has relatively limited resolving capabilities, however, by linking radio dishes together in an interferometer, the resolving capability is greatly increased.

13. d (p. 89; video lesson; objective 6)

14. d (p. 92; objective 7)

15. b (p. 93; objective 7)

Short-Answer Questions

1. X rays, unlike light waves, are so energetic they penetrate through glass that normally reflects light. Ground-based optical telescopes reflect light to a focus where an instrument analyzes the energy. The Chandra telescope uses highly polished mirrors as well; however, these mirrors are shaped like cylinders with the inside being highly polished. The X rays skim the inside of these mirrors and are focused onto detectors. (p. 95; objective 7)

2. Far infrared, far ultraviolet, X rays, gamma rays, and some radio waves cannot be observed from Earth's surface. Water vapor in Earth's atmosphere absorbs infrared wavelengths, the ozone layer absorbs far ultraviolet wavelengths, and layers in the upper atmosphere absorb X rays and gamma rays. (pp. 74–76, 92–93; objectives 1 & 7)

3. The most energetic forms of electromagnetic energy have the shortest wavelengths. Gamma rays are the most energetic with the shortest wavelengths. In order of decreasing energy and increasing wavelength the forms of electromagnetic energy are as follows: gamma rays, X rays, ultraviolet, visible light, infrared, microwave, radio waves. (pp. 75–76; video lesson; objective 1)

4. The magnifying power of a telescope determines by how much the image can be made bigger. It is calculated by dividing the focal length of the objective by the focal length of the eyepiece. The light-gathering power of a telescope is determined by the diameter of the telescope. The larger the telescope the better its light-gathering power and its resolving power. That is why in optical and radio interferometers, the diameter of the equivalent telescope equals the distance between the individual telescopes and therefore improves resolution. (pp. 78–80; objective 4)

5. The spectrograph spreads out light produced by the star or galaxy to create a spectrum of the object. By examining this light carefully, we notice that that light contains many lines (spectral lines). These lines are the signature of the atoms producing the light. The grating is the actual component of the spectrograph that spreads the light out. It consists of thousands of microscope grooves producing the phenomenon called diffraction. The comparison spectrum is used to identify the spectral lines in the light of the object and hence the atoms producing the light. This comparison spectrum is generated by the scientists studying the light. (pp. 86–87; video lesson; objective 5)

Applications

1. The light-gathering power of the 10-meter Keck is 1.56 times greater than the 8-meter Gemini. Divide the square of the diameter of the Keck by the square of the diameter of the Gemini. (p. 84, Reasoning with Numbers 5-1; objective 4)

2. The resolving power of the 40-inch Yerkes is 10 times better than a 4-inch refractor. Use the formula found on page 84 of the *Horizons* textbook for resolving power. (p. 84, Reasoning with Numbers 5-1; objective 4)

3. The magnifying power of a telescope is 40 times. Divide the focal length of the 1-meter objective (equal to 1,000 mm) by the focal length of the eyepiece (25 mm). (p. 86, Reasoning with Numbers 5-1; objective 4)

4. The energy of a 700 nm photon is 2.84×10^{-19} joules. The energy of a 400 nm photon is 4.97×10^{-19} joules. The violet photon has 1.75 times energy as the red photon. Use the formula on page 75 of the *Horizons* textbook. Remember to convert the units of nm (nanometers) to meters before dividing. (pp. 74–75; objective 1)

5. The energy of the 21-cm wavelength of cool hydrogen is 9.47×10^{-25} joules. Dividing the energy of a red photon by the energy of the 21-cm wavelength demonstrates that the energy of the red photon is almost 3×10^5 times greater than the 21-cm wavelength. Be sure to convert the unit cm (centimeters) to meters before calculating the energy. (pp. 74–75; objective 1)

Lesson Review

Lesson 5: Astronomical Tools

PLEASE NOTE: Use this matrix to guide your study and achieve the learning objectives of this lesson. It will also help you to view the video, which defines and demonstrates important concepts and principles as they relate to everyday life and actual case studies.

Learning Objective	Textbook	Student Guide
1. Describe the nature of electromagnetic radiation, explain its characteristics at different wavelengths, and discuss how this influences the way astronomers make observations.	pp. 74–76	Key Terms: 1, 2, 3, 4, 5; Matching I: 1, 2, 3; Matching II: 3, 9; Completion: 1; Self-Test: 1, 2, 3; Short-Answer: 2, 3; Applications: 4, 5.
2. Compare and contrast the different optical telescope designs, including their relative advantages and disadvantages.	pp. 76–78, 82–83	Key Terms: 7, 8, 9, 10, 11, 12, 13, 14, 16, 21, 22, 23, 24, 25, 26, 27, 28, 29; Matching I: 5, 6, 7, 8, 9, 10, 11, 12, 13, 15, 16, 18; Matching II: 1, 10, 11, 12, 13, 14, 15; Completion: 2, 3; Self-Test: 4, 5.
3. Discuss how astronomers use technology to enhance the performance of optical telescopes.	pp. 81–86	Key Terms: 30, 31, 33; Matching I: 14, 21; Matching II: 2, 6; Self-Test: 7, 8.
4. Describe the factors that determine the light-gathering, magnifying, and resolving powers of a telescope, noting the effects of Earth's atmosphere.	pp. 78–80	Key Terms: 6, 13, 14, 15, 16, 17, 18, 19, 20; Matching I: 4, 9, 17, 18, 19, 20; Matching II: 1, 5, 8; Completion: 2; Self-Test: 6, 9; Short-Answer: 4; Applications: 1, 2, 3.

Learning Objective	Textbook	Student Guide
5. Describe the instrumentation used for photography and spectroscopy with a telescope.	pp. 86–88	Key Terms: 35, 36, 37; Matching I: 21, 22, 23; Matching II: 4; Self-Test: 10, 11; Short-Answer: 5.
6. Explain the operation, advantages, and limitations of radio telescopes, including the method of radio interferometry.	pp. 88–91	Key Terms: 32, 38; Matching I: 24; Matching II: 7; Completion: 4; Self-Test: 12, 13.
7. Describe the advantages of placing telescopes in space and the types of observations they are designed to make.	pp. 91–95	Self-Test: 14, 15; Short-Answer: 1, 2.

LESSON
6

The Science of Starlight

Checklist

For the most effective study of this lesson, complete the following activities in this sequence.

Before Viewing the Video

❑ Read the Preview, Learning Objectives, and Viewing Notes below.

❑ Read Chapter 6, "Starlight and Atoms," pages 98–117, in the *Horizons* textbook.

What to Watch

❑ After reading the textbook chapter, watch the video for Lesson 6, *The Science of Starlight.*

After Viewing the Video

❑ Briefly note your answers to questions listed at the end of the Viewing Notes.

❑ Review the Summary below.

❑ Review all reading assignments for this lesson, especially the Chapter 6 summary on page 116 in *Horizons* and the Viewing Notes in this lesson.

❑ Write brief answers to the review questions at the end of Chapter 6 in *Horizons*.

❑ Complete the Review Exercises below. Check your answers with the Answer Key and review when necessary.

❑ Use the Lesson Review matrix found at the end of this lesson to review and assess your knowledge of each Learning Objective.

❑ As assigned by your instructor, complete the Applications activities and any additional activities for this lesson.

Preview

Until the early nineteenth century, humans knew little about the composition of the sun and other stars in our universe. When scientists began to study light and the solar spectrum, they realized that the patterns that they saw in the spectrum were related to the various atoms in the sun. It was then that they began to understand the true nature of a star—atoms leave their fingerprints on light.

In Lesson 5, you learned that modern astronomers use a spectrograph—a device that separates light by wavelengths—to analyze the electromagnetic radiation coming from celestial objects. The spectrum of electromagnetic radiation is often referred to as light, but it includes much more than visible light. By studying the light coming from a celestial object such as a star, we can determine its temperature, chemical composition, and its motion relative to Earth.

Analyzing light—and how it interacts with matter—is the key to understanding our universe. In this lesson, you'll gain a basic understanding of the structure of an atom. Studying the infinitesimally small atom can help us unlock the secrets of the infinitely large universe. This will also help you comprehend the concepts introduced in upcoming lessons: how stars produce energy, how they live, and how they die.

Concepts to Remember

- Recall from Lesson 5 that *electromagnetic radiation* is the changing electric and magnetic fields that travel through space and transfer energy from one place to another. Examples include visible light, radio waves, and ultraviolet rays (pp. 78–79 in this guide).

- In Lesson 5, you also learned that a *photon* is a quantum of electromagnetic energy. It carries an amount of energy that depends inversely on its wavelength. Photons can be emitted or absorbed by matter (p. 79 in this guide).

Learning Objectives

After you complete this lesson, you should be able to:

1. Describe the basic structure of an atom, including different kinds of atoms, and the nature of electron shells. HORIZONS TEXTBOOK PAGES 100–102.

2. Explain the interaction of light and matter, including the excitation of atoms and radiation from a heated object, and describe the relationship between temperature, wavelength of maximum intensity, and total radiated energy. HORIZONS TEXTBOOK PAGES 102–106.

3. Describe the appearance, origin, and characteristics of continuous, emission, and absorption spectra, including the visible and nonvisible spectrum of hydrogen. *HORIZONS* TEXTBOOK PAGES 106–109.

4. Explain spectral classification and how the appearance of a spectrum indicates stellar temperature and composition. *HORIZONS* TEXTBOOK PAGES 107, 110–111.

5. Explain what spectra indicate about the speed and direction of stellar motions as revealed by the Doppler effect. *HORIZONS* TEXTBOOK PAGES 111–114.

At this point, read Chapter 6, "Starlight and Atoms," pages 98–117.

Look at the stars! look, look up at the skies!
O look at all the fire-folk sitting in the air!
—**Gerard Manley Hopkins**
The Starlight
Night (1918)

Viewing Notes

In the video program, you'll discover how astronomers can analyze light to learn more about the universe, including the chemical composition of stars, their temperature, and their motion relative to Earth.

The video program contains the following segments:

⚙ Interaction of Light & Matter

⚙ Atomic Spectra

⚙ Temperature & Stellar Classification

⚙ Light in Motion

The following information will help you better understand the video program:

The video program discusses how the process of spectroscopy—analyzing a spectrum—can help astronomers learn more about the chemical composition of celestial objects. Remember that light and matter interact to form a spectrum—electrons in an atom can change **energy levels** and absorb or emit photons in the process. Different atoms have different energy levels so their spectra contain different patterns called **spectral lines**.

Kirchhoff's laws describe the different types of spectra: continuous, emission, and absorption.

In a **continuous spectrum**, a solid, liquid, or dense gas is excited to emit electromagnetic radiation at all wavelengths. This produces a spectrum with no spectral lines. A hot solid object like a tungsten filament in an incandescent lightbulb produces a continuous spectrum.

Astronomers can point their spectrographs to the gases near a star to study an **emission spectrum**—also known as a **bright-line spectrum**. In this type of spectrum, a low-density gas near a continuous light source is excited to emit light at certain wavelengths. There must be a source to excite the atoms in the gases; otherwise the gases would not be seen in the visible portion of the spectrum.

An object like a star produces an **absorption spectrum**—or a **dark-line spectrum**. Astronomers can point their spectrographs directly at the star to observe this type of spectrum; the star's deeper layers are shining and emitting photons of all wavelengths. When those photons pass through the cooler, thinner gas above, the gas absorbs some of them.

The video program mentions that the seven main **spectral classes** or **types** are from hottest to coolest: O, B, A, F, G, K, and M. Remember from the textbook that each spectral classification is further broken down into ten subclasses: from 0 to 9. For example, spectral class B can be further subdivided from B0 to B9.

QUESTIONS TO CONSIDER

- What are the basic parts of an atom?

- How does an electron absorb or emit a photon of light?

- What are the three types of spectra?

- How is each of the spectra produced?

- How can the analysis of a star's spectrum indicate its temperature?

- How can the spectrum indicate a star's motion?

Watch the video for Lesson 6, *The Science of Starlight*.

Key Terms and Concepts

Page references are keyed to the *Horizons* textbook.

1. **spectral line:** A line in a spectrum at a specific wavelength produced by the absorption or emission of light by certain atoms. (p. 100; objectives 1 & 3)

2. **nucleus:** The central core of an atom containing protons and neutrons; carries a net positive charge. (p. 100; video lesson; objective 1)

3. **proton:** A positively charged atomic particle contained in the nucleus of an atom. The nucleus of a hydrogen atom. (p. 100; video lesson; objective 1)

4. **neutron:** An atomic particle with no charge and about the same mass as a proton. (p. 100; video lesson; objective 1)

5. **electron:** Low-mass atomic particle carrying a negative charge. (p. 100; video lesson; objective 1)

6. **isotope:** An atom that has the same number of protons but a different number of neutrons. (p. 101; objective 1)

7. **ionization:** The process in which atoms lose or gain electrons. (p. 101; objective 1)

8. **ion:** An atom that has lost or gained one or more electrons. (p. 101; objective 1)

9. **molecule:** Two or more atoms bonded together. (p. 101; objective 1)

10. **Coulomb force:** The electrostatic force of repulsion or attraction between charged bodies. (p. 101; objective 2)

11. **binding energy:** The energy needed to pull an electron away from its atom. (p. 101; objective 2)

12. **quantum mechanics:** The study of the behavior of atoms and atomic particles. (p. 101; objective 2)

13. **permitted orbit:** One of the energy levels in an atom that an electron may occupy. (p. 101; video lesson; objective 2)

14. **energy level:** One of a number of states an electron may occupy in an atom, depending on its binding energy. (p. 103; video lesson; objective 2)

15. **excited atom:** An atom in which an electron has moved from a lower to a higher energy level. (p. 103; video lesson; objective 2)

16. **ground state:** The lowest permitted electron energy level in an atom. (p. 103; video lesson; objective 2)

17. **heat:** The flow of thermal energy. (p. 104; objective 2)

18. **temperature:** A measure of the agitation among the atoms and molecules of a material. The intensity of the atomic motion. (p. 104; objective 2)

19. **Kelvin temperature scale:** A temperature scale using Celsius degrees and based on zero at absolute zero. (p. 104; video lesson; objective 2)

20. **absolute zero:** The theoretical lowest possible temperature at which a material contains no extractable heat energy; established as zero on the Kelvin temperature scale. (p. 104; video lesson; objective 2)

21. **black body radiation:** Radiation emitted by a hypothetical perfect radiator. The spectrum is continuous, and the wavelength of maximum emission depends on the body's temperature. (p. 104; video lesson; objective 2)

22. **wavelength of maximum intensity:** The wavelength at which a perfect radiator emits the maximum amount of energy. Depends only on the object's temperature. (p. 104; objective 2)

23. **joule:** A unit of energy equivalent to a force of 1 newton acting over a distance of 1 meter. One joule per second equals 1 watt of power. (p. 105; objective 2)

24. **continuous spectrum:** A spectrum in which there are no absorption or emission lines. (p. 108; video lesson; objective 3)

25. **absorption spectrum:** A spectrum that contains absorption lines caused by photons being absorbed by atoms or molecules. (p. 108; video lesson; objective 3)

26. **emission spectrum:** A spectrum containing emission lines. (p. 108; objective 3)

27. **emission line:** A bright line in a spectrum caused by the emission of photons from atoms. (p. 108; objective 3)

28. **absorption line:** A dark line in a spectrum; is produced by the absence of photons absorbed by atoms or molecules. (p. 108; video lesson; objective 3)

29. **Kirchhoff's laws:** A set of laws that describe the absorption and emission of light by matter. (p. 108; video lesson; objective 3)

30. **transition:** The movement of an electron from one atomic energy level to another. (p. 109; objectives 2 & 3)

31. **Lyman series:** Spectral lines in the ultraviolet spectrum of hydrogen produced by transitions whose lowest energy level is the ground state. (p. 109; objective 3)

32. **Balmer series:** A series of spectral lines produced by hydrogen in the near-ultraviolet and visible parts of the spectrum. (p. 109; video lesson; objective 3)

33. **Paschen series:** Spectral lines in the infrared spectrum of hydrogen produced by transitions whose lowest energy level is the third. (p. 109; objective 3)

34. **spectral class or type:** A star's position in the temperature classification O, B, A, F, G, K, or M; is based on the appearance of the star's spectrum. (p. 110; video lesson; objective 4)

35. **spectral sequence:** The arrangement of spectral classes ranging from hot to cool. (p. 110; objective 4)

36. **L dwarf:** A main sequence star cooler than an M star. (p. 110; objective 4)

37. **T dwarf:** A very cool, low-mass star or brown dwarf located below the L stars on the main sequence. (pp. 110–111; objective 4)

38. **Doppler effect:** The change in the wavelength of radiation due to the relative radial motion of source and observer. (p. 111; video lesson; objective 5)

39. **blueshift:** A Doppler shift toward shorter wavelengths caused by a velocity of approach. (p. 112; video lesson; objective 5)

40. **redshift:** A Doppler shift toward longer wavelengths caused by a velocity of recession. (p. 112; video lesson; objective 5)

41. **radial velocity (V_r):** A component of an object's velocity directed away from or toward Earth. (p. 113; objective 5)

Summary

In order to appreciate the nature of a star and its spectrum, it helps to understand the basic structure of an atom and how atoms interact with light.

Atoms

An atom has a positively charged **nucleus** that is surrounded by an **electron shell** or cloud, but it is mostly empty space. The nucleus contains two kinds of particles: a positively charged **proton** and a neutrally charged **neutron**. The electron shell contains negatively charged electrons that whirl around the nucleus. The nucleus has a net positive charge and the electron shell has a net negative charge. In a normal atom, the electrons and protons balance each other out so the atom has a neutral charge.

There are more than one hundred different types of atoms known as chemical elements—substances that cannot be decomposed by chemical reaction into a simpler substance. The number of protons in the nucleus determines which element the atom is: hydrogen has one proton, helium has two, lithium has three, and so forth. The number of protons in an element is fixed and corresponds to its atomic number, but the number of neutrons and electrons can vary without significantly altering the element's chemical properties.

Atoms that have the same number of protons—the same atomic number—but have a different number of neutrons are called different **isotopes**. Isotopes of an element have different atomic weights and different nuclear properties. Atoms can also vary in the number of electrons they have. Because the electrons are loosely bound in the electron shell, they can be removed by a process called **ionization**. Likewise, an atom can gain extra electrons under certain circumstances. An atom that has lost or gained an electron and is no longer neutral is called an **ion**.

When two or more atoms collide, they may form bonds by exchanging or sharing electrons. These bonded atoms are called **molecules**. For instance, carbon monoxide (CO) is a molecule made from one carbon atom and one oxygen atom; carbon dioxide (CO_2) is a made from one carbon atom and *two* oxygen atoms. This type of chemical bonding requires gentle collisions in cool matter; in the next lesson you'll learn what happens when atoms collide violently in a process called nuclear fusion.

The negative electrons are bound to the positive nucleus by **binding energy**; this attraction between an electron and the nucleus is called the **Coulomb force**. As the electrons whirl around the nucleus, they do so in **permitted orbits** or **energy levels**; an atom can have only a certain amount of binding energy and this determines its permitted orbits. The sizes of the orbits depend on the energy that binds the atom together. An electron can occupy only a permitted orbit—not the space in between. The number of energy levels, the number of electrons permitted in each energy level, and how closely they orbit the nucleus depend on the atom.

The Interaction of Light and Matter

Every atom has a **ground state**—the lowest permitted electron energy level—and the atom is most stable in this state. An electron can move from one energy level to another if energy of the right wavelength is added to it. When an electron is moved from one energy level to a higher energy level, the atom is an **excited atom**.

An atom can be excited by collision—if two atoms collide violently, the electrons in one or both atoms may move to a higher energy level. An atom can also be excited when it absorbs a photon of the appropriate wavelength. Only a photon with the exact amount of energy can move an electron from one energy level to another—and this varies by the type of atom. Different energy levels within an atom can absorb photons of different wavelengths.

An atom can remain in its excited state for a very short period of time before quickly returning to its ground state. When the electron in an excited atom drops from a higher energy level to a lower one, it can emit the excess energy as a photon. Depending on the energy that excited it, an electron can move through a combination of energy levels and each transition emits a single photon. For example, if the ground state of an electron is the first energy level, and it gets excited to the fourth energy level, it may come back down to the first energy level directly, or it may make a stop at the third or the second energy level. Because each type of atom has a unique set of energy levels, each type absorbs and emits photons of different wavelengths. We can therefore identify the elements in certain types of matter by studying the wavelengths of the photons that are emitted or absorbed by its atoms. This is how atoms leave their fingerprints on light.

Molecules and atoms in an object are in perpetual motion—they are constantly colliding with one another. In a hot object, they move faster than in a cool object; this motion is known as thermal energy or **heat**. **Temperature** refers to the average speed of the particles in an object.

Astronomers prefer to use the **Kelvin temperature scale** to measure the temperature of celestial objects because the scale has no negative values. The temperature at which an object contains no heat energy is known as **absolute zero** or zero degrees Kelvin. Absolute zero is theoretically impossible to obtain—even if we could stop the motion in atoms caused by heat, there still would be vibratory motion. Any object that has a temperature greater than zero degrees Kelvin is considered to be heated—its particles are moving. The hotter an object is, the faster the particles move. When an object becomes hotter, its particles become more agitated and collide with electrons. When the electrons are accelerated, they emit radiation and part of the energy is emitted as photons.

The radiation that's emitted by a heated object is known as **black body radiation**. A black body is a theoretical object that absorbs 100 percent of the radiation that hits it. Therefore it reflects no radiation and appears perfectly black; at a particular temperature the black body would emit the maximum amount of energy possible for that temperature. The hotter the object is, the more black body radiation it emits because the agitated atoms collide more frequently and violently with electrons. Even cold objects—anything with a temperature over zero degrees Kelvin (0 K)—emit black body radiation.

Not only is there a relationship between an object's temperature and the amount of black body radiation it emits, there is a relationship between the temperature of the object and the wavelengths of the photons it emits. Hotter objects emit more blue light and cooler objects emit more red light. The wavelength at which an object emits the most radiation is known as its **wavelength of maximum intensity (λmax)**. An object's wavelength of maximum intensity depends on its temperature.

Stellar Spectra

The wavelengths of the photons emitted by an excited atom are determined by the energy levels in an atom. Whether an atom is emitting a photon or absorbing one, the energy levels in the atom are the same and the wavelengths are the same. Different elements have different energy levels so their spectra contain different patterns called **spectral lines**. There are three different types of spectra: a continuous spectrum, an emission spectrum, and an absorption spectrum. **Kirchhoff's laws** describe them.

In a **continuous spectrum**, a solid, liquid, or dense gas is excited to emit electromagnetic radiation at all wavelengths and produces a spectrum of all colors with no spectral lines. The benefit of a continuous spectrum is that its black body radiation can reveal the temperature of an object. For example, a blue star is hot and a red star is cool. But there are no lines to reveal chemical composition.

In an **emission spectrum**—also known as a **bright-line spectrum**—a low-density gas surrounding a light source is excited to emit light at certain wavelengths. If we observe the gas surrounding the light source, not the source itself, we can detect emission lines. The benefit of the emission spectrum is that it reveals the chemical composition of the gas, but does not directly reveal its temperature. For example, neon is orange because it's neon, not because it's cool.

In an **absorption spectrum**—also known as a **dark-line spectrum**— continuous light coming from an object passes through a cooler, thinner gas that absorbs some of the photons. Atoms in the gas absorb photons of certain wavelengths, depending on the gas. These wavelengths are missing from the spectrum and you see their positions as dark absorption lines against the continuous background. For example, if the gas surrounding the light source is made up of hydrogen atoms, it will absorb different wavelengths than a gas comprised of helium atoms and therefore produce a different pattern of spectral lines. The benefit of the absorption spectrum is that it reveals both the temperature (from the peak color of the continuous background) and the composition (from the pattern of dark lines).

When the electrons in a hydrogen atom are in the second energy level, they can produce Balmer absorption lines and astronomers can study them to determine the temperature of a star. Both hot and cool stars produce weak Balmer absorption lines. In a cool star, the electrons in a hydrogen atom are not excited and are mostly in the ground state orbiting in the first energy level. In a hot star, the hydrogen atoms are undergoing violent collisions and the electrons have been knocked out of most atoms, so few remain in the second energy level. Only medium-temperature stars of about 10,000 K have enough hydrogen atoms with electrons in their second energy level to produce strong Balmer lines.

You might wonder how studying hydrogen Balmer lines can help astronomers determine the temperature of a star when they are only strong in medium-temperature stars. The absorption lines created by other atoms are similarly affected by the temperature of the source. By combining the information gathered by the Balmer absorption lines and the absorption lines of other atoms, we are able to more accurately measure the temperature of a star. For example, the absorption lines of helium are strongest at 20,000 K, those of ionized iron are strongest at 5800 K, and those of titanium oxide are strongest at 3000 K.

Most stars are nearly identical in their composition, being mostly hydrogen and helium. The different patterns in their spectra are actually caused by having different surface temperatures. Therefore, we can surmise that all stars of a particular temperature will have the same spectral lines. After studying the spectra of thousands of stars, astronomers have developed a classification system. The seven **spectral classes** or **types** are from hottest to coolest: O, B, A, F, G, K, and M. Each spectral classification is further broken down into ten subclasses: from 0 to 9. For example, spectral class A is broken down into A0 (the hottest within the A class), A1, A2, and so forth. This allows astronomers to gauge the temperature of a star with accuracy within about 5 percent. It may be difficult to remember this seemingly random classification, but if you can remember the phrase "Oh Boy, An F Grade Kills Me," then you can remember the order of classification from hottest to coolest.

Not only can we determine the composition and temperature of a star by studying its spectrum, we can also establish if it's moving toward or away from Earth. We can do this by studying the **Doppler effect** in the star's spectrum—the apparent change in the wavelength of radiation caused by relative motion of the source.

You are most likely familiar with the Doppler effect when you hear a car with a siren pass you by. Although sound is not electromagnetic radiation, it travels as a wave and is therefore subject to the Doppler effect. Sounds with short wavelengths are higher-pitched while sounds with longer wavelengths are lower-pitched. As the car approaches you, its sound is shifted to shorter wavelengths and therefore sounds higher-pitched. As the car moves away from you, its sound is shifted to longer wavelengths and sounds lower-pitched (see Figure 6-11 on p. 114 in the *Horizons* textbook).

The Doppler effect reveals relative motion—it doesn't matter if the source of the sound is moving toward or away from us or if we're moving toward or away from it. Furthermore, it only indicates the component of the object's velocity that is coming toward you or away from you known as **radial velocity** (V_r); it does not indicate motion that's perpendicular to you.

When analyzing the motion of a star, astronomers can study the shift in the spectral lines to identify its relative motion and its speed. If astronomers detect a shift in the spectral lines toward the shorter wavelength—the blue end of the spectrum—they know the star is moving toward Earth (or Earth toward it). This is known as **blueshift**. Conversely, if astronomers detect a shift in the spectral lines toward the longer wavelength—the red end of the spectrum—they know the star is moving away from Earth. This is known as **redshift**.

By measuring how much the spectral lines shift, astronomers can determine the star's radial velocity—a slow-moving star has a smaller Doppler effect and a smaller shift in the spectral lines than a fast-moving star. An object that is rotating is, in a sense, moving toward us and away from us at the same time. If the absorption lines were broadened, we know that a star is rotating quite rapidly.

Studying the interaction of light and atoms and celestial objects' spectra provides the key that unlocks the universe. The information gathered can help us identify the composition, the temperature, and the motion of objects in the universe. In the next few lessons you'll learn more about how the atoms in the sun and other stars produce energy.

Review Exercises

Matching I

Match each term with the appropriate definition or description.

1. _____ spectral line		6. _____ isotope
2. _____ nucleus		7. _____ ionization
3. _____ proton		8. _____ ion
4. _____ neutron		9. _____ molecule
5. _____ electron		10. _____ Coulomb force

    a. Two or more atoms bonded together.

    b. The central core of an atom containing protons and neutrons; carries a net positive charge.

    c. The electrostatic force of repulsion or attraction between charged bodies.

    d. A line in a spectrum at a specific wavelength produced by the absorption or emission of light by certain atoms.

    e. The process in which atoms lose or gain electrons.

    f. A positively charged atomic particle contained in the nucleus of an atom.

    g. Low-mass atomic particle carrying a negative charge.

    h. An atom that has the same number of protons but a different number of neutrons.

i. An atomic particle with no charge and about the same mass as a proton.

j. An atom that has lost or gained one or more electrons.

Matching II

Match each term with the appropriate definition or description.

1. _____ binding energy	7. _____ heat
2. _____ quantum mechanics	8. _____ temperature
3. _____ permitted orbit	9. _____ Kelvin temperature scale
4. _____ energy level	10. _____ absolute zero
5. _____ excited atom	11. _____ black body radiation
6. _____ ground state	

a. One of a number of states an electron may occupy in an atom, depending on its binding energy.

b. The flow of thermal energy.

c. One of the energy levels in an atom that an electron may occupy.

d. The study of the behavior of atoms and atomic particles.

e. A measure of the agitation among the atoms and molecules of a material; the intensity of the atomic motion.

f. An atom in which an electron has moved from a lower to a higher energy level.

g. A temperature scale using Celsius degrees and based on zero at absolute zero.

h. The lowest permitted electron energy level in an atom.

i. The energy needed to pull an electron away from its atom.

j. The theoretical lowest possible temperature at which a material contains no extractable heat energy.

k. Radiation emitted by a hypothetical perfect radiator.

Matching III

Match each term with the appropriate definition or description.

1. _____ wavelength of maximum intensity	7. _____ emission line
	8. _____ Kirchhoff's laws
2. _____ joule	9. _____ transition
3. _____ continuous spectrum	10. _____ Lyman series
4. _____ absorption spectrum	11. _____ Balmer series
5. _____ absorption line	12. _____ Paschen series
6. _____ emission spectrum	

a. A dark line in a spectrum; is produced by the absence of photons absorbed by atoms or molecules.

b. The movement of an electron from one atomic energy level to another.

c. A spectrum containing emission lines.

d. A series of spectral lines produced by hydrogen in the near-ultraviolet and visible parts of the spectrum.

e. Spectral lines in the infrared spectrum of hydrogen produced by transitions whose lowest energy level is the third

f. Spectral lines in the ultraviolet spectrum of hydrogen produced by transitions whose lowest energy level is the ground state.

g. A bright line in a spectrum caused by the emission of photons from atoms.

h. A spectrum in which there are no absorption or emission lines.

i. A set of laws that describe the absorption and emission of light by matter.

j. A spectrum that contains absorption lines caused by photons being absorbed by atoms or molecules.

k. The wavelength at which a perfect radiator emits the maximum amount of energy; depends only on the object's temperature.

l. A unit of energy equivalent to a force of 1 newton acting over a distance of 1 meter.

Matching IV

Match each term with the appropriate definition or description.

1. _____ spectral class or type	5. _____ Doppler effect
2. _____ spectral sequence	6. _____ blueshift
3. _____ L dwarf	7. _____ redshift
4. _____ T dwarf	8. _____ radial velocity

a. The arrangement of spectral classes ranging from hot to cool.

b. A Doppler shift toward shorter wavelengths caused by a velocity of approach.

c. A very cool, low-mass star located below the L stars on the main sequence.

d. A star's position in the temperature classification; based on the appearance of the star's spectrum.

e. A component of an object's velocity directed away from or toward Earth.

f. The change in the wavelength of radiation due to the relative radial motion of source and observer.

g. A Doppler shift toward longer wavelengths caused by a velocity of recession.

h. A main sequence star cooler than an M star.

Completion

Fill each blank in the sentences below with the most appropriate term from the list of completion answers that follow. A term may be used once, more than once, or not at all. Check your answers with the Answer Key and review when necessary.

absolute zero	longer	size
Balmer	Lyman	T
ion	Q	temperature
isotope	R	triple point
L	shorter	

1. An atom that has the same number of protons but a different number of neutrons is called a(n) _____.

2. An atom that has the same number of protons but a different number of electrons is called a(n) _____.

3. Long after the spectral sequence was created, astronomers found stars at temperatures even cooler than the M stars. These are called _____ and _____ stars.

4. The sun is approximately a black body. This means that the radiation emitted depends on its _____.

5. The hydrogen lines that are in the visible region of the electromagnetic region are _____ series.

6. Zero on the Kelvin temperature scale is also called _____.

7. A horseshoe is being heated in a forge. As it heats up, it changes from red to white, as the atoms become more excited and cause more collisions and radiate heat at _____ wavelengths.

Self-Test

Select the best answer.

1. Because of the Doppler effect, the spectral lines of a luminous object moving away from Earth will appear
   a. shifted slightly toward the blue.
   b. shifted slightly toward the red.
   c. split into several adjacent lines.
   d. thicker than normal.

2. The chemical element found in the greatest abundance in all normal stars is
   a. iron.
   b. carbon.
   c. helium.
   d. hydrogen.

3. Which of the following spectral classes has stars with the hottest temperature?
   a. A
   b. F
   c. G
   d. O

4. The sun produces
   a. no spectrum.
   b. an emission spectrum.
   c. an absorption spectrum.
   d. a continuous spectrum.

5. If an atom were 4.5 football fields long, the nucleus would be the size of
   a. a grape seed.
   b. a watermelon.
   c. a car.
   d. a house.

6. How many elements have been found in the sun using the solar spectrum?
   a. 3
   b. 5
   c. 10
   d. more than 90

7. Who personally inspected and classified the spectra of over a quarter of a million stars?
   a. Edwin Hubble
   b. Annie Jump Cannon
   c. Fred Whipple
   d. Jocelyn Bell

8. Which spectral class has strong hydrogen Balmer lines and weak ionized calcium spectral lines?
   a. O
   b. A
   c. G
   d. M

9. Without any additional interactions, an excited atom will
   a. drop to a lower energy level.
   b. jump to a higher energy level.
   c. do nothing.
   d. ionize.

10. Light passing through a cool, thin gas will produce
    a. a continuous spectrum.
    b. an emission spectrum.
    c. an absorption spectrum.
    d. no spectrum.

11. Light produced in a hot, thin gas will produce
    a. a continuous spectrum.
    b. an emission spectrum.
    c. an absorption spectrum.
    d. no spectrum.

12. Light emitted from a hot dense object will produce
    a. a continuous spectrum.
    b. an emission spectrum.
    c. an absorption spectrum.
    d. no spectrum.

Short-Answer Questions

1. How do permitted energy levels of the electron allow us to identify a specific element?

2. What is the difference between heat and temperature?

3. Why are titanium oxide bands produced in stars cooler than 3000 K?

4. Why does the Doppler effect only find the radial velocity?

5. How can you find the temperature of a star by its color?

6. What is an ion and how is it formed?

Applications

1. What would be the maximum intensity wavelength of a star that has a temperature of 10,000 K?

2. If a star emits its most intense light at a wavelength of 650 nm, how hot is it?

3. How much more energy does a 9000 K star emit compared to a 3000 K star (three times the absolute temperature)?

4. What is your maximum intensity wavelength and in what region of the spectrum does it fall?

5. Explain how emission, absorption and continuous spectra are produced.

Answer Key

Matching I

1. d (p. 100; objectives 1 & 3)
2. b (p. 100; video lesson; objective 1)
3. f (p. 100; video lesson; objective 1)
4. i (p. 100; video lesson; objective 1)
5. g (p. 100; video lesson; objective 1)
6. h (p. 101; objective 1)
7. e (p. 101; objective 1)
8. j (p. 101; objective 1)
9. a (p. 101; objective 1)
10. c (p. 101; objective 2)

Matching II

1. i (p. 101; objective 2)
2. d (p. 101; objective 2)
3. c (p. 101; video lesson; objective 2)
4. a (p. 103; video lesson; objective 2)
5. f (p. 103; video lesson; objective 2)
6. h (p. 103; video lesson; objective 2)
7. b (p. 104; objective 2)
8. e (p. 104; objective 2)
9. g (p. 104; video lesson; objective 2)
10. j (p. 104; video lesson; objective 2)
11. k (p. 104; video lesson; objective 2)

Matching III

1. k (p. 104; objective 2)
2. l (p. 105; objective 2)
3. h (p. 108; video lesson; objective 3)
4. j (p. 108; video lesson; objective 3)
5. a (p. 108; video lesson; objective 3)

6. c (p. 108; objective 3)

7. g (p. 108; objective 3)

8. i (p. 108; video lesson; objective 3)

9. b (p. 109; objectives 2 & 3)

10. f (p. 109; objective 3)

11. d (p. 109; video lesson; objective 3)

12. e (p. 109; objective 3)

Matching IV

1. d (p. 110; video lesson; objective 4)

2. a (p. 110; objective 4)

3. h (p. 110; objective 4)

4. c (pp. 110–111; objective 4)

5. f (p. 111; video lesson; objective 5)

6. b (p. 112; video lesson; objective 5)

7. g (p. 112; video lesson; objective 5)

8. e (p. 113; objective 5)

Completion

1. isotope (p. 101; objective 1)

2. ion (p. 101; objective 1)

3. L, T [in any order] (pp. 110–111; objective 4)

4. temperature (p. 104; objective 2)

5. Balmer (p. 109; video lesson; objective 3)

6. absolute zero (p. 104; video lesson; objective 2)

7. shorter (pp. 104–105; video lesson; objective 2)

Self-Test

1. b (p. 112; video lesson; objective 5)

2. d (p. 115; objective 4)

3. d (p. 110; objective 4)

4. c (pp. 106–110; video lesson; objective 4)

5. a (p. 100; objective 1)

6. d (p. 114; objective 4

7. b (p. 110; video lesson; objective 4)

8. b (p. 110; objective 4)

9. a (p. 103; video lesson; objective 2)

10. c (p. 108; video lesson; objective 3)

11. b (p. 108; video lesson; objective 3)

12. a (p. 108; video lesson; objective 3)

Short-Answer Questions

1. The transitions for the electron energy levels produce specific wavelengths to identify specific elements, like a fingerprint. (pp. 103–104, 109; video lesson; objective 2)

2. Temperature is a measure of hotness or coldness. Heat is a transfer of energy. (p. 104; objective 2)

3. If it were any hotter, the temperatures would not allow the molecules to stay bonded. (pp. 106–107; objective 4)

4. You cannot use the Doppler effect to detect any part of the velocity that is perpendicular to your line of sight. It is only sensitive to the part of the velocity directed away from you or toward you. (p. 113; objective 5)

5. For black bodies, the maximum intensity wavelength (color) changes with temperature. (pp. 104–105; objective 2)

6. An ion is an atom with an electric charge. The atom can either gain or lose one or more electrons. (p. 101; objective 1)

Applications

1. The answer is 300 nm. 3,000,000/10,000 = 300 (p. 105; objective 2)

2. The answer is 4615 degrees Kelvin. 3,000,000/650 = 4615 K (p. 105; objective 2)

3. The answer is 81 times. Energy is related to temperature to the fourth power, and $3^4 = 81$ (p. 105; objective 2)

4. The answer is 9,677 nm, infrared. 98.6° F = 37° C = 310 K; and 3,000,000/310 = 9,677 (p. 105; objective 2)

5. An absorption spectrum results when radiation passes through a cool gas. An emission spectrum is produced by photons emitted by an excited gas. A hot dense object produces a continuous spectrum. (p. 108; video lesson; objective 3)

Lesson Review

Lesson 6: The Science of Starlight

PLEASE NOTE: Use this matrix to guide your study and achieve the learning objectives of this lesson. It will also help you to view the video, which defines and demonstrates important concepts and principles as they relate to everyday life and actual case studies.

Learning Objective	Textbook	Student Guide
1. Describe the basic structure of an atom, including different kinds of atoms, and the nature of electron shells.	pp. 100–102	Key Terms: 1, 2, 3, 4, 5, 6, 7, 8, 9; Matching I: 1, 2, 3, 4, 5, 6, 7, 8, 9; Completion: 1, 2; Self-Test: 5; Short-Answer: 6.
2. Explain the interaction of light and matter, including the excitation of atoms and radiation from a heated object, and describe the relationship between temperature, wavelength of maximum intensity, and total radiated energy.	pp. 102–106	Key Terms: 10, 11, 12, 13, 14, 15, 16, 17, 18, 19, 20, 21, 22, 23, 29; Matching I: 10; Matching II: 1, 2, 3, 4, 5, 6, 7, 8, 9, 10, 11; Matching III: 1, 2, 9; Completion: 4, 6, 7; Self-Test: 9; Short-Answer: 1, 2, 5; Applications: 1, 2, 3, 4.
3. Describe the appearance, origin, and characteristics of continuous, emission, and absorption spectra, including the visible and nonvisible spectrum of hydrogen.	pp. 106–109	Key Terms: 1, 24, 25, 26, 27, 28, 29, 30, 31, 32; Matching I: 1; Matching III: 3, 4, 5, 6, 7, 8, 9, 10, 11, 12; Completion: 5; Self-Test: 10, 11, 12; Applications: 5.
4. Explain spectral classification and how the appearance of a spectrum indicates stellar temperature and composition.	pp. 107, 110–111	Key Terms: 33, 34, 35, 36; Matching IV: 1, 2, 3, 4; Completion: 3; Self-Test: 2, 3, 4, 6, 7, 8; Short-Answer: 3.

Learning Objective	Textbook	Student Guide
5. Explain what spectra indicate about the speed and direction of stellar motions as revealed by the Doppler effect.	pp. 111–114	Key Terms: 37, 38, 39, 40; Matching IV: 5, 6, 7, 8; Self-Test: 1; Short-Answer: 4.

Notes:

LESSON
7

The Sun—Our Star

Checklist

For the most effective study of this lesson, complete the following activities in this sequence.

Before Viewing the Video

❑ Read the Preview, Learning Objectives, and Viewing Notes below.

❑ Read Chapter 7, "The Sun," pages 118–141, in the *Horizons* textbook.

What to Watch

❑ After reading the textbook chapter, watch the video for Lesson 7, *The Sun—Our Star.*

After Viewing the Video

❑ Briefly note your answers to questions listed at the end of the Viewing Notes.

❑ Review the Summary below.

❑ Review all reading assignments for this lesson, especially the Chapter 7 summary on pages 139–140 in *Horizons* and the Viewing Notes in this lesson.

❑ Write brief answers to the review questions at the end of Chapter 7 in *Horizons.*

❑ Complete the Review Exercises below. Check your answers with the Answer Key and review when necessary.

❑ Use the Lesson Review matrix found at the end of this lesson to review and assess your knowledge of each Learning Objective.

❑ As assigned by your instructor, complete the Applications activities and any additional activities for this lesson.

Preview

Think about how the sun affects us here on Earth; it warms our planet and provides us with energy. Through photosynthesis, the sun fuels the cycle of life on our planet. The sun is similar to millions of other stars in the universe. But, because it's so much closer, we can discover more about it. And, because it's like so many other stars in the universe, we can learn about them, too.

As you discovered in Lesson 6, we can learn a lot about a star by its spectrum. By studying the sun's spectrum, astronomers have been able to understand how it produces energy and how that energy is transmitted to Earth.

Like any other star, the sun has a dense core, a visible surface, and an atmosphere. Energy is created in the sun's core by a series of nuclear reactions and that energy flows outward to the surface continuing through the solar atmosphere. This energy reaches Earth in the form of heat and light. The sun's strong magnetic field can also cause climate changes and can interfere with the electricity that powers our computers and cell phones.

CONCEPTS TO REMEMBER

- Recall from Lesson 6 that an *absorption spectrum* is a spectrum that is produced when radiation passes through a cooler gas. Like other stars, our sun emits an absorption spectrum as radiation emerges from its interior (p. 104 in this guide).

- In Lesson 6, you learned that *ionization* is a process in which atoms lose or gain electrons and become positively or negatively charged. The ions that result from this process are good conductors of electricity and can trace the sun's magnetic fields (p. 103 in this guide).

- In Lesson 2, you also learned that a *scientific model* is a carefully devised mental conception of how something works. Such models help scientists think about how some aspect of nature, like how the sun creates its energy, or why solar activity occurs (p. 16 in this guide).

Learning Objectives

After you complete this lesson, you should be able to:

1. Identify and describe the major layers of the solar interior and solar atmosphere, and discuss how they are observed. *HORIZONS* TEXTBOOK PAGES 120–125; VIDEO PROGRAM.

2. Describe the behavior and suspected cause of the solar magnetic cycle and its effects on sunspots, prominences, flares, coronal activity, solar wind, and Earth. *HORIZONS* TEXTBOOK PAGES 131–138.

3. Describe nuclear fusion and explain how the proton–proton chain works to generate most of the sun's energy. *HORIZONS* TEXTBOOK PAGES 126–128.

4. Describe observations of solar neutrinos and explain why they are important to astronomers. *HORIZONS* TEXTBOOK PAGES 128–129.

At this point, read Chapter 7, "The Sun," pages 118–141.

> All cannot live on the piazza, but everyone may enjoy the sun.
>
> **—Italian proverb**

Viewing Notes

The video for this lesson illustrates how the sun—a star like so many others in the universe—creates vast amounts of energy. Some of that energy reaches Earth to sustain life.

The video program contains the following segments:

- ✪ Nuclear Fusion in the Sun
- ✪ The Solar Surface & Atmosphere
- ✪ Solar Activity
- ✪ Effects on Earth

The following information will help you better understand the video program:

The sun produces energy by fusing atomic nuclei in a nuclear reaction called the **proton–proton chain**. When this nuclear reaction occurs, excess energy is released in the form of gamma rays, positrons, and neutrinos. You'll recall from Lesson 5 that a *gamma ray* is a form of electromagnetic radiation. A *positron* is a positively charged electron, and a **neutrino** is a neutrally charged, weakly interacting particle. In the proton–proton chain, the gamma rays are absorbed in the sun's interior which is *opaque*—impenetrable by light. It's so crowded because of the collision between photons and free electrons that the rays can't escape. But their energy eventually reaches the photosphere as heat and escapes primarily as visible light.

The video program shows how regions known as "granules" appear in the sun's photosphere. These granules are large cells of gas caused by **convection**—a circulation of fluid driven by heat. The heated gas that's swirling around in the sun's atmosphere is called *plasma*, which is a collection of *ions*. Plasma is an excellent conductor of electricity because the negative and positive charges are free to move around and trace the magnetic field in the gas.

The sun's magnetic field causes solar activity. The video discusses the fact that this activity is related to the *magnetic dynamo*—this is the same thing as the *dynamo effect* mentioned in the *Horizons* textbook (p. 134). The **dynamo effect** occurs when a rapidly rotating conductor—in this case, the plasma—is stirred by convection to produce a magnetic field.

QUESTIONS TO CONSIDER

- What scientific evidence supported the theory that the sun was a sphere of gas?

- What discovery supported the theory that the sun produced energy by nuclear fusion?

- How do astronomers know that convection takes place within the sun?

- How does the sun's strong magnetic field cause solar activity?

- How does solar activity affect Earth?

- How does the study of our sun help us understand other stars?

- How does the study of other stars help us understand our sun?

Watch the video for Lesson 7, *The Sun—Our Star*.

Key Terms and Concepts

Page references are keyed to the *Horizons* textbook.

1. **sunspots:** Relatively dark spots on the sun that contain intense magnetic fields. (p. 120; video lesson; objective 1)

2. **granulation:** The fine structure of bright grains covering the sun's surface. (p. 121; video lesson; objective 1)

3. **convection:** Circulation in a fluid driven by heat. (p. 121; objective 1)

4. **filtergram:** A photograph (usually of the sun) taken in the light of a specific region of the spectrum. (p. 122; objective 1)

5. **filaments:** Solar prominences seen from above silhouetted against the bright photosphere. (p. 123; objective 1)

6. **spicules:** Small, flamelike projections in the chromosphere of the sun. (p. 123; objective 1)

7. **supergranules:** Very large convective features in the sun's surface. (p. 121; video lesson; objective 1)

8. **magnetic carpet:** The network of small magnetic loops that covers the solar surface. (p. 123; objective 1)

9. **solar wind:** Rapidly moving atoms and ions that escape from the solar corona and blow outward through the solar system. (p. 124; objective 1)

10. **helioseismology:** The study of the interior of the sun by the analysis of its modes of vibration. (p. 124; objective 1)

11. **Maunder butterfly diagram:** A graph showing the latitude of sunspots versus time. (p. 132; objective 2)

12. **Zeeman effect:** The splitting of spectral lines into multiple components when the atoms are in a magnetic field. (p. 133; video lesson; objective 2)

13. **Maunder minimum:** A period of less numerous sunspots and other solar activity between 1645 and 1715. (p. 133; video lesson; objective 2)

14. **active regions:** Magnetic regions on the solar surface that include sunspots, prominences, flares, and similar features. (p. 133; objective 2)

15. **differential rotation:** The rotation of a body in which different parts of the body have different periods of rotation. (p. 134; objective 2)

16. **dynamo effect:** The process by which a rotating, convecting body of conducting matter can generate a magnetic field. (p. 134; objective 2)

17. **Babcock model:** A model of the sun's magnetic cycle in which the differential rotation of the sun winds up and tangles the solar magnetic field in a 22-year cycle. (p. 134; video lesson; objective 2)

18. **weak force:** One of the four forces of nature; is responsible for some forms of radioactive decay. (p. 126; objective 3)

19. **strong force:** One of the four forces of nature; binds protons and neutrons together in atomic nuclei. (p. 126; objective 3)

20. **nuclear fission:** Reactions that break the nuclei of atoms into fragments. (p. 126; video lesson; objective 3)

21. **nuclear fusion:** Reactions that join the nuclei of atoms to form more massive nuclei. (p. 126; video lesson; objective 3)

22. **prominence:** A looping eruption on the solar surface. (p. 136; video lesson; objective 2)

23. **flares:** Violent eruptions on the sun's surface. (p. 137; objective 2)

24. **reconnection:** On the sun, the merging of magnetic fields to release energy in the form of flares. (p. 137; objective 2)

25. **auroras:** The glowing light displays that result when a planet's magnetic field guides charged particles so that they strike the upper atmosphere and excite atoms to emit photons. (p. 137; objective 2)

26. **coronal holes:** Areas of the solar surface that are dark at X-ray wavelengths; may be the source of solar wind. (p. 137; objective 2)

27. **coronal mass ejections (CMEs):** Matter ejected from the sun's corona in powerful surges guided by magnetic fields. (p. 137; objective 2)

28. **Coulomb barrier:** The electrostatic force of repulsion between bodies of like charge. (p. 127; objective 3)

29. **proton–proton chain:** A series of three nuclear reactions that builds a helium atom by adding together protons. (p. 127; video lesson; objective 3)

30. **deuterium:** An isotope of hydrogen in which the nucleus contains one proton and one neutron. (p. 127; video lesson; objective 4)

31. **neutrino:** A neutral atomic particle that travels at or nearly at the speed of light. (p. 127; video lesson; objective 4)

Summary

The sun is like many other stars in the universe. It is a huge ball of gas held together by its own gravity. Scientists have studied the sun for hundreds of years. But not until the 1930s did astronomers understand how the sun makes its energy.

Nuclear Fusion in the Sun

The sun's energy is generated by **nuclear fusion**, which takes place near the center of the sun where it is hot and dense. Because conditions near the center of the sun make it so hot, the atoms are ionized and collide violently with one another. In the **proton–proton chain**, a series of three nuclear reactions combines four of these hydrogen nuclei in the gas to build one helium nucleus.

When these particles combine, there's a little bit of mass left over and that mass is converted into energy. The energy is released in the form of gamma rays, positrons, and neutrinos. The positrons and gamma rays help keep the star hot. The neutrinos, on the other hand, almost never interact with other particles and eventually make their way to Earth, allowing astronomers to study them. Neutrinos provided evidence that the sun makes its energy by nuclear fusion.

It's important to note that the way the sun makes energy is different than the way a nuclear power plant makes energy. The sun makes energy by fusion (combining of atoms) while a nuclear power plant makes energy by fission (splitting of uranium atoms).

The Solar Atmosphere

The sun's atmosphere is comprised of three layers: the *photosphere*, the *chromosphere*, and the *corona* (introduced in Lesson 3, p. 36 of this guide). We can tell a lot about these layers by their spectra.

The photosphere is the thin layer of gas closest to the sun from which Earth receives most of its light. Although it appears to be a solid surface, it is not.

It is less than 500 km deep and has an average temperature of about 5800 K. Astronomers have learned about the photosphere by studying its *absorption spectrum*. The lower layers of the photosphere are dense and produce light in a continuous spectrum, but the gases in the upper layers absorb photons of specific wavelengths.

The chromosphere is the layer just above the photosphere and has an average depth of less than Earth's diameter. Because the chromosphere is faint in comparison to the photosphere, it can only be detected with the unaided eye during a total solar eclipse. Astronomers have learned about the chromosphere by studying its *emission spectrum*—a spectrum that is produced by photons emitted by an excited gas (Lesson 6, p. 104 of this guide). The presence of an emission spectrum tells you that the chromosphere must contain an excited, low-density gas.

The corona is the outermost layer of the sun's atmosphere. The density of the corona is very low and does not emit much radiation. It contains very hot gas, but it's not very bright. You can see the corona during a total solar eclipse and it doesn't even shine as brightly as the full moon.

The temperature of the corona rises with altitude and reaches more than 1 million degrees Kelvin at its outermost regions. You might wonder why the gas gets hotter the farther away it is from the sun's core. Astronomers believe that this is caused by the very strong magnetic fields that extend up through the photosphere into the corona and churn the gas and heat it. This hot, ionized gas follows the magnetic fields away from the sun in a solar wind that blows past Earth. The **solar wind** is composed of rapidly moving atoms and ions.

You've probably heard of a phenomenon called the Aurora Borealis, or Northern Lights, or you may have heard of its equivalent in the southern hemisphere called the Aurora Australis. An **aurora** is the beautiful, irregular light displayed in the night sky. It occurs when Earth's magnetic field guides the charged particles in the solar wind toward the north and south magnetic poles. The atoms and ions carried by the solar wind strike oxygen and nitrogen molecules in Earth's atmosphere, causing them to glow in various colors.

Auroras are related to the sun's magnetic cycle and are most frequent during sunspot maximum. An aurora can have a significant impact on radio communication, radar, and power systems on Earth. Auroras may be seen from a very large geographic area, but the closer you are to Earth's polar regions, the more likely you are to see one. In northern Alaska, auroras can be seen more than 200 nights a year. In the southern regions of the United States, they are much less frequent and only seen every 10–30 years.

Solar Activity

Because the sun is gaseous, different parts of it can rotate at different speeds. The photosphere rotates faster at the equator than at the higher latitudes and the deeper layers of the sun rotate at different speeds than the outer layer. This phenomenon is called **differential rotation**.

The **Babcock model** explains the sun's magnetic cycle as a progressive tangling of the solar magnetic field. The sun's highly ionized gas is locked into its magnetic field and as the gas moves, the magnetic field moves with it. Convection in the sun's atmosphere and the sun's differential rotation cause the magnetic field to become tangled and unstable. This instability manifests itself in the forms of solar activity, such as sunspots, prominences, and flares.

Sunspots are areas on the sun that appear dark and the sun's strong magnetic fields are believed to cause them. Rising and sinking currents of ionized gas twist the magnetic field into loops; where the loops of tangled magnetic field rise through the surface, sunspot pairs occur. Convection is reduced in certain areas and the gas remains cooler and therefore appears darker. The sunspot cycle is 11 years long and is related to the tangling of the magnetic field. After about 11 years of tangling, the magnetic field becomes so complex that it corrects itself. The newly corrected magnetic field is reversed, and the next sunspot cycle begins with magnetic north replaced by magnetic south. The complete magnetic cycle is 22 years long—11 years to tangle the magnetic field in one direction until it corrects itself and 11 years to tangle in the opposite direction until it corrects itself again.

A **prominence** is a cloud of gas that appears as a magnetic arch rising through the photosphere and chromosphere and into the corona. The ionized gas orients itself along the magnetic lines of the sunspot pairs. The prominence can erupt and send large amounts of material outward into space; this is called a **coronal mass ejection**, which can trigger a solar flare.

Solar **flares** are massive eruptions on the sun's surface. They occur in active regions where oppositely directed magnetic fields meet and cancel each other out in a process called **reconnection** (also known as recombination). When the magnetic fields cancel each other out, excess energy is thrown off in the form of a solar flare. The X rays and ultraviolet photons emitted from the flare travel to Earth at the speed of light.

Astronomers have learned about the sun—and its different layers—by observing the absorption and emission spectra. Likewise, because astronomers are not able to see the details of the solar interior, they rely on measuring the vibrations of the sun in a process called **helioseismology**. Just like geologists can learn about Earth by studying the vibrations in its surface (through a process called seismology), astronomers can learn about the sun by studying the vibrations on *its* surface. Helioseismology has allowed astronomers to determine the temperature, density and rate of rotation inside the sun.

Review Exercises

Matching I

Match each term with the appropriate definition or description.

1. _____ sunspots	7. _____ supergranules
2. _____ granulation	8. _____ magnetic carpet
3. _____ convection	9. _____ solar wind
4. _____ filtergram	10. _____ helioseismology
5. _____ filaments	11. _____ Maunder butterfly diagram
6. _____ spicules	

    a. Solar prominences seen from above silhouetted against the bright photosphere.
    b. The network of small magnetic loops that covers the solar surface.
    c. Relatively dark spots on the sun that contain intense magnetic fields.
    d. Rapidly moving atoms and ions that escape from the solar corona and blow outward through the solar system.
    e. Very large convective features in the sun's surface.
    f. A graph showing the latitude of sunspots versus time.
    g. The study of the interior of the sun by the analysis of its modes of vibration.
    h. A photograph (usually of the sun) taken in the light of a specific region of the spectrum.
    i. Small, flamelike projections in the chromosphere of the sun.
    j. The fine structure of bright grains covering the sun's surface.
    k. Circulation in a fluid driven by heat.

Matching II

Match each term with the appropriate definition or description.

1. _____ Zeeman effect	6. _____ Babcock model
2. _____ Maunder minimum	7. _____ weak force
3. _____ active regions	8. _____ strong force
4. _____ differential rotation	9. _____ nuclear fission
5. _____ dynamo effect	10. _____ nuclear fusion

    a. A period of less numerous sunspots and other solar activity between 1645 and 1715.

b. The rotation of a body in which different parts of the body have different periods of rotation.

c. Reactions that join the nuclei of atoms to form more massive nuclei.

d. A model of the sun's magnetic cycle in which the differential rotation of the sun winds up and tangles the solar magnetic field in a 22-year cycle.

e. The process by which a rotating convecting body of conducting matter can generate a magnetic field.

f. Magnetic regions on the solar surface that include sunspots, prominences, and flares.

g. The splitting of spectral lines into multiple components when the atoms are in a magnetic field.

h. One of the four forces of nature; is responsible for some forms of radioactive decay.

i. One of the four forces of nature; binds protons and neutrons together in atomic nuclei.

j. Reactions that break the nuclei of atoms into fragments.

Matching III

Match each term with the appropriate definition or description.

1. _____ prominence	6. _____ coronal mass ejections
2. _____ flares	7. _____ Coulomb barrier
3. _____ reconnections	8. _____ proton–proton chain
4. _____ auroras	9. _____ deuterium
5. _____ coronal holes	10. _____ neutrino

a. A neutral atomic particle that travels at or nearly at the speed of light.

b. An isotope of hydrogen in which the nucleus contains one proton and one neutron.

c. An area of the solar surface that is dark at X-ray wavelengths; may be the source of solar wind.

d. The electrostatic force of repulsion between bodies of like charge.

e. A looping eruption on the solar surface.

f. Matter ejected from the sun's corona in powerful surges guided by magnetic fields.

g. The glowing light display that results when a planet's magnetic field guides charged particles so that they strike the upper atmosphere and excite atoms to emit photons.

h. Violent eruptions on the sun's surface.

i. On the sun, the merging of magnetic fields to release energy in the form of flares.

j. A series of three nuclear reactions that builds a helium atom by adding together protons.

Completion

Fill each blank in the sentences below with the most appropriate term from the list of completion answers that follow. A term may be used once, more than once, or not at all. Check your answers with the Answer Key and review when necessary.

auroras   100 times   prominences
deuterium   1,000 times   spicules
fission   positrons   twice
fusion

1. The main mechanism for producing energy in the sun is nuclear

   _____.

2. In the proton–proton chain, energy appears in the form of gamma rays, neutrinos, and _____.

3. In the first reaction of the proton–proton chain, two protons combine to form a heavy hydrogen nucleus called _____.

4. An average sunspot is _____ the size of Earth.

5. Ionized gas trapped in a magnetic field causes arched shapes that occur in the chromosphere called _____.

6. Energy in the solar wind guided by Earth's magnetic field excites gases in the upper atmosphere and produces _____.

Self-Test

Select the best answer.

1. Which of the following describes the sequence of the layers of the solar atmosphere from the inside out?
   a. photosphere, corona, chromosphere
   b. corona, photosphere, chromosphere
   c. chromosphere, photosphere, corona
   d. photosphere, chromosphere, corona

2. Which of the following describes the sequence of the layers of the interior of the sun from the inside out?
   a. core, radiative zone, convective zone
   b. core, convective zone, radiative zone
   c. convective zone, radiative zone, core
   d. radiative zone, core, conductive zone

3. The Babcock model explains the magnetic cycles as
   a. an effect of the helioseismic vibrations in the sun.
   b. a progressive tangling of the solar magnetic field.
   c. a result of sunspot activity.
   d. being caused by neutrino production.

4. During the Maunder minimum, Europe and North America experienced a period of unusually
   a. wet weather.
   b. warm weather.
   c. cool weather.
   d. dry weather.

5. Granules just below the photosphere are caused by what kind of heating?
   a. radiative
   b. convective
   c. conductive
   d. chemical

6. When referring to the absorption lines in a spectrum, the photons of specific wavelength are absorbed in the sun's
   a. core.
   b. convective zone.
   c. photosphere.
   d. radiative zone.

7. Astronomers use the Zeeman effect to measure the
   a. rotation rate of the sun.
   b. speed of the solar wind.
   c. length of the sunspot cycle.
   d. strength of the magnetic field.

8. Auroras are most frequent
   a. at sunspot maximum.
   b. at sunspot minimum.
   c. during the Maunder minimum.
   d. at varied times—not related to the sunspot cycle.

9. Flares produce sudden eruptions of X rays and UV photons that reach Earth
   a. instantaneously.
   b. in 8 minutes.
   c. in a few hours.
   d. in a few days.

10. The sun rotates
    a. the slowest near the poles.
    b. the slowest near the mid-latitudes.
    c. the slowest near the equator.
    d. equally slow at all latitudes.

11. The four fundamental forces are
    a. gravity, electromagnetic, fission, and fusion.
    b. gravity, electromagnetic, weak nuclear, and strong nuclear.
    c. conduction, convection, radiation, and plasma.
    d. kinetic, potential, chemical, and radiation.

Short-Answer Questions

1. Why are sunspots cooler than the surrounding surface?

2. Why must a gas be incredibly hot for nuclear fusion to occur?

3. Explain why the sunspot cycle could be considered 11 or 22 years long.

4. Explain what causes the dynamo effect.

5. Why would astronauts on the moon or Mars be concerned about solar flares?

6. Explain why the "missing" solar neutrinos are no longer missing.

7. Use the Maunder butterfly diagram to describe the change in position (latitude) of sunspots through one cycle.

8. Explain how scientists probe the sun's interior by detecting neutrinos.

9. Describe how the Babcock model explains the magnetic cycle.

10. Explain how the nature of sunspots led astronomers to the currently accepted theory for the sun's magnetic cycle.

Applications

1. How much mass would have to be converted to energy to get the energy of one megaton bomb of TNT? One megaton of TNT = 4×10^{15} Joules.

2. If the sun releases 3.826×10^{26} Joules of energy every second, how many fusion reactions does it take to produce this each second?

3. If this many reactions (described in Question 2 above) occur each second, how many reactions occur over the ten-billion-year lifespan of the sun?

4. If 0.048×10^{-27} kg of mass is converted to energy in a single hydrogen fusion reaction, how much mass is lost over the lifespan of the sun? (Hint: Multiply by the answer from Question 3 above.)

5. What is the percentage of mass lost through fusion reactions compared to the total mass of the sun? Does this correspond with the value given in the reading on page 126 of the *Horizons* textbook?

6. How might the sun be different if it rotated as a solid sphere rather than with differential rotation?

7. Explain the process and products in each of the three reactions in the proton–proton chain.

Answer Key

Matching I

1. c (p. 120; video lesson; objective 1)
2. j (p. 121; video lesson; objective 1)
3. k (p. 121; video lesson; objective 1)
4. h (p. 122; objective 1)
5. a (p. 123; objective 1)
6. i (p. 123; objective 1)
7. e (pp. 121–122; video lesson; objective 1)
8. b (p. 123; objective 1)
9. d (p. 124; objective 1)
10. g (p. 124; objective 1)
11. f (p. 132; objective 2)

Matching II

1. g (p. 133; video lesson; objective 2)
2. a (p. 133; video lesson; objective 2)
3. f (p. 133; objective 2)
4. b (p. 134; objective 2)
5. e (p. 134; objective 2)
6. d (p. 134; video lesson; objective 2)
7. h (p. 126; objective 3)
8. i (p. 126; objective 3)
9. j (p. 126; objective 3)
10. c (p. 126; video lesson; objective 3)

Matching III

1. e (p. 136; video lesson; objective 2)
2. h (p. 137; video lesson; objective 2)
3. i (p. 137; objective 2)
4. g (p. 137; objective 2)
5. c (p. 137; objective 2)
6. f (p. 137; objective 2)
7. d (p. 127; objective 3)
8. j (p. 127; video lesson; objective 3)
9. b (p. 127; video lesson; objective 4)
10. a (p. 127; video lesson; objective 4)

Completion

1. fusion (p. 126; video lesson; objective 3)
2. positrons (p. 127; objective 3)
3. deuterium (p. 127; video lesson; objective 3)
4. twice (p. 132; objective 2)
5. prominences (p. 136; video lesson; objective 2)
6. auroras (p. 137; objective 2)

Self-Test

1. d (pp. 120–123; objective 1)
2. a (pp. 127–129; video lesson; objective 1)
3. b (pp. 134–135; video lesson; objective 2)
4. c (p. 133; video lesson; objective 2)
5. b (p. 121; video lesson; objective 1)
6. c (p. 121; objective 1)
7. d (p. 133; video lesson; objective 2)
8. a (p. 137; Background Notes 7; objective 2)
9. b (p. 137; video lesson; objective 2)
10. a (p. 134; objective 2)
11. b (p. 126; objective 3)

Short-Answer Questions

1. The strong fields are believed to inhibit gas motion below the photosphere; consequently, convection is reduced below the sunspot, and the surface is cooler. (p. 133; objective 2)

2. In a hotter gas, the particles move faster. In order for fusion to occur, they must collide violently by overcoming the repulsive Coulomb barrier. (pp. 126–127; objective 3)

3. The complete magnetic field cycle (sunspots switching from N-S back to N-S) is 22 years long and the sunspot cycle is 11 years long. (p. 135; objective 2)

4. A magnetic field is produced when a rapidly rotating conductor is stirred by convection. (p. 134; video lesson; objective 2)

5. The moon and Mars lack significant magnetism and would not protect astronauts from many of the particles released by the sun during a flare. (p. 137; video lesson; objective 2)

6. Only one-third of predicted neutrinos were observed. They are no longer missing because it was determined that neutrinos can be found in three different forms, or flavors, and scientists were only detecting one of them. (pp. 128–129; video lesson; objective 4)

7. Early in the cycle, the sunspots appear in the higher latitudes. Later in the cycle, the sunspots appear closer to the equator. (p. 132; objective 2)

8. If you can detect neutrinos, you can compare the number of them with the products of theorized thermonuclear fusion. (pp. 128–129; video lesson; objective 4)

9. The magnetic field gets captured in the gas and as the sun rotates with differential rotation, the magnetic field gets tangled. When it gets so tangled

that it breaks through the surface, you see sunspots. (pp. 134–135; video lesson; objective 2)

10. Sunspots always come in pairs and the polarity (N-S) reverses in each solar sunspot maximum. Using the Zeeman effect, we could tell that sunspots were magnetic. (pp. 131–133; video lesson; objective 2)

Applications

1. $m = 0.044$ kg or 44 grams (p. 127; video lesson; objective 3).

   $E = mc^2$

   $m = E/c^2 = (4 \times 10^{15}) / (9 \times 10^{16})$

2. The answer is 8.9×10^{37} reactions (pp. 126–127; objective 3).

   Total Energy / Energy in one reaction = Total reactions

   $(3.826 \times 10^{26}) / (4.32 \times 10^{-12})$

3. The answer is 2.8×10^{55} reactions (objective 3).

   Reactions per second × seconds in 10 billion years = Total reactions

   $(8.9 \times 10^{37}) \times (3.15 \times 10^7$ seconds in one year$) \times (1 \times 10^{10}$ years$)$

4. The answer is 1.35×10^{27} kg (pp. 126–127; objective 3).

   Mass loss per reaction × total reactions = Total mass lost

   $(0.048 \times 10^{-27}) \times (2.8 \times 10^{55})$

5. The answer is 0.068% (p. 126; objective 3).

   (Mass lost / Mass of the sun) × 100

   $((1.35 \times 10^{27}) / (1.989 \times 10^{30})) \times 100$

6. The magnetic field might not get twisted and there might not be a solar cycle. Flares, prominences, and sunspots might not occur. The sun might not have a magnetic field and a high temperature gas above its photosphere. (pp. 134–138; objective 2)

7. In the first reaction, two hydrogen nuclei combine to form deuterium and emits a positron and a neutrino. In the second reaction, the heavy hydrogen nucleus absorbs another proton and emits a gamma ray. In the last reaction, two light helium nuclei combine to form a normal helium nucleus and two protons. (p. 127; objective 3)

Lesson Review

Lesson 7: The Sun—Our Star

PLEASE NOTE: Use this matrix to guide your study and achieve the learning objectives of this lesson. It will also help you to view the video, which defines and demonstrates important concepts and principles as they relate to everyday life and actual case studies.

Learning Objective	Textbook	Student Guide
1. Identify and describe the major layers of the solar interior and solar atmosphere, and discuss how they are observed.	pp. 120–125	Key Terms: 1, 2, 3, 4, 5, 6, 7, 8, 9, 10; Matching I: 1, 2, 3, 4, 5, 6, 7, 8, 9, 10; Self-Test: 1, 2, 5, 6.
2. Describe the behavior and suspected cause of the solar magnetic cycle and its effects on sunspots, prominences, flares, coronal activity, solar wind, and Earth.	pp. 131–138	Key Terms: 11, 12, 13, 14, 15, 16, 17, 22, 23, 24, 25, 26, 27; Matching I: 11; Matching II: 1, 2, 3, 4, 5, 6; Matching III: 1, 2, 3, 4, 5, 6; Completion: 4, 5, 6; Self-Test: 3, 4, 7, 8, 9, 10; Short-Answer: 1, 3, 4, 5, 7, 9, 10; Applications: 6.
3. Describe nuclear fusion and explain how the proton–proton chain works to generate most of the sun's energy.	pp. 126–128	Key Terms: 18, 19, 20, 21, 28, 29; Matching II: 7, 8, 9, 10; Matching III: 7, 8; Completion: 1, 2, 3; Self-Test: 11; Short-Answer: 2; Applications: 1, 2, 3, 4, 5, 7.
4. Describe observations of solar neutrinos and explain why they are important to astronomers.	pp. 128–129	Key Terms: 30, 31; Matching III: 9, 10; Short-Answer: 6, 8.

Notes:

LESSON
8

The Family of Stars

Checklist

For the most effective study of this lesson, complete the following activities in this sequence.

Before Viewing the Video

❑ Read the Preview, Learning Objectives, and Viewing Notes below.

❑ Read Chapter 8, "The Family of Stars," pages 142–167, in the *Horizons* textbook.

What to Watch

❑ After reading the textbook chapter, watch the video for Lesson 8, *The Family of Stars*.

After Viewing the Video

❑ Briefly note your answers to questions listed at the end of the Viewing Notes.

❑ Review the Summary below.

❑ Review all reading assignments for this lesson, especially the Chapter 8 summary on page 166 in *Horizons* and the Viewing Notes in this guide.

❑ Write brief answers to the review questions at the end of Chapter 8 in *Horizons*.

❑ Complete the Review Exercises below. Check your answers with the Answer Key and review when necessary.

❑ Use the Lesson Review matrix found at the end of this lesson to review and assess your knowledge of each Learning Objective.

❑ As assigned by your instructor, complete the Applications activities and any additional activities for this lesson.

Preview

As you look up into the night sky, the universe appears to be two-dimensional—it looks like all of the stars are equidistant and simply vary in brightness. But, the universe is three-dimensional and two stars that appear to be equally bright can vary greatly in their size, mass, and how much energy they emit.

In Lesson 7, you learned about the sun—a typical star in the universe. We are able to learn a lot about the sun because we are able to observe it closely. Trying to understand a distant star can be quite challenging, but we can begin by studying our sun and other stars with similar properties.

Knowing the distance to a star is the first piece of information needed to understand the characteristics of a star. In this lesson, you will first learn how to calculate the distance from Earth to a star. Once you have that piece of information, you can gather more clues to determine a star's size, mass, and how much energy it emits.

Concepts to Remember

- Recall from Lesson 1 that an *astronomical unit (AU)* is the average distance from the Earth to the sun; 1.5×10^8 km, or 93×10^6 mi. Astronomers use AU as a standard unit of measurement (p. 4 in this guide).

- In Lesson 2, you learned that *apparent visual magnitude (m_v)* is the brightness of a star as seen by human eyes on Earth. This can help astronomers classify stars by their apparent brightness (p. 16 in this guide).

- In Lesson 4, you learned that *parallax* is the apparent change in position of an object due to a change of position of the observer. Astronomers can measure the parallax of a star to determine its distance from Earth (p. 55 in this guide).

- In Lesson 6, you learned that a *spectral class or type* describes a star's position in the temperature classification system (O, B, A, F, G, K, M) based on the appearance of the star's spectrum. By observing a star's spectrum, astronomers can approximate its temperature (p. 104 in this guide).

Learning Objectives

After you complete this lesson, you should be able to:

1. Define stellar parallax and the parsec, and describe the use of parallax to determine stellar distances. *HORIZONS* TEXTBOOK PAGES 144–146.

2. Define absolute visual magnitude and luminosity and explain how they relate to distance and apparent magnitude. *HORIZONS* TEXTBOOK PAGES 146–148.

3. Describe the purpose and design of the Hertzsprung–Russell (H–R) diagram, identify the locations of different types of stars on the diagram, and explain how it reveals the diameters of stars. *HORIZONS* TEXTBOOK PAGES 149–151.

4. Describe luminosity classification and how it can yield additional information about stellar diameters and distances. *HORIZONS* TEXTBOOK PAGES 151–154.

5. Describe the behavior and observation of visual, spectroscopic, and eclipsing binary star systems and explain how they are used to determine stellar masses and diameters. *HORIZONS* TEXTBOOK PAGES 156–160.

6. Describe the mass–luminosity relation and the relative abundance of stars of different spectral types. *HORIZONS* TEXTBOOK PAGES 161–166.

At this point, read Chapter 8, "The Family of Stars," pages 142–167.

Jim he allowed [the stars] was made, but I allowed they happened. Jim said the moon could'a laid them; well, that looked kind of reasonable, so I didn't say nothing against it, because I've seen a frog lay most as many, so of course it could be done.

—Mark Twain
The Adventures of Huckleberry Finn

Viewing Notes

Knowing the distance to a star in our three-dimensional universe unlocks its secrets. Once we know the distance from Earth to a star, we can determine its size and its luminosity.

The video program contains the following segments:

- ⦿ The Third Dimension
- ⦿ Surveying the Stars
- ⦿ Intrinsic Brightness
- ⦿ Diameters of Stars
- ⦿ The Masses of Stars

The following information will help you better understand the video program:

Light always obeys the "inverse square relation." By measuring the apparent brightness of a star and its distance, we can calculate the star's **luminosity**— how much energy it emits. Remember that luminosity is determined by a star's temperature and surface area.

We can use the **Hertzsprung–Russell (H–R) diagram** to help us understand the properties of stars not only by comparing them to other stars but also by separating the effects of temperature and surface area on luminosity. The H–R diagram organizes stars according to their temperature (or spectral type) and luminosity. For main sequence stars, we can apply the **mass–luminosity relation** to infer a star's diameter and mass by comparing it to other stars with similar properties. The mass–luminosity relation states that the more luminous a main-sequence star is, the more massive it must be.

The video program concludes with a discussion on a star's life cycle. You will learn more about how stars form and how they die in Lessons 9 and 10.

QUESTIONS TO CONSIDER

- What are two ways you can estimate the surface temperature of a star?
- What two properties affect the luminosity of a star?
- What property can cause two stars of the same temperature to have different luminosities?
- In a main sequence star, how are mass and luminosity related?

Watch the video for Lesson 8, *The Family of Stars.*

Key Terms and Concepts

Page references are keyed to the *Horizons* textbook.

1. **flux:** A measure of the flow of energy out of a surface, usually applied to light. (p. 147; objective 2)

2. **absolute visual magnitude (M_v):** Intrinsic brightness of a star; the apparent visual magnitude the star would have if it were 10 parsecs away. (p. 147; video lesson; objective 2)

3. **luminosity (L):** The total amount of energy a star radiates in one second. (p. 148; video lesson; objective 2)

4. **Hertzsprung–Russell (H–R) diagram:** A plot of intrinsic brightness vs. surface temperature of stars. (pp. 149–150; video lesson; objective 3)

5. **main sequence:** The region of the H–R diagram running from upper left to lower right. (p. 150; video lesson; objective 3)

6. **giant:** Large, cool, highly luminous star in the mid-to-upper right of the H–R diagram. (p. 150; video lesson; objective 3)

7. **supergiant:** Exceptionally luminous star; diameter is 10 to 1,000 times that of the sun. (p. 150; video lesson; objective 3)

8. **red dwarf:** A faint, cool, low-mass, main-sequence star. (p. 150; video lesson; objective 3)

9. **white dwarf:** Dying star that has collapsed to the size of Earth and is slowly cooling off, found in lower left of the H–R diagram. (p. 150; video lesson; objective 3)

10. **stellar parallax:** A way to measure stellar distance by how far it shifts against the background stars. (pp. 144–145; objective 1)

11. **parsec (pc):** The distance to a star whose parallax is one second of arc. (p. 145; video lesson; objective 1)

12. **luminosity classes:** Categories of stars of similar luminosity, determined by the widths of lines in their spectra. (p. 152; objective 4)

13. **spectroscopic parallax:** The method of determining a star's distance by comparing its apparent magnitude with its absolute magnitude as estimated from its spectrum. (p. 154; objective 4)

14. **binary stars:** Pairs of stars that orbit around their common center of mass. (p. 154; video lesson; objective 5)

15. **visual binary system:** A binary star system in which the two stars are separately visible in the telescope. (p. 156; objective 5)

16. **spectroscopic binary system:** A star system in which the stars are too close together to be visible separately, but we determine that there are two stars by seeing two spectra. (p. 156; objective 5)

17. **eclipsing binary system:** A binary star system in which the stars eclipse each other. (p. 159; objective 5)

18. **light curve:** A graph of brightness vs. time. (p. 160; objective 5)

19. **mass–luminosity relation:** The relationship which states that the more massive a main-sequence star is, the more luminous it is. (p. 161; video lesson; objective 6)

Summary

In this lesson, you learned how to determine certain properties of stars—how much energy they emit, how big they are, and how much mass they contain. As you saw in the video program, the universe is three-dimensional and it's difficult to determine a star's size if you don't know how far away it is. We can understand much more about the properties of a star once we know the distance between it and Earth.

Measuring Distance to Stars

Scientists can approximate the distance to a star by measuring its parallax by using a method called triangulation. When thinking about the triangulation method, envision a triangle. We can establish one point of the triangle as the star and establish the other two points of the triangle as the two opposite ends of Earth's orbit. When measuring the distance to a star, we use the diameter of Earth's orbit (or 2 AU) as the baseline of the triangle.

To measure parallax and thereby calculate the distance from Earth to a star, astronomers can take a picture of a star at one end of Earth's orbit and then wait 6 months and take another picture when Earth is at the opposite side of its orbit. By taking pictures of a star at either end of Earth's orbit, astronomers can determine how far the star appears to shift. **Stellar parallax** is half the total shift of the star measured in seconds of an arc. One second of arc is 1/3,600 of a degree.

The distance to even the nearest star is so great that it's not convenient to measure it using astronomical units (AU). When we measure the distance to a star by measuring its parallax, we use the unit of distance called the **parsec (pc)**. One parsec equals the distance to an imaginary star that has a parallax of 1 second of arc. A parsec is 206,265 AU, or 3.26 light-years (ly) away.

Knowing the angle at the far end of the triangle (by measuring the stellar parallax) and its baseline (the diameter of Earth's orbit), we can calculate the distance to the star. To calculate the distance to a star (d), we divide one parsec by the star's parallax (p), as in this formula:

$$d = 1/p$$

For example, in Reasoning with Numbers 8-1 on p. 145 of the *Horizons* textbook, you learned that the star Altair has a parallax of 0.20 seconds of arc. We can divide 1 by 0.20 to determine the star is 5 parsecs (or about 16.3 ly) away.

The blur of Earth's atmosphere can make it difficult to measure stellar parallax. As you learned in the video program, the satellite HIPPARCOS (the **HI**gh **P**recision **PAR**allax **CO**llecting **S**atellite) was launched to measure the parallax of about 2,000 stars. Once we can determine the distance to a star by measuring its parallax, we can learn so much more about it, including its luminosity.

Absolute Visual Magnitude

In Lesson 2, you learned about the apparent magnitude scale as it refers to stellar brightness. *Apparent visual magnitude* indicates how bright a star *appears* to be here on Earth, not how bright it actually is. In our three-dimensional universe, a small, cool star that's nearby can appear to be just as bright as a large, hot star that is farther away.

In order to compare the brightness of stars that are of varying distances, astronomers calculate the brightness based on a standard distance of 10 parsecs. This would be like lining up all the stars in the universe to compare them side by side. By standardizing the distance, we take into account the difference in apparent brightness caused by distance. Astronomers express the intrinsic brightness as **absolute visual magnitude (M_v)**.

The video program provided a good example of the difference between apparent visual magnitude and absolute visual magnitude. If you view a 100-watt lightbulb that is close to you, it looks much brighter than if the same lightbulb is across the street. The lightbulb throws off the same amount of light no matter what the distance, which is similar to the absolute visual magnitude of a star. However, the lightbulb looks much dimmer when viewed from across the street, which is similar to a star's apparent visual magnitude. The light that you see is inversely proportional to the square of the distance to the source.

To calculate absolute visual magnitude, we start by using the apparent magnitude scale and use the distance from Earth to the star, as in this formula:

$$m_v - M_v = -5 + 5 \log_{10}(d)$$

Let's continue with our study of the star Altair. Astronomers have assigned an apparent visual magnitude (m_v) of 0.76 to it by using the apparent visual magnitude scale. We already know the distance to Altair ($d = 5$ pc), which we determined by measuring its parallax. A calculator tells us that $\log_{10}(5)$ is equal to 0.7. Now, we can use the formula to determine the absolute visual magnitude of Altair.

$$0.76 - M_v = -5 + 5(0.7)$$
$$M_v = 2.26$$

The absolute visual magnitude (M_v) of Altair is 2.26 compared to the sun's absolute visual magnitude of 4.78. The difference between the sun's magnitude and Altair's is 2.52. You'll remember from Lesson 2 that magnitudes are on a logarithmic scale and the lower the magnitude, the greater the intensity. If we refer to Reasoning with Numbers 2-1 (p. 15 in the *Horizons* textbook), we can take the difference in magnitudes between the sun and Altair to determine that Altair's intrinsic brightness is about 10 times greater than the sun ($2.512^{2.52}$).

Luminosity

Now that we know how to measure the distance to a star, we can focus on the second goal of this lesson and determine how much energy it emits, or its luminosity. Where absolute visual magnitude refers to light or brightness, **luminosity** refers to how much total energy a star emits. This includes energy that we cannot see, like ultraviolet and infrared radiation.

The luminosity of a star depends on two things: its size and its temperature. If a star has a very large surface area, it is much more luminous than a smaller star with the same temperature.

We can determine a star's luminosity by starting with its absolute visual magnitude and adjusting it for the radiation the star emits that we cannot see. Astronomers express luminosity in one of two ways: either in solar luminosities or in joules per second (J/s). One joule per second is equal to one watt of power. Based on Altair's absolute visual magnitude, we can express its luminosity as 10 solar luminosities or 4×10^{27} J/s.

The Diameters of Stars

The second goal in this lesson is to determine the diameter of a star, which is twice its radius. Once we know a star's luminosity and temperature, we are able to calculate its radius (see Reasoning with Numbers 8-3 on p. 149 in the *Horizons* textbook). You remember from Lesson 6 that we can estimate a star's temperature by determining its spectral class (see Table 6-1 on p. 105 in the *Horizons* textbook).

In the formula below, L is the luminosity, R is radius, and T is the temperature of the star that we are studying and $L_\odot$, $R_\odot$, and $T_\odot$ are that of the sun.

$$L/L_\odot = (R/R_\odot)^2\ (T/T_\odot)^4$$

If we continue with our example of Altair, we can calculate its radius if we know its luminosity and temperature. By looking at a star catalog, we can identify Altair's spectral type of A7, which we can use to determine its approximate temperature of 7500 K and the sun's temperature is 6000 K (see Appendix A in the *Horizons* textbook). When we calculated Altair's luminosity, we discovered that it is 10 times more luminous than the sun. Using the formula, we can solve for the radius.

$$10/1 = (R/R_\odot)^2\ (7500/6000)^4$$
$$10 = (R/R_\odot)^2\ (2.4414)$$
$$4.1 = (R/R_\odot)^2$$
$$R = 2.0$$

Just by knowing Altair's temperature and its luminosity, we can determine that its radius is about twice that of the sun. Knowing the radius and therefore the diameter of a star is necessary to calculate its surface area, which can tell us about the life of the star.

Hertzsprung–Russell (H–R) Diagram

The **Hertzsprung–Russell (H–R) diagram** is a graph that helps astronomers separate the effects of temperature and surface area and compare stars according to their luminosity and size. The diagram has luminosity or absolute visual magnitude on the vertical axis and spectral type or temperature on the horizontal axis. A point on the graph represents a star. It's important to note that the position of the star on the graph doesn't represent its position in the sky; it represents the effect of a star's temperature and surface area on luminosity.

Since luminosity is on the vertical axis of the graph, points near the top of the diagram represent very luminous stars and points near the bottom represent stars that have low luminosity. Likewise, since temperature (or spectral type) is on the horizontal axis (with the hottest on the left), points to the left of the graph represent very hot blue stars and points to the right represent cooler redder stars.

Roughly 90 percent of all normal stars are "main-sequence" stars, or stars that fall within the main sequence of the diagram. Those that fall outside the main sequence are white dwarfs, giants, and supergiants. White dwarfs are very hot stars that have low luminosity because of their small size. Giants and supergiants are cooler stars but are highly luminous because of their large size. In Lessons 9 and 10, you will learn that a star's position on the H–R diagram can help astronomers determine where it is in its life cycle.

Luminosity Classes

There are several different ways that astronomers can classify stars. In Lesson 6, you learned about spectral classes (O, B, A, F, G, K, M). Another way for astronomers to classify stars is based on their luminosities. Luminosity classes (Ia, Ib, II, III, IV, V) are also determined from spectral lines, and therefore are related to spectral classes, but are not the same thing. Spectral classes categorize stars based on the patterns of their spectral lines and indicate temperature; luminosity classes categorize stars based on the widths of their lines, which vary with diameter.

Since a star's luminosity is a function of its size and temperature, a star's luminosity classification can help us determine how large and hot it is. It can also help us determine the distance to a star that is too far away to have measurable parallax.

Spectroscopic parallax is the estimation of the distance to a star based on its spectral type, luminosity class, and apparent magnitude. The luminosity type appears after the spectral type—for example, Altair is classified as A7 V. This tells us that Altair has a temperature of approximately 7500 K (A7) and is a main-sequence star (V).

The Masses of Stars

The third goal of this lesson is to determine the mass of a star and we can do this by studying its gravitational field. Remember from Lesson 4 that all matter produces a gravitational field; gravitation is mutual and two bodies of different mass balance at the center of mass, which is proportionately closer to the more massive object. To determine a star's mass, we must study **binary stars**—stars that orbit one another. When two stars orbit each other, their mutual gravitation makes them each follow an orbit around a single point (the center of mass) between the two stars. By studying binary stars, we can generalize to determine the properties of stars that are not in a binary system.

To find the mass of a single star in a binary system, we must first find the total mass of the binary star system. The total mass of the binary system is related to the average distance (a) between the two stars and their orbital period (P). In this formula, M_A and M_B represent the individual mass of each star in solar masses, a is expressed in astronomical units (AU), and P is in years.

$$M_A + M_B = a^3/P^2$$

For example, we can observe two different binary star systems. One system has an average distance between the stars (a) of 16 AU and an orbital period (P) of 32 years. The total mass of the system is 16^3 divided by 32^2 or 4 solar masses. A different system also has an average distance between the stars of 16 AU but has an orbital period of 15 years. The total mass of the system is 16^3 divided by 15^2 or 18.2 solar masses. The greater the total mass of the system, the faster both stars have to move to stay in orbit.

To find the mass of the individual star in a binary system, we can study the relationship between the size of each star's orbit. If one star has more mass than its companion, it orbits the center of mass more closely and has a smaller orbit. The less massive star would be further from the center of mass and therefore have a larger orbit.

The masses of the individual stars in a binary system have an inverse relationship to the size of their orbits ($M_A/M_B = r_B/r_A$), where M_A and M_B represent the individual mass of each star in solar masses, and r_A and r_B represent their radii. In other words, if one star has an orbit twice as large as the other star's orbit, we can deduce that it is half as massive as its companion.

More than half of all stars are in a binary system. There are many different types of binary star systems, but astronomers focus on three types to help determine stellar masses: visual, spectroscopic, and eclipsing binary star systems.

In a **visual binary star system**, each star is visible in a telescope because the orbits are large, the stars are far enough away from one another to appear as separate points of light, and the system is relatively close to Earth.

In a **spectroscopic binary system**, the stars are so close together that they look like a single point of light in a telescope. By observing the spectrum, astronomers can determine that there are actually two stars present. The stars orbiting each other produce spectral lines with Doppler shifts. You learned about the Doppler effect in Lesson 6—it is the change in wavelength of radiation because of the relative radial motion of source and observer.

Calculating the total mass in a spectroscopic binary system can be difficult because we can't see the stars or the inclinations of the orbits—how the planes are tipped—and correct for them. But, we can use the Doppler shifts in the spectral lines to determine the orbital period and minimum orbital velocity and use those figures to calculate the minimum mass of the system.

When the stars are in a position in their orbit where they are both perpendicular to our line of sight, there are no Doppler shifts in the spectral lines. When one star is approaching Earth, it reveals a blueshift and as it is receding from Earth it shows a redshift. By watching the Doppler shifts in the spectral lines, we can determine the orbital period by waiting until the Doppler shifts return to their normal positions (see Figure 8-13 on p. 157 in the *Horizons* textbook). We can also determine the orbital velocities for each respective star by measuring the size of the Doppler shifts. Once we have those two pieces of information, we can calculate the circumference of the orbit and derive the mass of the binary system. But, if the orbits are tilted somewhat from our point of view, we would underestimate the velocities and mass by an unknown amount.

In an **eclipsing binary system**, the plane of the orbits is nearly edge-on to Earth and the stars cross in front of each other as observed from Earth. We can determine the orbital period and orbital velocity of each of the stars by watching the variation in light: when one star crosses in front of the other, it blocks some light and the total brightness of the system decreases. The resulting variation in the brightness of the system is shown on a **light curve**—a graph of brightness versus time.

We can easily determine not just the masses but also the diameters by observing how long it takes for a small star to cross in front of its larger companion star. We can determine the orbital velocities for each respective star by measuring the size of the Doppler shifts, just like in the spectroscopic binary system. Then we multiply the smaller star's orbital velocity by the time it takes to cross in front of the larger star to obtain the diameter of the larger star. We can even determine the diameter of the smaller star by observing how long it takes to disappear behind the edge of the larger star. Just like in other binary star systems, once we know the size of the orbit and the orbital period, we can calculate the mass of the binary system. In an eclipsing binary system, we know the orbit is not tipped or there wouldn't be an eclipse.

Mass, Luminosity, and Density

By looking at stars on the H–R diagram, we are able to detect patterns in how a star's size is related to its luminosity and mass. In the main-sequence stars, the most massive stars are the hottest and the lower-mass stars are the coolest. Only the main-sequence stars obey this mass–luminosity relation; giants, supergiants, and white dwarfs do not have this relationship.

If you know only the mass of a main-sequence star, you can estimate its luminosity by comparing it to the mass of the sun. This mass–luminosity relation can help you determine the mass or the luminosity of a main-sequence star based on the properties of similar stars. This relationship can be expressed in this formula, where L is luminosity in solar luminosities and M is mass in solar masses: $L = M^{3.5}$

For example, if we know that the mass of Altair is 1.94 solar masses, then we can estimate its luminosity: $L = (1.94)^{3.5} = 10$ solar luminosities.

Also, if you know any star's mass and diameter, you can calculate its average density by dividing its mass by its volume. In general, larger stars are less dense and smaller stars are more dense.

As you can see, once astronomers discover a few pieces of information about a star, they are able to learn a lot more about its properties and life cycle. By detecting patterns in mass and luminosity, they are also able to learn more about other stars like it. In upcoming lessons you will discover that stars go through a life cycle and their densities, temperatures, and luminosities reflect the different stages of evolution of the stars.

Review Exercises

Matching I

Match each term with the appropriate definition or description.

1. _____ flux	5. _____ main sequence	
2. _____ absolute visual magnitude (M_v)	6. _____ giant	
	7. _____ supergiant	
3. _____ luminosity	8. _____ red dwarf	
4. _____ Hertzsprung–Russell (H–R) diagram	9. _____ white dwarfs	

a. Exceptionally luminous star whose diameter is 10 to 1,000 times greater than that of the sun.

b. The total amount of energy a star radiates in one second.

c. Intrinsic brightness of a star; the apparent visual magnitude the star would have if it were 10 parsecs away.

d. Large, cool, highly luminous star in the mid-to-upper right of the H–R diagram.

e. Dying star that has collapsed to the size of Earth and is slowly cooling off; found in lower left of the H–R diagram.

f. A faint, cool, low-mass, main-sequence star.

g. The region of the H–R diagram running from upper left to lower right.

h. A plot of intrinsic brightness vs. surface temperature of stars.

i. A measure of the flow of energy out of a surface, usually applied to light.

Matching II

Match each term with the appropriate definition or description.

1. _____ stellar parallax		6. _____ visual binary system	
2. _____ parsec (pc)		7. _____ spectroscopic binary system	
3. _____ luminosity classes		8. _____ eclipsing binary system	
4. _____ spectroscopic parallax		9. _____ light curve	
5. _____ binary stars		10. _____ mass–luminosity relation	

a. The distance to a star whose parallax is one second of arc.

b. The relationship which states that the more massive a main-sequence star is, the more luminous it is.

c. Categories of stars of similar luminosity, determined by the widths of lines in their spectra.

d. A graph of brightness vs. time.

e. A way to measure stellar distance by how far it shifts against the background stars.

f. Pairs of stars that orbit around their common center of mass.

g. Method of determining a star's distance by comparing its apparent magnitude with its absolute magnitude as estimated from its spectrum.

h. A binary star system in which the stars eclipse each other.

i. A binary star system in which the two stars are separately visible in the telescope.

j. A star system in which the stars are too close together to be visible separately, but we determine that there are two stars by seeing two spectra.

Completion

Fill each blank in the sentences below with the most appropriate term from the list of completion answers that follow. A term may be used once, more than once, or not at all. Check your answers with the Answer Key and review when necessary.

eclipsing	lower left-hand	temperature (or
greater	lower right-hand	spectral type)
high	luminosity	upper left-hand
lesser	spectroscopic	upper right-hand
low		visual

1. The H–R diagram is a plot of the _____ vs. the _____ of stars.

2. You see a single point of light, but can detect the spectra of two stars in a _____ binary system.

3. The large size of the giants and supergiants means their spectra have sharper spectra lines and their atmospheres have _____ densities.

4. For main-sequence stars the greater the luminosity, the _____ the mass.

5. Red dwarfs are found in the _____ part of the H–R diagram.

6. White dwarfs are found in the _____ part of the H–R diagram.

Self-Test

Select the best answer.

1. The main sequence includes roughly what percentage of all normal stars?
   a. 10 percent
   b. 25 percent
   c. 50 percent
   d. 90 percent

2. The brightest luminosity class is
   a. Ia.
   b. Ib.
   c. IV.
   d. V.

3. If you moved three times farther from a star, the star would appear to be how many times fainter?
   a. 3
   b. 6
   c. 9
   d. 27

4. A light curve measures
   a. brightness vs. distance.
   b. mass vs. luminosity.
   c. brightness vs. time.
   d. curvature vs. temperature.

5. Which best describes the percentage of stars that are in binary systems?
   a. less than 10 percent
   b. less than 25 percent
   c. greater than 50 percent
   d. greater than 90 percent

6. If you measure apparent visual magnitude and you know absolute visual magnitude, you can solve directly for the
   a. age.
   b. distance.
   c. mass.
   d. distance to its companion.

7. A parsec is the
   a. average distance between Earth and the sun.
   b. the distance that light travels in one year.
   c. the number of seconds in a year.
   d. the distance to a star with a parallax of one second of arc.

8. The most common spectral class of main-sequence stars is
   a. A.
   b. O.
   c. G.
   d. M.

9. The most luminous main-sequence stars are spectral class
   a. A.
   b. O.
   c. G.
   d. M.

10. The most massive main-sequence stars are spectral class
    a. A.
    b. O.
    c. G.
    d. M.

11. A star has a parallax of 0.01 arc seconds. How far is it?
    a. 0.01 parsecs
    b. 0.1 parsecs
    c. 10 parsecs
    d. 100 parsecs

12. Which star has the largest diameter?
    a. G2 Ib
    b. G2 II
    c. G2 III
    d. G2 IV

13. A star's classification is K5 III. What is the temperature of this star?
    a. 3000 K
    b. 8000 K
    c. 10,000 K
    d. 20,000 K

14. A star's classification is K5 III. What type of star is this?
    a. supergiant
    b. giant
    c. subgiant
    d. main sequence

Short-Answer Questions

1. How can you tell which star is hotter by looking at an eclipsing binary light curve?

2. What is the difference between flux, luminosity, and absolute visual magnitude?

3. Why would it be easier to take parallax measurements if we were on a planet farther from the sun than Earth?

4. Describe the process of using spectroscopic parallax to determine the distance to a star.

5. How can we tell that two stars in a spectroscopic binary are in gravitational orbit around a common center of mass?

6. Why do extremely cool stars look fainter than you might expect with their luminosities and distances?

7. What is the difference between spectral class and luminosity class?

8. How would the spectra from a G2 V star differ from a G2 III?

Applications

1. If you observe a binary system with an average separation of 100 AU and a period of 400 years, what is the total mass of the system?

2. τ Ceti is 11.9 ly away. How many arc seconds of parallax will it have?

3. How far would the sun (absolute visual magnitude = 4.83) have to be to be barely visible to the unaided eye (magnitude 6)?

4. A candle has a luminosity of 1 watt which corresponds to an absolute visual magnitude of 71. How many kilometers away could this candle still be seen by the Hubble Space Telescope (limit 28.0 apparent visual magnitude)?

5. How much more luminous is a 3 solar mass main-sequence star than the sun?

6. A main-sequence star is 20 times more luminous than the sun. How massive is it?

7. If a star has twice the temperature of the sun and twice the radius, how many times would its luminosity be compared to the sun?

Answer Key

Matching I

1. i (p. 147; objective 2)

2. c (p. 147; video lesson; objective 2)

3. b (p. 148; video lesson; objective 2)

4. h (pp. 149–150; video lesson; objective 3)

5. g (p. 150; video lesson; objective 3)

6. d (p. 150; video lesson; objective 3)

7. a (p. 150; video lesson; objective 3)

8. f (p. 150; video lesson; objective 3)

9. e (p. 150; video lesson; objective 3)

Matching II

1. e (pp. 144–145; objective 1)

2. a (p. 145; video lesson; objective 1)

3. c (p. 152; objective 4)

4. g (p. 154; objective 5)

5. f (p. 154; video lesson; objective 5)

6. i (p. 156; objective 5)

7. j (p. 156; objective 5)

8. h (p. 159; objective 5)

9. d (p. 160; objective 5)

10. b (p. 161; video lesson; objective 6)

Completion

1. luminosity, temperature (or spectral type) (pp. 148–150; video lesson; objective 3)

2. spectroscopic (p. 156; objective 5)

3. low (pp. 151–153; objective 3)

4. greater (pp. 161–162; objective 6)

5. lower right-hand (p. 151; objective 3)

6. lower left-hand (p. 151; objective 3)

Self-Test

1. d (p. 150; objective 3)
2. a (p. 152; objective 6)
3. c (p. 147; video lesson; objective 2)
4. c (pp. 159–160; objective 5)
5. c (p. 158; objective 5)
6. b (p. 147; objective 2)
7. d (p. 145; video lesson; objective 1)
8. d (p. 165; objective 6)
9. b (p. 152; objective 3)
10. b (p. 163; objective 6)
11. d (p. 145; objective 1)
12. a (pp. 151–153; objective 4)
13. a (p. 152; objective 3)
14. b (p. 152; objective 4)

Short-Answer Questions

1. You can look at the light curve and point to the deeper of the two eclipses and say, "That is where the hotter star is behind the cooler star." (pp. 159–160; objective 5)

2. Flux is the energy in Joules per second falling on one square meter. Luminosity is how much energy per second a star emits. Absolute visual magnitude is the apparent visual magnitude a star would have if it were 10 parsecs away. (pp. 147–148; objective 2)

3. The orbit is bigger for planets further from the sun, the baseline would be wider and the shifts would be larger and easier to measure. (pp. 144–145; objective 1)

4. To determine spectroscopic parallax, record the spectrum of a star. This will help you determine its spectral class and that tells you its horizontal location on the H–R diagram. You can also determine its luminosity class by looking at the widths of the spectral lines. Once you plot the point on the H–R diagram, you can determine its absolute magnitude. You can find the distance to a star if you know the absolute visual magnitude and the apparent visual magnitude. (pp. 153–154; objective 5)

5. The Doppler shifts of the spectrum lines can tell that one star is blue shifted (moving toward us) and the other is red shifted (moving away from us). As the stars orbit, the spectral lines would move back and forth across each other and would alert you that you were observing a spectroscopic binary system. (pp. 156–159; objective 5)

6. The coolest stars radiate the vast majority of their photons in the infrared, which you can't see. (p. 148; objective 2)

7. Spectral class is based mainly on temperature. Luminosity class is dependent mainly on size. Larger stars of the same temperature give off more luminosity than smaller stars. (pp. 149–153; objectives 3 & 4)

Applications

1. The answer is 6.25 solar masses. (p. 155; objective 5)

$$M_A + M_B = a^3/P^2$$

$$M_A + M_B = 100^3/400^2 = 1,000,000/160,000 = 6.25$$

2. The answer is 0.27 arc seconds. (p. 145; objective 1)

11.9 ly $\times$ (1 parsec/3.26 ly) = 3.65 parsecs

$p = 1/d = 1/3.65 = 0.27$ arc seconds

3. The answer is 17 parsecs. (p. 147; objective 2)

$$m_v - M_v = -5 + 5 \log_{10}(d)$$

$$6 - 4.83 = -5 + 5 \log_{10}(d)$$

$$1.17 = -5 + 5 \log_{10}(d)$$

$$6.17 = 5 \log_{10}(d)$$

$$0.234 = \log_{10}(d)$$

$$d = 10^{(0.234)} = 17.13 \text{ parsecs}$$

4. 7.7×10^5 km. [about twice the distance to the moon] (p. 147; objective 2)

$$m_v - M_v = -5 + 5 \log_{10}(d)$$

$$28 - 71 = -5 + 5 \log_{10}(d)$$

$$-43 = -5 + 5 \log_{10}(d)$$

$$-38 = 5 \log_{10}(d)$$

$$-7.6 = \log_{10}(d)$$

$$d = 10^{(-7.6)} = 2.5 \times 10^{-8} \text{ parsecs}$$

2.5×10^{-8} parsecs $\times (3.085 \times 10^{13}$ km/1 parsec) = 771,250 km

5. The answer is 47 times. (p. 162; objective 6)

$$L = M^{3.5} = 3^{3.5} = 47$$

6. The answer is 2.4 solar masses. (p. 162; objective 6)

$$20 = M^{3.5}$$

$$20^{(1/3.5)} = M$$

7. The answer is 64 times. (p. 149; objective 2)

$$L/L_\odot = (R/R_\odot)^2 (T/T_\odot)^4$$

$$L/L_\odot = (2)^2 (2)^4 = 4 \times 16 = 64$$

Lesson Review

Lesson 8: The Family of Stars

PLEASE NOTE: Use this matrix to guide your study and achieve the learning objectives of this lesson. It will also help you to view the video, which defines and demonstrates important concepts and principles as they relate to everyday life and actual case studies.

Learning Objective	Textbook	Student Guide
1. Define stellar parallax and the parsec, and describe the use of parallax to determine stellar distances.	pp. 144–146	Key Terms: 10, 11; Matching II: 1, 2; Self-Test: 7, 11; Short-Answer: 3; Applications: 2.
2. Define absolute visual magnitude and luminosity and explain how they relate to distance and apparent magnitude.	pp. 146–148	Key Terms: 1, 2, 3; Matching I: 1, 2, 3; Self-Test: 3, 6; Short-Answer: 2, 6; Applications: 3, 4, 7.
3. Describe the purpose and design of the Hertzsprung–Russell (H–R) diagram, identify the locations of different types of stars on the diagram, and explain how it reveals the diameters of stars.	pp. 149–151	Key Terms: 4, 5, 6, 7, 8, 9; Matching I: 4, 5, 6, 7, 8, 9; Completion: 1, 3, 5, 6; Self-Test: 1, 9, 13; Short-Answer: 7.
4. Describe luminosity classification and how it can yield additional information about stellar diameters and distances.	pp. 151–154	Key Terms: 12, 13; Matching II: 3; Self-Test: 12, 14; Short-Answer: 7.
5. Describe the behavior and observation of visual, spectroscopic, and eclipsing binary star systems and explain how they are used to determine stellar masses and diameters.	pp. 156–160	Key Terms: 14, 15, 16, 17, 18; Matching II: 4, 5, 6, 7, 8, 9; Completion: 2; Self-Test: 4, 5; Short-Answer: 1, 4, 5; Applications: 1.

Learning Objective	Textbook	Student Guide
6. Describe the mass–luminosity relation and the relative abundance of stars of different spectral types.	pp. 161–166	Key Terms: 19; Matching II: 10; Completion: 4; Self-Test: 2, 8, 10; Applications: 5, 6.

LESSON
9

Stellar Births

Checklist

For the most effective study of this lesson, complete the following activities in this sequence.

Before Viewing the Video

- ❑ Read the Preview, Learning Objectives, and Viewing Notes below.
- ❑ Read Chapter 9, "The Formation and Structure of Stars," pages 168–195, in the *Horizons* textbook.

What to Watch

- ❑ After reading the textbook chapter, watch the video for Lesson 9, *Stellar Births*.

After Viewing the Video

- ❑ Briefly note your answers to questions listed at the end of the Viewing Notes.
- ❑ Review the Summary below.
- ❑ Review all reading assignments for this lesson, especially the Chapter 9 summary on page 194 in *Horizons* and the Viewing Notes in this guide.
- ❑ Write brief answers to the review questions at the end of Chapter 9 in *Horizons*.
- ❑ Complete the Review Exercises below. Check your answers with the Answer Key and review when necessary.
- ❑ Use the Lesson Review matrix found at the end of this lesson to review and assess your knowledge of each Learning Objective.
- ❑ As assigned by your instructor, complete the Applications activities and any additional activities for this lesson.

Preview

The life of a star contains an amazing series of events. Swirling clouds of cool dust and gas coalesce under the force of gravity to form a new star. The star can live a relatively ordinary life and when its fuel is exhausted it can become unstable. A star's life can end in a violent explosion with its matter scattered throughout the galaxy. The remnants of the star can become incorporated into new celestial objects.

Studying the formation of stars even within our own galaxy has proven to be a difficult task. In previous lessons, you learned how a star's spectrum and position on the H–R diagram can tell you more about it. You also learned about different types of telescopes that can detect visible light as well as other forms of electromagnetic radiation that our eyes cannot detect. Now it's time to put those tools to use and understand how stars are born and how they live. In Lesson 10, you'll discover how they die.

Concepts to Remember

- Recall from Lesson 8 that the concept of *mass–luminosity relation* explains that the more massive a main-sequence star is, the more luminous it is (p. 147 in this guide).

- In Lesson 8, you also learned that *Hertzsprung–Russell (H–R) diagram* is a plot of intrinsic brightness versus the surface temperature of stars. A star's position on the H–R diagram can help you understand its evolution (p. 146 in this guide).

Learning Objectives

After you complete this lesson, you should be able to:

1. Describe the composition and distribution of the interstellar medium and how its contents are detected at visible and nonvisible wavelengths. HORIZONS TEXTBOOK PAGES 170–175.

2. Outline the theory of star formation, from molecular cloud to main sequence. HORIZONS TEXTBOOK PAGES 175–179.

3. Describe the observational evidence we have of main-sequence star formation at both visible and nonvisible wavelengths. HORIZONS TEXTBOOK PAGES 180–181, 190–192.

4. Describe different types of thermonuclear fusion in main-sequence stars, and explain how these reactions are governed by a pressure–temperature thermostat. HORIZONS TEXTBOOK PAGES 182–183.

5. State the four laws of stellar structure and describe how they can be used to create a model of a star's interior. *HORIZONS* TEXTBOOK PAGES 183–187.

6. Use the mass-luminosity relation to explain the stability of main-sequence stars. *HORIZONS* TEXTBOOK PAGES 187–188.

7. Explain the changes a star undergoes during the main-sequence phase, discuss the relationship between mass and lifespan, and estimate the life expectancy of a star from its mass. *HORIZONS* TEXTBOOK PAGES 188–189, 192.

At this point, read Chapter 9, "The Formation and Structure of Stars," pages 168–195.

Unknowingly, we plow the dust of stars, blown about us by the wind, and drink the universe in a glass of rain.

—Ihab Hassan
Egyptian-born
author and critic

Viewing Notes

Early astronomers believed that stars were eternal. But modern astronomers know that stars are born from clouds of dust and gas distributed through space.

The video program contains the following segments:

- ☸ The Interstellar Medium
- ☸ Nebulae
- ☸ Stellar Formation
- ☸ The Core Ignites
- ☸ Main Sequence

The following information will help you better understand the video program:

A star is born in the cool, dark recesses of space. The gas and dust in the **interstellar medium** coalesce to form dark clouds called nebulae. The force of gravity compresses the gases in nebulae to form a cluster of young stars. Once formed, the hotter stars can cause the surrounding nebulae to glow visibly.

In the video, you will see several examples of the three types of nebulae: dark nebulae, reflection nebulae, and emission nebulae. An **emission nebula** is produced when a hot star excites the gas near it to produce an emission spectrum. You learned about the emission spectrum in Lesson 6—the excited gas produces bright lines in its spectrum. A **reflection nebula** is produced when starlight scatters from a dusty nebula—it is just the reflected absorption spectrum of starlight. A **dark nebula** is a dense cloud of gas and dust that obstructs the view of more distant stars.

The video mentions that the shock wave theory of star formation is not widely accepted, which seems to contradict the textbook. But, in the video Dr. Basri is referring only to the shock wave produced by an exploding supernova. All star formation probably involves some kind of shock wave in order to compress the gas. For example, the interstellar medium is spread out between the stars, but it is denser in some places than in others—like in the spiral arms of a galaxy. You learned about the spiraling arms of a galaxy in Lesson 1 and many astronomers believe that these arms are long, curved shock waves travelling through the interstellar medium. The collision of an interstellar cloud with a spiral arm could trigger the formation of new stars.

QUESTIONS TO CONSIDER

- How did astronomers come to the conclusion that dark nebulae were not holes in space but clouds of opaque dust?

- Why do stars form only from cold, dark clouds? What types of external events can trigger such a cloud to form stars?

- In what ways can we observe protostars at different stages of their development?

- What are the different types of thermonuclear fusion that occur in main-sequence stars?

- What is the purpose of a stellar model? How can we use it to understand the evolution of a star?

- Why are main-sequence stars so stable? Why are massive stars more luminous and short-lived?

Watch the video for Lesson 9, *Stellar Births*.

Key Terms and Concepts

Page references are keyed to the *Horizons* textbook.

1. **interstellar medium:** The dust and gas distributed between the stars. (p. 170; video lesson; objective 1)

2. **nebula:** A glowing cloud of gas or a cloud of dust reflecting or obscuring the light of nearby stars. (p. 171; video lesson; objective 1)

3. **interstellar reddening:** The process in which dust scatters blue light out of starlight and makes the stars look redder. (p. 171; video lesson; objective 1)

4. **interstellar dust:** Microscopic solid grains in the interstellar medium. (p. 170; video lesson; objective 1)

5. **emission nebula:** A cloud of glowing gas excited by ultraviolet radiation from hot stars. (p. 172; video lesson; objective 1)

6. **HII region:** A region of ionized hydrogen around a hot star. (p. 172; video lesson; objective 1)

7. **reflection nebula:** A nebula produced by starlight reflecting off dust particles in the interstellar medium. (p. 172; video lesson; objective 1)

8. **dark nebula:** A cloud of gas and dust of any size seen silhouetted against a brighter nebula or star field. (p. 173; video lesson; objective 1)

9. **molecular cloud:** A dense interstellar gas cloud in which atoms are able to link together to form molecules such as H_2 and CO. (p. 175; video lesson; objective 2)

10. **shock wave:** A sudden change in pressure that travels as an intense sound wave. (p. 175; video lesson; objective 2)

11. **association:** Group of widely scattered stars (10 to 100) moving together through space. (p. 176; objective 2)

12. **protostar:** A collapsing cloud of gas and dust destined to become a star. (p. 177; video lesson; objective 2)

13. **evolutionary track:** A path a star follows on the H–R diagram as it gradually changes its surface temperature and luminosity. (p. 177; objective 2)

14. **birth line:** In the H–R diagram, the line above the main sequence where protostars first become visible. (p. 178; objective 3)

15. **CNO cycle:** A series of nuclear reactions that use carbon as a catalyst to combine four hydrogen atoms to make one helium atom plus energy. (p. 182; video lesson; objective 4)

16. **T-Tauri star:** A young star surrounded by gas and dust, believed to be contracting toward the main sequence. (p. 180; objective 3)

17. **Bok globule:** Small dark cloud only about 1 ly in diameter that contains 10 to 1,000 solar masses of gas and dust. (p. 180; video lesson; objective 3)

18. **Herbig–Haro object:** A small nebula that varies irregularly in brightness, believed to be associated with star formation. (p. 181; objective 3)

19. **bipolar flow:** Jets of gas flowing away from a central object in opposite directions, usually applied to protostars. (p. 181; objective 3)

20. **triple-alpha process:** The nuclear fusion process that combines three helium nuclei to make one carbon nucleus. (p. 182; objective 4)

21. **conservation of mass:** One of the basic laws of stellar structure that states the total mass of the star must equal the sum of the masses of the shells. (p. 183; objective 5)

22. **conservation of energy:** One of the basic laws of stellar structure that states the amount of energy flowing out of the top of a shell must equal the amount coming in at the bottom plus whatever energy is generated within the shell. (p. 183; objective 5)

23. **hydrostatic equilibrium:** The balance between the weight of the material pressing downward on a layer in a star and the pressure in that layer. (p. 184; objective 5)

24. **energy transport:** Flow of energy from hot regions to cooler regions by one of three methods: conduction, convection, or radiation. (p. 185; objective 5)

25. **opacity:** The resistance of a gas to the passage of radiation. (p. 185; objective 5)

26. **stellar model:** A table of numbers representing the conditions in various layers within a star. (p. 185; objective 5)

27. **brown dwarf:** A star whose mass is too low to ignite nuclear fusion. (p. 187; video lesson; objective 6)

28. **zero-age main sequence (ZAMS):** The location in the H–R diagram where stars first reach stability as hydrogen-burning stars. (p. 189; objective 7)

29. **pressure–temperature thermostat:** The relation between gas pressure and temperature, the stability of a star depends upon this. (video lesson; objective 4)

Summary

For thousands of years astronomers noticed dark patches between the stars. They didn't realize that space is filled with clouds of dust and gas called the interstellar medium. As the electromagnetic radiation from distant stars travels through the interstellar medium toward Earth, the medium scatters certain wavelengths and makes the stars behind it virtually undetectable. It wasn't until astronomers turned their infrared telescopes toward the sky that they were able to detect the interstellar medium and the stars it obscured.

In some cases, the interstellar medium can be easily seen as dense clouds of gas and dust—called nebulae—with thin gas and dust interspersed between them. There are three kinds of nebulae: emission nebulae, reflection nebulae, and dark nebulae.

By studying the spectrum of the interstellar medium—and those of the stars it obscures—astronomers were able to determine its composition. This interstellar medium is primarily made up of about 75 percent hydrogen and about 25 percent helium gas, with traces of carbon, nitrogen, oxygen, calcium, sodium and heavier atoms. About 1 percent of this mass is in the form of dust made of carbon and silicates mixed or coated with frozen water. As you've learned in previous lessons, these are the elements that stars are made of.

You might think that gravity alone would cause the gas and dust in the interstellar medium to form stars. Although the temperature of the medium is relatively cool—about 10 K—the rapidly moving atoms in the gas are hot enough to resist collapse. In the densest of the interstellar clouds, hydrogen can exist as molecules (H_2) rather than as atoms. Most molecular clouds in the interstellar medium are relatively stable, but when the cloud becomes gravitationally unstable, stars can form.

The Birth of Stars

A molecular cloud can become unstable when it collides with a **shock wave**—a sudden change in pressure that travels as an intense sound wave. Shock waves can be triggered by a supernova, by the collision of two interstellar clouds, or by the birth of a very hot star releasing a sudden blast of radiation. Inside a giant molecular cloud, a few massive stars can drive a continuing cycle of star formation.

Once a portion of a cloud is triggered to collapse, gravity draws each atom towards its center. The contraction of the cloud heats its gas by converting gravitational energy into thermal energy. About half of this thermal energy dissipates into space and the remaining energy raises the temperature of the cloud. As the cloud contracts, it becomes more dense and hot, and a **protostar**—an object that will eventually become a star—is formed.

When the center of the protostar gets hot enough, nuclear reactions begin to take place in its core and it generates energy. This hydrogen burning increases until the star stops contracting and becomes stable, and it can then be classified as a main-sequence star, which you studied in Lesson 8. The length of this process can take anywhere from 160,000 to 1 billion years depending on the star's mass. A more massive star has greater gravitational forces that make it collapse and form more rapidly than a lower-mass star.

Of course, humans have not been observing space for a billion years, so astronomers must develop theories about how stars are formed and rely on their observations to validate them. You'll recall from Lesson 8 that a star's position on the H–R diagram can provide clues as to where it is in its life cycle. The **birth line** on the H–R diagram indicates where a contracting protostar first becomes visible. A protostar in the early stages of evolution can only be detected in the infrared. When it becomes hot and luminous, it crosses the birth-line and can be detected by optical telescopes.

Astronomers are always searching for evidence to help them understand how stars are formed; studying star clusters and nebulae can enlighten them. Because the stars in a cluster are approximately the same age, composition, and distance away from Earth, astronomers can study them to find evidence of how stars form.

When astronomers detect Bok globules inside a nebula, they can surmise that the interstellar medium is about to contract to form protostars. **Bok globules** are dark nebulae that are very cold at their centers. They are dense, relatively small, and circular and the sun is thought to have come from an object such as this. When astronomers detect **Herbig–Haro** objects—small nebulae that fluctuate in brightness and are produced by energetic jets—they can confirm that a star has recently been born because young stars are thought to produce them. When astronomers find a nebula with young stars, it usually contains a number of **T-Tauri** stars—stars surrounded by gas and dust that are believed to be contracting toward the main sequence.

Fusion in Stars

In Lesson 6, you discovered that the sun produces energy by fusing hydrogen into helium in a series of nuclear reactions called the proton–proton chain. All main-sequence stars produce energy by nuclear fusion, but the elements it combines and how it combines them depend on how massive it is and where it is in its life cycle.

Main-sequence stars that are more massive than the sun fuse hydrogen into helium using the **CNO cycle**. Temperatures greater than 16,000,000 K are required to make this process work, so it only happens in stars hotter and more massive than the sun. The CNO cycle involves a carbon nucleus combining one at a time with four hydrogen nuclei. It changes from carbon to nitrogen to oxygen and back to carbon plus a helium nucleus. So, the carbon is a catalyst which is not consumed. The net result is four protons becoming helium, just as in the proton–proton chain (see Figure 9-9 on p. 182 in the *Horizons* textbook).

In older stars, the hydrogen fuel has been exhausted and a star may fuse other elements, such as helium—using the **triple-alpha process**—or carbon, to create energy. These elements require much hotter temperature to fuse than required to fuse hydrogen: helium fusion requires a temperature of 100 million K and carbon fusion requires 600 million K. Nuclear reactions at these temperatures can convert heavy atoms such as magnesium, aluminum, and silicon into even heavier atoms. Understanding these processes will be important when you study the death of massive stars in Lesson 10.

The relationship between the gas pressure and the temperature acts as a thermostat that keeps the nuclear reactions in a main sequence star under control. If left unchecked, gravity would continue to pull the gases into the star and it would continue to get hotter and denser until it would ultimately collapse. But, the fuel burning inside the star increases its pressure, which counterbalances the gravity that's trying to collapse it; the star maintains this delicate balance throughout most of its life.

Stellar Structure

In Lesson 6, you learned about the structure of the sun—it has a core, radiative and convective inner layers, and a photosphere "surface" at the base of its atmosphere. When studying a star's interior structure, it helps to visualize the star having shells—or layers—similar to those of an onion. These imaginary layers of the star can vary in composition, temperature, density and pressure.

The structure of a star depends on how it generates its energy, what it is made of, and on four simple laws of structure. The four laws of structure include the conservation of mass, the conservation of energy, hydrostatic equilibrium, and energy transport.

The law of **conservation of mass** states that the total mass of a star must equal the total masses of its shells. This is a simple concept that indicates if you were able to determine the mass of each of the imaginary shells individually, their collective mass would equal the total mass of the star.

The law of **conservation of energy** states that the amount of energy flowing out of the top of a layer must be equal to the amount of energy coming in at the bottom of the layer plus whatever energy is generated within the layer.

The third law of stellar structure is that of **hydrostatic equilibrium**. This is closely related to the gas pressure-temperature thermostat that keeps a star stable. It states that the weight of each layer is balanced by the pressure in that layer. The pressure of the gas depends on its temperature and density. This tells you that a star must be hot inside to maintain the pressure needed to support its own weight. In order to balance its own weight, a massive star has to be hotter than a less massive star.

The fourth law is that of **energy transport**. This states that energy must flow from hot regions to cooler regions by conduction, convection, or radiation. In Lesson 6, you learned how the sun transports its energy from the solar interior to the solar atmosphere. Similar processes happen in other stars as described below.

In a star like a white dwarf, conduction is a significant source of heat transfer because it is extremely dense. *Conduction* is a form of heat flow that requires close contact between the particles—heat energy is passed through particles that are touching one another. However, most other stars are much less dense and don't rely on conduction to transfer heat.

In a star like the sun, radiation is the primary method of heat transfer. Photons are absorbed and re-emitted in random directions as they work their way outward from the star's interior. The flow of energy by radiation depends on the density of the gas and the density of the gas depends on its temperature. In a hot, thin gas, photons can move about more freely than in a cooler, denser gas.

If a star's layer is so cool for its density that the heat cannot escape via radiation, it builds up in lower layers of the star. The gas begins to churn and hot gas rises while cooler gas sinks in a process called *convection*. Convection mixes the gas in the layers—like convection mixes cool cream into hot coffee—and equally distributes it.

By knowing what a star is made of, how it generates energy, and by applying the four laws of stellar structure, astronomers can build a stellar model to better understand its properties and its life cycle. A **stellar model** is a mathematical model in the form of a table that indicates the temperature, density, mass, and luminosity of each one of the star's layers or shells. This allows astronomers to understand something that is difficult to study in reality and lets them look into a star's past and future by comparing it to similar stars.

Main-Sequence Stars

When you studied the family of stars, you learned about the *H–R diagram* and how a main-sequence star's mass is related to its luminosity. A star's *luminosity* depends on its temperature and its surface area: a large, hot star is more luminous than a small, cool star.

Main-sequence stars live a long and stable life because of hydrostatic equilibrium—the star's internal pressure balances its weight. Massive stars are more luminous because they have to make more energy to balance their weight. Conversely, because a star on the lower main sequence weighs less, it cannot raise its internal temperature or pressure without having all of its gases dispersed through space.

All main sequence stars obey the mass-luminosity relation. A star's position on the H–R diagram can help you understand where it is in its life cycle. A star begins its life on the lower edge of the main-sequence line of the H–R diagram. As the star begins to use up its fuel, the balance between its gas pressure and gravity changes. The gradual change in luminosity and surface temperature moves the position of the star up and to the right of the diagram. Remember, the change in position on the diagram doesn't mean the star is actually moving—it's simply a graphical representation of the effect of luminosity and temperature on the star. As the star uses up more and more of its fuel, it reaches the upper limit of the main sequence, where its surface is larger, cooler, and more luminous than when it first formed.

The number of years a star spends on the main sequence depends on how massive it is. High-mass stars consume fuel more rapidly and are short-lived. By contrast, low-mass stars are cooler and burn their fuel more slowly and can shine for trillions of years. You can estimate a star's life expectancy—the number of years it will be on the main sequence—by estimating the amount of fuel it has and the rate at which it consumes it. The amount of fuel a star has is proportional to its mass; the rate at which it uses its fuel is proportional to its luminosity. See Reasoning with Numbers 9-1 on p. 192 in the *Horizons* textbook.

Once a star moves out of the main sequence, it is providing us with evidence that it is beginning to die. It is exhausting its fuel supply and gravity begins to win the battle of hydrostatic equilibrium. A star can die a quiet death, or it can end in a massive explosion, which you will learn about in Lesson 10.

Review Exercises

Matching I

Match each term with the appropriate definition or description.

1. _____ interstellar medium	6. _____ HII region
2. _____ nebula	7. _____ reflection nebula
3. _____ interstellar reddening	8. _____ dark nebula
4. _____ interstellar dust	9. _____ molecular cloud
5. _____ emission nebula	

   a. A cloud of gas and dust of any size seen silhouetted against a brighter nebula.

   b. A cloud of glowing gas excited by ultraviolet radiation from hot stars.

   c. A glowing or dark cloud of gas or a cloud of dust reflecting the light of nearby stars.

   d. A nebula produced by starlight reflecting off dust particles in the interstellar medium.

   e. Microscopic solid grains in the interstellar medium.

   f. The process in which dust scatters blue light out of starlight and makes the stars look redder.

   g. A dense interstellar gas cloud in which atoms are able to link together to form molecules such as H_2 and CO.

   h. A region of ionized hydrogen around a hot star.

   i. The dust and gas distributed between the stars.

Matching II

Match each term with the appropriate definition or description.

1. _____ shock wave		6. _____ CNO cycle	
2. _____ association		7. _____ T-Tauri star	
3. _____ protostar		8. _____ Bok globule	
4. _____ evolutionary track		9. _____ Herbig–Haro object	
5. _____ birth line		10. _____ bipolar flow	

a. A young star surrounded by gas and dust, believed to be contracting toward the main sequence.

b. In the H–R diagram, the line above the main sequence where protostars first become visible.

c. A series of nuclear reactions that use carbon as a catalyst to combine four hydrogen atoms to make one helium atom plus energy.

d. A sudden change in pressure that travels as an intense sound wave.

e. Group of widely scattered stars (10 to 1,000) moving together through space.

f. Jets of gas flowing away from a central object in opposite directions; usually applied to protostars.

g. Small dark cloud only about 1 ly in diameter that contains 10 to 1,000 solar masses of gas and dust.

h. A small nebula that varies irregularly in brightness, believed to be associated with star formation.

i. A path a star follows in the H–R diagram as it gradually changes its surface temperature and luminosity.

j. A collapsing cloud of gas and dust destined to become a star.

Matching III

Match each term with the appropriate definition or description.

1. _____ triple-alpha process		6. _____ opacity	
2. _____ conservation of mass		7. _____ stellar model	
3. _____ conservation of energy		8. _____ brown dwarf	
4. _____ hydrostatic equilibrium		9. _____ zero-age main sequence	
5. _____ energy transport		10. _____ pressure–temperature thermostat	

a. One of the basic laws of stellar structure that states the total mass of the star must equal the sum of the masses of the shells.

b. Flow of energy from hot regions to cooler regions by one of three methods: conduction, convection, or radiation.

c. The location in the H–R diagram where stars first reach stability as hydrogen-burning stars.

d. The resistance of a gas to the passage of radiation.

e. The nuclear fusion process that combines three helium nuclei to make one carbon nucleus.

f. One of the basic laws of stellar structure that states the amount of energy flowing out of the top of a shell must equal the amount coming in at the bottom plus whatever energy is generated within the shell.

g. The balance between the weight of the material pressing downward on a layer in a star and the pressure in that layer.

h. A table of numbers representing the conditions in various layers within a star.

i. A star whose mass is too low to ignite nuclear fusion.

j. The relation between gas pressure and temperature, the stability of a star depends upon this.

Completion

Fill each blank in the sentences below with the most appropriate term from the list of completion answers that follow. A term may be used once, more than once, or not at all. Check your answers with the Answer Key and review when necessary.

birth line	most	T-Tauri
Herbig–Haro objects	1	25,000
infrared	10	ultraviolet
least	10,000	zero-age main sequence

1. The wavelength region of the electromagnetic spectrum in which protostars are observable is the _____.

2. Interstellar dust makes up roughly _____ percent of the mass of the interstellar medium.

3. Emission nebulae are produced when a hot star excites the gas near it to produce an emission spectrum. To do this, the star must be hotter than _____ K.

4. On the H–R diagram, protostars remain hidden in visible wavelengths until they cross the _____.

5. Main-sequence stars that produce energy through the CNO cycle are the _____ massive.

6. When powerful jets from a newborn star strike the interstellar medium, they produce _____.

7. Stars that fluctuate in brightness and appear to be newborn stars just blowing away their dust cocoons are known as _____.

Self-Test

Select the best answer.

1. Reflection nebulae appear to be what color?
   a. pink
   b. red
   c. white
   d. blue

2. Stars that contract the slowest to land on the zero-age main sequence line on the H–R diagram
   a. are the most massive.
   b. are the least massive.
   c. are not dependent on mass.
   d. are neither the most nor least massive.

3. To begin to collapse into stars, an interstellar cloud must be
   a. cold and thin.
   b. cold and dense.
   c. hot and thin.
   d. hot and dense.

4. Starlight passing through interstellar dust becomes
   a. fainter and redder.
   b. brighter and bluer.
   c. fainter and bluer.
   d. brighter and redder.

5. Main-sequence stars of which spectral class use the CNO cycle?
   a. A
   b. G
   c. K
   d. M

6. Bok globules contain
   a. 1–2 solar masses.
   b. 10–1,000 solar masses.
   c. 10,000–1,000,000 solar masses.
   d. more than 1,000,000,000 solar masses.

7. Which of the following is NOT one of the four laws of stellar structure?
   a. Conservation of mass
   b. Conservation of energy
   c. Conservation of charge
   d. Hydrostatic equilibrium

8. If an object formed from a cloud of dust and gas is less than 0.08 solar masses, it forms a
   a. yellow dwarf.
   b. black dwarf.
   c. white dwarf.
   d. brown dwarf.

9. The major gas component in the interstellar medium is
   a. helium.
   b. hydrogen.
   c. carbon.
   d. silicon.

10. What two characteristics are balanced in hydrostatic equilibrium?
    a. positive and negative charges
    b. weight and pressure
    c. mass and energy
    d. gravity and magnetism

11. When a star begins its stable life, it begins on the
    a. birth line.
    b. zero-age main sequence.
    c. asymptotic giant branch.
    d. turnoff point.

12. The average star spends _____ of its life on the main sequence.
    a. 10 percent
    b. 25 percent
    c. 50 percent
    d. 90 percent

Short-Answer Questions

1. Describe the process by which stars form from molecular cloud to main sequence.

2. Why does the fusion of heavier atoms require higher temperatures than the fusion of hydrogen?

3. How can very cold clouds make very hot stars?

4. What would happen if the sun stopped generating energy?

5. Describe the difference between a reflection nebula, an emission nebula, and a dark nebula and what spectra would they show, if any.

6. What are the four laws of stellar structure and what do they state?

7. What are the different kinds of thermonuclear fusion in main-sequence stars and at what temperatures do they occur?

8. How does the Hubble photo of the Eagle Nebula (on p. 181 in *Horizons*) show stars in the act of forming?

Applications

1. What is the life expectancy of a 2-solar-mass star?

2. If a star's lifespan is 1.1×10^7 years, what is its mass and spectral type?

3. Describe which reactions would have to occur to get from carbon-12 (^{12}C) to fluorine-18 (^{18}F).

4. What would be the fewest number of reactions to get from carbon-12 (^{12}C) to silicon-28 (^{28}Si)?

5. Using the stellar model in Figure 9-14 on page 187 in *Horizons*, approximately what portion of the sun's radius has a density greater than water (1 g/cm^3)?

6. Explain why the sky is blue and how that relates to reflection nebulae.

7. Where would brown dwarfs fall on the H–R diagram if it were expanded to include L stars and T stars?

Answer Key

Matching I

1. i (p. 170; video lesson; objective 1)

2. c (p. 171; video lesson; objective 1)

3. f (p. 171; video lesson; objective 1)

4. e (p. 170; video lesson; objective 1)

5. b (p. 172; video lesson; objective 1)

6. h (p. 172; video lesson; objective 1)

7. d (p. 172; video lesson; objective 1)

8. a (p. 173; video lesson; objective 1)

9. g (p. 175; video lesson; objective 2)

Matching II

1. d (p. 175; video lesson; objective 2)

2. e (p. 176; objective 2)

3. j (p. 177; video lesson; objective 2)

4. i (p. 177; objective 2)

5. b (p. 178; objective 3)

6. c (p. 182; video lesson; objective 4)

7. a (p. 180; objective 3)

8. g (p. 180; video lesson; objective 3)

9. h (p. 181; objective 3)

10. f (p. 181; objective 3)

Matching III

1. e (p. 182; objective 4)

2. a (p. 183; objective 5)

3. f (p. 183; objective 5)

4. g (p. 184; objective 5)

5. b (p. 185; objective 5)

6. d (p. 185; objective 5)

7. h (p. 185; objective 5)

8. i (p. 187; video lesson; objective 6)

9. c (p. 189; objective 7)

10. j (video lesson; objective 4)

Completion

1. infrared (p. 171; video lesson; objective 1)

2. 1 (p. 170; objective 1)

3. 25,000 (p. 172; objective 1)

4. birth line (p. 178; objective 3)

5. most (p. 182; objective 4)

6. Herbig–Haro objects (p. 181; objective 3)

7. T-Tauri (p. 180; objective 3)

Self-Test

1. d (p. 172; video lesson; objective 1)

2. b (pp. 177–178; objective 2)

3. b (p. 177; video lesson; objective 2)

4. a (p. 171; video lesson; objective 1)

5. a (p. 182; video lesson; objective 4)

6. b (p. 180; video lesson; objective 3)

7. c (pp. 183–185; objective 5)

8. d (p. 187; video lesson; objective 6)

9. b (p. 170; objective 1)

10. b (p. 184; objective 5)

11. b (p. 189; objective 7)

12. d (p. 189; objective 7)

Short-Answer Questions

1. When the densest part of the molecular gas cloud becomes unstable and contracts under the influence of its own gravity, it becomes a protostar. It grows hotter but is hidden in a dusty cocoon. The newborn star may have bipolar flows and be very active producing Herbig-Haro objects. When it settles into stable hydrostatic equilibrium, it falls on the zero-age main sequence line on the H–R diagram. (pp. 175–179, 188–189; video lesson; objective 2)

2. Heavier atomic nuclei have higher positive charges so their Coulomb barrier is higher. This requires that the gas be hotter so the particles move faster and the collisions will be more violent. (p. 182; objective 4)

3. Gravity causes the cold gas to contract to the point that it turns on its thermonuclear processes. (p. 175; objective 2)

4. With no pressure to push outward against gravity, it would collapse. (pp. 183–184; objective 5)

5. A reflection nebula is when starlight is scattered in a dusty nebula. It reflects the absorption spectrum of the star. An emission nebula is produced when a hot star excites the gas to produce an emission spectrum. A dark nebula is a dense cloud of dust and gas that obstructs the view of more distant stars. It produces no spectra. (pp. 172–173; video lesson; objective 3)

6. Conservation of mass: One of the basic laws of stellar structure that states the total mass of the star must equal the sum of the masses of the shells.

   Conservation of energy: One of the basic laws of stellar structure that states the amount of energy flowing out of the top of a shell must equal the amount coming in at the bottom plus whatever energy is generated within the shell.

   Hydrostatic equilibrium: The balance between the weight of the material pressing downward on a layer in a star and the pressure in that layer.

   Energy transport: Flow of energy from hot regions to cooler regions by one of three methods: conduction, convection, or radiation. (pp. 183–185; objective 5)

7. The proton–proton chain occurs over 10,000,000 K and the CNO cycle occurs over 16,000,000 K. (pp. 127, 182; objective 4)

8. Radiation is evaporating the dust and driving away the gas to expose small globules of denser dust and gas. At least one of these globules seems to contain a new star. (pp. 180–181; objective 3)

Applications

1. The answer is 0.18 solar lifetimes. (p. 192; objective 7)

   For a 2-solar-mass star, $\tau = 1/M^{2.5} = 1/2^{2.5} = 0.177$

2. The star's mass is 15 solar masses; using Table 9-2 on page 189 in *Horizons*, its spectral class is B0. (pp. 189, 192; objective 7)

3. $^{12}C \rightarrow {}^{13}C \rightarrow {}^{16}O \rightarrow {}^{17}O \rightarrow {}^{14}N \rightarrow {}^{18}F$

   (p. 182; objective 4)

4. The answer is three. (p. 182; objective 4)

   $^{12}C \rightarrow {}^{23}Na \rightarrow {}^{24}Mg \rightarrow {}^{28}Si$ or $^{12}C \rightarrow {}^{20}Ne \rightarrow {}^{24}Mg \rightarrow {}^{28}Si$

5. The answer is a little more than half. $R/R_o = 0.50$

   (p. 187; objective 5)

6. The sky is blue because of the scattering of blue light in the atmosphere. In both cases, short wavelengths (blue light) scatters more easily than long wavelengths (red light). (p. 172; video lesson; objective 1)

7. They would fall to the lower right. (p. 189; objective 6)

Notes:

Lesson Review

PLEASE NOTE: Use this matrix to guide your study and achieve the learning objectives of this lesson. It will also help you to view the video, which defines and demonstrates important concepts and principles as they relate to everyday life and actual case studies.

Learning Objective	Textbook	Student Guide
1. Describe the composition and distribution of the interstellar medium and how its contents are detected at visible and nonvisible wavelengths.	pp. 170–175	Key Terms: 1, 2, 3, 4, 5, 6, 7, 8, 9; Matching I: 1, 2, 3, 4, 5, 6, 7, 8; Completion: 1, 2, 3; Self-Test: 1, 4, 9; Applications: 6.
2. Outline the theory of star formation, from molecular cloud to main sequence.	pp. 175–179	Key Terms: 10, 11, 12, 13, 14; Matching I: 9; Matching II: 1, 2, 3, 4; Self-Test: 2, 3; Short-Answer: 1, 3.
3. Describe the observational evidence we have of main-sequence star formation at both visible and nonvisible wavelengths.	pp. 180–181, 190–192	Key Terms: 15, 17, 18, 19, 20; Matching II: 7, 8, 9, 10; Completion: 4, 6, 7; Self-Test: 6; Short-Answer: 5, 8.
4. Describe different types of thermonuclear fusion in main-sequence stars, and explain how these reactions are governed by a pressure–temperature thermostat.	pp. 182–183	Key Terms: 16, 21, 30; Matching II: 6; Matching III: 1, 10; Completion: 5; Self-Test: 5; Short-Answer: 2, 7; Applications: 3, 4.

Learning Objective	Textbook	Student Guide
5. State the four laws of stellar structure and describe how they can be used to create a model of a star's interior.	pp. 183–187	Key Terms: 22, 23, 24, 25, 26, 27; Matching III: 2, 3, 4, 5, 6, 7; Self-Test: 7, 10; Short-Answer: 4, 6; Applications: 5.
6. Use the mass–luminosity relation to explain the stability of main-sequence stars.	pp. 187–188	Key Terms: 28; Matching III: 8; Self-Test: 8; Applications: 7.
7. Explain the changes a star undergoes during the main-sequence phase, discuss the relationship between mass and lifespan, and estimate the life expectancy of a star from its mass.	pp. 188–189, 192	Key Terms: 29; Matching III: 9; Self-Test: 11, 12; Applications: 1, 2.

LESSON
10

Stellar Deaths

Checklist

For the most effective study of this lesson, complete the following activities in this sequence.

Before Viewing the Video

- ❑ Read the Preview, Learning Objectives, and Viewing Notes below.
- ❑ Read Chapter 10, "The Deaths of Stars," pages 196–223, in the *Horizons* textbook.

What to Watch

- ❑ After reading the textbook chapter, watch the video for Lesson 10, *Stellar Deaths*.

After Viewing the Video

- ❑ Briefly note your answers to questions listed at the end of the Viewing Notes.
- ❑ Review the Summary below.
- ❑ Review all reading assignments for this lesson, especially the Chapter 10 summary on page 222 in *Horizons* and the Viewing Notes in this guide.
- ❑ Write brief answers to the review questions at the end of Chapter 10 in *Horizons*.
- ❑ Complete the Review Exercises below. Check your answers with the Answer Key and review when necessary.
- ❑ Use the Lesson Review matrix found at the end of this lesson to review and assess your knowledge of each Learning Objective.
- ❑ As assigned by your instructor, complete the Applications activities and any additional activities for this lesson.

Preview

The death of a star provides life in the universe. In previous lessons, you followed the life cycle of a star in which gravity plays an integral part. Stars are born in the nurseries of the interstellar medium where gravity causes a cloud of dust and gas to collapse and form a star. A typical star lives its life on the main sequence; throughout its life, the star's internal pressure counterbalances the force of gravity that would otherwise make it collapse. When the star begins to run out of fuel, it evolves as the battle between gravity and the internal pressure continues. The star eventually runs out of fuel and it dies.

You can trace the evolution of a star on the *H–R diagram*, which you learned about in Lesson 8. How long a star will live and how it dies depend on its mass. A massive star burns its fuel at a furious rate and lives a relatively short time until it dies a violent death. A less massive star burns fuel slowly and can live for billions of years before dying a quiet death. The remnants of the star are mixed in with the interstellar medium to form new clouds of dust and gas that contribute to new life in the universe.

Concepts to Remember

- Recall from Lesson 8 that *binary systems* are pairs of stars that orbit around their common center of mass. Stars that are in a binary system can evolve differently than those that are not (p. 147 in this guide).

- In Lesson 8, you also learned that the *Hertzsprung–Russell (H–R) diagram* is a plot of intrinsic brightness versus the surface temperature of stars. A star's position on the H–R diagram can help you understand where it is in its life cycle (p. 146 in this guide).

- In Lesson 9, you learned that the *interstellar medium* is the gas and dust distributed between the stars. Dense clouds of gas within the medium are where stars are formed (p. 168 in this guide).

- In Lesson 7, you learned that *nuclear fusion* describes reactions that join the nuclei of atoms to form more massive nuclei. Stars produce energy by fusing atoms together (p. 127 in this guide).

Learning Objectives

After you complete this lesson, you should be able to:

1. Describe the internal processes that cause an aging main-sequence star to expand into a giant. *HORIZONS* TEXTBOOK PAGES 198–203.

2. Explain how studies of star clusters are critical to understanding the evolution of stars. *HORIZONS* TEXTBOOK PAGES 202–205.

3. Describe the deaths of low- and medium-mass stars, including the formation of planetary nebulae and white dwarfs, and discuss the effects that our sun's death will have on Earth. HORIZONS TEXTBOOK PAGES 203, 206–210.

4. Describe the evolution of close binary stars, including possible effects of mass transfer such as novae and type I supernovae. HORIZONS TEXTBOOK PAGES 211–214.

5. Describe the deaths of the most massive stars, including type II supernovae and supernova remnants. HORIZONS TEXTBOOK PAGES 214–221.

At this point, read Chapter 10, "The Deaths of Stars," pages 196–223.

Our birth is nothing but our death begun.
—Edward Young
English poet
(1683–1765)

Viewing Notes

During the life of a star, there is a delicate balance between gravity and pressure. But, once a star's fuel begins to run out, this balance is upset and the death of the star is imminent.

The video program contains the following segments:

- Expansion into a Giant
- The Deaths of Medium-Mass Stars
- Low-Mass Stars
- The Death of Massive Stars
- The Evolution of Binary Systems
- Evidence of Stellar Evolution

The following information will help you better understand the video program:

The video illustrates that a main-sequence star evolves into a giant as it exhausts its fuel and expands. As you recall from Lesson 8, a "main-sequence star" is one that falls within the main sequence of the H–R diagram. The position of the star on the graph doesn't represent its position in the sky; it represents the effect of a star's temperature and surface area on luminosity. As the star ages, the point on the graph that represents the star moves to depict the change in the star's surface area and temperature. The stars on the middle- and upper-main sequence have enough mass and heat to expand into a giant; stars on the lower-main sequence don't expand into giants.

You will see how matter can be transferred from one star in a binary system to another. This transfer of mass can cause a nova explosion to occur—hydrogen gas accumulates on the surface of a white dwarf in a binary system and causes an explosion.

The video also illustrates a **supernova**—a violent collision triggered by the death of certain types of stars. You'll recall from the textbook that there are three types of supernovae: type Ia, type Ib, and type II. **Type II supernovae** are rare and occur when a massive star, with hydrogen-rich outer layers, develops an iron core that collapses to cause a violent explosion. In a **type Ia supernova**, a white dwarf that no longer contains hydrogen gains mass from a binary companion until it can no longer support itself and detonates. A **type Ib supernova** is a supernova explosion that is just like a type II supernova except that the massive star has lost its atmosphere.

QUESTIONS TO CONSIDER

- What is the sequence of events leading to the death of a main-sequence star?

- How can studying a star cluster help astronomers understand the evolution of stars?

- How does the death of a low-mass star like a red dwarf differ from the death of a medium-mass star like our sun?

- Why do certain stars die and produce planetary nebulae instead of ending in a violent explosion?

- How do stars in a binary system transfer matter? How does this play a part in the evolution of each of the stars?

- How does the death of a massive star differ from the death of a medium-mass star like our sun?

- What types of supernovae can occur in a binary system?

Watch the video for Lesson 10, *Stellar Deaths*.

Key Terms and Concepts

Page references are keyed to the *Horizons* textbook.

1. **nova:** A sudden brightening of a star making it appear as a new star in the sky (from the Latin, meaning "new"). Believed to be associated with eruptions on white dwarfs in binary systems. (p. 196; video lesson; objective 4)

2. **supernova:** The explosion of a star in which it increases its brightness by a factor of about a million. (p. 196; objective 5)

3. **degenerate matter:** Extremely high-density matter in which pressure no longer depends on temperature caused by quantum mechanical effects. (p. 200; objective 1)

4. **helium flash:** The explosive ignition of helium burning that takes place in some giant stars. (p. 201; objective 1)

5. **planetary nebula:** An expanding shell of gas ejected from a star during the latter stages of its evolution. (p. 206; video lesson; objective 3)

6. **open cluster:** A cluster of between 100 and 1,000 stars with an open, transparent appearance. The stars are not tightly grouped, and they are usually relatively young and located in the disk of the galaxy. (p. 204; video lesson; objective 2)

7. **globular cluster:** A star cluster containing 100,000 to 1 million stars in a sphere about 75 ly in diameter. Stars in such clusters are generally old, metal-poor, and found in the spherical component of the galaxy. (p. 204; objective 2)

8. **turnoff point:** The point in an H–R diagram at which a cluster's stars turn off the main sequence and move toward the red giant region, revealing the age of the cluster. (p. 204; objective 2)

9. **horizontal branch:** In the H–R diagram, stars fusing helium in a shell and evolving back toward the giant region. (p. 205; objective 2)

10. **black dwarf:** The end state of a white dwarf star that has cooled to a low temperature. (p. 207; objective 3)

11. **Chandrasekhar limit:** The maximum mass of a white dwarf, about 1.4 solar masses. (p. 207; video lesson; objective 3)

12. **Roche lobe:** The volume of space a star controls gravitationally within a binary system. (p. 211; objective 4)

13. **Roche surface:** The dumbbell-shaped surface that encloses the Roche lobes around a close binary star. (p. 211; objective 4)

14. **Lagrangian points:** Points of gravitational stability in the orbital plane of a binary star system or of a planet and its moon. (p. 211; objective 4)

15. **inner Lagrangian point:** The point of gravitational equilibrium between two orbiting stars through which matter can flow from one star to the other. (p. 211; objective 4)

16. **angular momentum:** A measure of the tendency of a rotating body to continue spinning. (p. 212; objective 4)

17. **accretion disk:** The whirling disk of gas that forms around a compact object, such as a white dwarf, neutron star, or black hole, as matter is drawn in. (p. 212; objective 4)

18. **type II supernova:** A supernova explosion caused by the collapse of a massive star. (p. 217; video lesson; objective 5)

19. **type Ia supernova:** A supernova explosion caused by the collapse of a white dwarf. (p. 217; objective 4)

20. **type Ib supernova:** A supernova explosion that is just like a type II supernova except the massive star has lost its atmosphere. (p. 217; objective 4)

21. **synchrotron radiation:** Radiation emitted when high-speed electrons move through a magnetic field. (p. 217; objective 5)

22. **supernova remnant:** The expanding shell of gas marking the site of a supernova explosion. (p. 219; video lesson; objective 5)

Summary

All stars evolve and eventually die; their matter is mixed in with the interstellar medium—the incubator of the universe. The way a star evolves and dies depends on its mass.

Giant Stars

A medium-mass star like our sun will live most of its life on the main sequence, but will eventually become a giant. It generates its energy by fusing hydrogen together in its core to make helium by the process of *nuclear fusion*. As the star uses up all of its hydrogen fuel, helium "ash" is left over in its core: a main-sequence star is hot enough to fuse hydrogen, but not hot enough to fuse helium to produce more energy. As the star runs out of hydrogen, gravity takes over and the core begins to contract.

As the core contracts, gravitational energy is converted into thermal energy; the core becomes hot enough to fuse what little hydrogen is left in a shell just outside the core. The energy created by the fused hydrogen pushes out toward the surface and causes the outer layers of the star to expand dramatically.

As the star's exterior expands, its low-density gas begins to cool. The star becomes a giant or supergiant with a large diameter and luminosity but low density. On the H–R diagram, the point that represents the star that was once on the main sequence quickly moves to the right and upward.

As the outer layers of the core expand, the helium ash core continues to contract and grow hotter. When the core eventually reaches a temperature of 100,000,000 K, it is finally hot enough to fuse the remaining helium to make carbon. How the star fuses helium depends on its mass—a massive star greater than 3 solar masses contracts quickly and begins to fuse helium, while a less massive star contracts and heats more slowly.

In a less massive star, the helium core takes so long to reach fusion temperature that the gas becomes **degenerate** first—the electrons are not free to change their energy because the gas is so dense. In this state, the pressure–temperature thermostat no longer regulates energy production. But because the gas is degenerate, it resists compression and the pressure doesn't increase. So the helium fusion begins suddenly in what is called a **helium flash** where the core of the star can generate more energy per second than does the rest of the entire galaxy. As a result of the released energy, the star becomes so hot it is no longer degenerate and the pressure–temperature thermostat quickly brings the helium fusion under control.

The ignition of helium in a star's core changes its structure: it now produces energy in its helium core and it also produces energy in its hydrogen shell. The core stops contracting and the outside of the star stops expanding and begins to contract. The point that represents the star on the H–R diagram moves downward and to the left as the star gets smaller and hotter.

The helium fusion in the star's core continues and produces carbon and oxygen atoms that require much hotter temperatures to fuse. Since the temperature is not hot enough to fuse carbon and oxygen, they accumulate in the star's core. When the helium in the core has been used up, the core begins to contract again and heats up even more. Eventually, a new helium-fusion shell ignites beneath the existing hydrogen fusion shell. As the star is creating energy in both of these shells, it once again expands and the resulting low-density gas is cooled. The point that represents the star on the H–R diagram moves back to the right and the giant continues to evolve.

The Deaths of Lower-Main-Sequence Stars

The death of a lower-main-sequence star differs from the death of massive upper-main-sequence star. As you recall, the more massive a star, the faster it burns its fuel. We can divide the lower-main-sequence stars into two groups based on their structural differences: very-low-mass red dwarfs and medium-mass stars like the sun. The death of stars in each of these two groups differs because the extent of their interior convection differs.

A very-low-mass red dwarf with a mass less than 0.4 solar masses can live a very long life as a result of two factors. First, it consumes its hydrogen very slowly and therefore can live a long life. Second, it is totally *convective*—its gases are stirred and hot gas rising from the interior is mixed with cool gas that is sinking from the exterior. Therefore, hydrogen and helium are equally distributed throughout the star and not just concentrated in its core. Since it is able to burn fuel throughout the star—not just in its core—it doesn't end up with an inert helium core when all of the hydrogen has been used up. Therefore, it doesn't expand and become a giant star.

A medium-mass star is not convective throughout and the hydrogen and helium are not equally distributed. Stars with solar masses of 1.1 or less are not mixed at all because they have virtually no convection near their center. Stars with solar masses of 1.1 or more have small convective zones near their centers that mix roughly 12 percent of their gases.

Because these medium-mass stars are not mixed thoroughly by convection, helium accumulates in the core and is surrounded by unprocessed hydrogen. The core contracts, the helium is ignited, and the cool exterior expands and causes the star to swell into a giant as explained above. As the star expands, it can expel its outer atmosphere to form a **planetary nebula**—a greenish-blue ionized gas that radiates from the dying star. The excited, low-density gases of the planetary nebula expand at 10 to 20 km/s and ultimately mix in with the interstellar medium.

When the dying star blows away its outer atmosphere, the hot interior of the star is exposed. It collapses into a small hot object containing a carbon and oxygen interior surrounded by hydrogen and helium fusion shells with a thin hydrogen atmosphere. As the fusion dies out, the core of the star evolves into the nucleus of the planetary nebula and eventually cools to become a white dwarf. A white dwarf is no longer a true star because it doesn't generate nuclear energy, is almost totally degenerate, and contains practically no normal gas. Therefore, we can refer to it as a "compact object" instead of a star. In Lesson 11, you will learn about two other compact objects: neutron stars and black holes.

As our sun is a medium-mass star, we can expect it to die a similar death. As it grows into a giant, its increased luminosity will evaporate Earth's oceans and atmosphere and vaporize its crust. Mathematical models suggest that the sun may survive about another 5 billion years.

The Evolution of Binary Systems

In order to understand the death of more massive stars, we first must study the evolution of binary systems. You'll recall from Lesson 8 that a binary system contains a pair of stars that orbit around their common center of mass. If two binary stars orbit far away from each other, they can evolve separately. But, if they are close together, when one of the stars in the system swells into a giant, it can affect the evolution of its companion star.

The gravitational fields of a binary system define a dumbbell-shaped area called the **Roche lobes**, and matter inside of each lobe is gravitationally bound to the star it surrounds. Stars that are close together can transfer matter through the **inner Lagrangian point**—the point where the two Roche lobes meet.

Matter can be transferred through the inner Lagrangian point in one of two ways. First, a star can produce a very strong stellar wind that blows through the inner Lagrangian point. Second, an evolving star expands so far that it fills its Roche lobe and the matter spills out into its companion's lobe.

Matter flowing from one star to another through the inner Lagrangian point does not become assimilated directly by its companion. Rather, the matter flows into a whirling disk around the star. Keeping its **angular momentum**—the tendency of a rotating object to continue rotating—the gas flows rapidly into an accretion disk that surrounds the star that's assimilating the mass.

This transfer of mass can cause a nova explosion to occur if the star receiving the material has become a white dwarf. The matter transferred from one star to the other can lose its angular momentum in the accretion disk and it settles onto the surface of the white dwarf. As the layer of mainly hydrogen gas accumulates on the white dwarf, it becomes denser and hotter and the hydrogen fuses in an explosion that blows the surface off of the white dwarf. The explosion appears in the sky and then fades away but it hardly disturbs the surface of the white dwarf. A nova can occur repeatedly on a white dwarf whenever the gas accumulates into an explosive layer.

The Deaths of Massive Stars

A massive star dies a much more dramatic death than other stars. Like a medium-mass star, a massive star consumes hydrogen in its core, ignites its hydrogen shell, and swells into a giant or a supergiant. As the core contracts, it fuses the helium that resulted from the hydrogen fusion—first in its core, and then in its expanding shell.

Unlike a medium-mass star, the extremely hot massive star is able to eventually ignite the carbon and oxygen that remains in its core and then in its shell. This process continues at increasing speed with each fuel that results from previous reactions until all fuel is exhausted and only iron remains. Iron cannot combine with any of the remaining atoms to produce energy. Rather, nuclear reactions involving iron tend to remove energy from the core, and release electrons and gamma rays. As a result, the core collapses even faster and draws on the energy stored in its gravitational field and continues to get hotter.

As the inner core continues to contract, a shock wave develops and moves outward propelled by two additional sources of energy: neutrinos and hot gas. The disruption of the iron nuclei in the core releases the neutrinos, and they carry large amounts of energy away from the collapsing core and heat the gas around it. The turbulence created by this released energy drives the shock wave outward and it bursts through the surface of the star, blowing it apart and creating a **supernova**—a violent collision triggered by the death of certain types of stars.

There are several different types of supernovae, which can be recognized by their spectra. In a **type II supernova**, a hydrogen-rich star's core collapses and then explodes. A **type I supernova** occurs in stars that no longer have hydrogen, and can be either a type Ia or type Ib supernova. In a type Ia supernova, a white dwarf gains mass in a binary system until it can no longer support itself and collapses; it is totally destroyed and doesn't leave a black hole or neutron star (you'll learn about black holes and neutron stars in Lesson 11). In a type Ib supernova, a massive star in a binary system transfers its hydrogen-rich outer layers to its companion; then an iron core develops and causes a violent explosion.

If you think about how long stars can live, it's no surprise that only a few supernovae have been seen with the naked eye in recorded history. A supernova explosion fades away in a year or two, but an expanding shell of gas remains. When the expanding gas collides with the interstellar medium, it can assimilate more gas and excite it to produce a supernova remnant. The **supernova remnant**—the nebulous remains of a supernova explosion—can compress the interstellar medium and can trigger new star formations.

Star Clusters

Studying star clusters can provide insight as to how stars are born, how they live, and how they die. All of the stars in a cluster are born about the same time from the same cloud of gas. Because stars in a cluster are the same age and began with the same chemical composition, the differences between them arise from differences in mass.

There are two types of star clusters: open clusters and globular clusters. An **open cluster** is a collection of 10 to 1,000 stars in a region about 25 pc in diameter. A **globular cluster** is nearly spherical and the stars are much closer together than those in an open cluster; it can contain 10^5 to 10^6 stars in a region 10–30 pc in diameter.

Studying the H–R diagram of a star cluster can help us understand the evolution of the stars in it. We can establish the age of a cluster by looking at its **turnoff point** on the H–R diagram—the point on the main sequence where stars evolve to the right to become giants. Because high-mass stars evolve quickly, they turn off the main sequence faster than the smaller, cooler, low-mass stars. Therefore, a higher turnoff point means a much younger cluster. Stars from older clusters have a much lower turnoff point.

As a star cluster ages, its main sequence grows shorter: the most massive stars have died off and no stars in the cluster are plotted on the upper left of the main sequence. The **horizontal branch stars** are those stars fusing helium first in the core and then in the shell and evolving back toward the giant region.

Review Exercises

Matching I

Match each term with the appropriate definition or description.

1. _____ nova	7. _____ globular cluster
2. _____ supernova	8. _____ turnoff point
3. _____ degenerate matter	9. _____ horizontal branch
4. _____ helium flash	10. _____ black dwarf
5. _____ planetary nebula	11. _____ Chandrasekhar limit
6. _____ open cluster	

a. A star cluster containing 100,000 to 1 million stars in a sphere about 75 ly in diameter.
b. The explosive ignition of helium burning that takes place in some giant stars.
c. A sudden brightening of a star making it appear as a new star in the sky.
d. The point in an H–R diagram at which a cluster's stars turn off the main sequence and move toward the red giant region.
e. In the H–R diagram, stars fusing helium in a shell and evolving back toward the giant region.
f. The maximum mass of a white dwarf, about 1.4 solar masses.
g. Extremely high-density matter in which pressure no longer depends on temperature because of quantum mechanical effects.
h. End state of a white dwarf star that has cooled to a low temperature.
i. The explosion of a star in which it increases its brightness by a factor of about a million.
j. A cluster of between 100 and 1,000 stars with an open, transparent appearance.
k. An expanding shell of gas ejected from a star during the latter stages of its evolution.

Matching II

Match each term with the appropriate definition or description.

1. _____ Roche lobe	7. _____ type II supernova
2. _____ Roche surface	8. _____ type Ia supernova
3. _____ Lagrangian points	9. _____ type Ib supernova
4. _____ inner Lagrangian point	10. _____ synchrotron radiation
5. _____ angular momentum	11. _____ supernova remnant
6. _____ accretion disk	

a. The point of gravitational equilibrium between two orbiting stars through which matter flows from one star to the other.

b. A supernova explosion caused by the collapse of a massive star.

c. Radiation emitted when high-speed electrons move through a magnetic field.

d. The dumbbell-shaped surface that encloses the Roche lobes around a close binary star.

e. The volume of space a star controls gravitationally within a binary system.

f. The whirling disk of gas that forms around a compact object, such as a white dwarf, neutron star, or black hole, as matter is drawn in.

g. The expanding shell of gas marking the site of a supernova explosion.

h. Points of gravitational stability in the orbital plane of a binary star system or of a planet and its moon.

i. A supernova explosion caused by the collapse of a white dwarf.

j. A measure of the tendency of a rotating body to continue spinning.

k. A supernova explosion that is just like a type II supernova except the massive star has lost its atmosphere.

Completion

Fill each blank in the sentences below with the most appropriate term from the list of completion answers that follow. A term may be used once, more than once, or not at all. Check your answers with the Answer Key and review when necessary.

Lagrangian point	red dwarf	type I
ninety	slowly	type II
older	ten	white dwarf
planetary nebula	three	younger
quickly		

1. The more massive stars will "burn" their fuel more

   _____.

2. The higher a cluster's turnoff point on the H–R diagram , the

   _____ it is.

3. After the red giant phase for a medium-mass star, it will throw off its outer layers as a _____.

4. After the red giant phase for a medium-mass star, its core will collapse into a _____.

5. A star in a binary system may expand to fill its Roche lobe and begin to accrete (or accumulate) material onto the other star by material passing through the inner _____.

6. When a solitary massive star dies, it dies spectacularly in a

   _____ supernova.

7. The amount of time a normal star spends as a red giant compared to its entire lifetime is _____ percent.

Self-Test

Select the best answer.

1. The last nuclear reaction in the core of massive stars that uses energy rather than producing energy fuses what element?
   a. helium
   b. carbon
   c. silicon
   d. iron

2. Of the following types of stars, the one that has never been and can never be a giant star is a
   a. red dwarf.
   b. white dwarf.
   c. black dwarf.
   d. yellow dwarf.

3. A shell of gas slowly ejected by a giant star is known as a
   a. supernova remnant.
   b. nova remnant.
   c. planetary nebula.
   d. reflection nebula.

4. A turnoff point corresponding to the youngest star cluster would occur at what spectral classification?
   a. B
   b. A
   c. F
   d. G

5. An explosive, although nondestructive, event on the surface of a white dwarf that is part of a binary system results in a
   a. planetary nebula.
   b. type I supernova.
   c. type II supernova.
   d. nova.

6. The blue glow from the Crab Nebula is produced by
   a. synchrotron radiation.
   b. reflection nebulae.
   c. a blue supergiant.
   d. Hawking radiation.

7. The conservation of angular momentum states that if a body contracts (gets smaller), its speed of rotation
   a. increases.
   b. decreases.
   c. stays the same.
   d. increases, then decreases.

8. The gravitationally bound volumes surrounding a binary system are called
   a. bipolar flows.
   b. Roche lobes.
   c. Einstein's volumes.
   d. Zeeman figures.

9. The horizontal branch stars are giants fusing what element in their cores and then their shells?
   a. hydrogen
   b. helium
   c. carbon
   d. silicon

10. A helium flash
    a. is sudden and powerful.
    b. destroys the star.
    c. is a slow expulsion of helium gas.
    d. occurs in very low mass stars.

11. Stars will never experience a helium flash if the mass is less than
    a. 3 solar masses.
    b. 1.4 solar masses.
    c. 1.1 solar masses.
    d. 0.4 solar masses.

12. The type of star that does not currently exist because our universe is not old enough is a
    a. red dwarf.
    b. white dwarf.
    c. brown dwarf.
    d. black dwarf.

Short-Answer Questions

1. What is the difference between a type Ia, a type Ib, and a type II supernova?

2. Briefly explain degenerate matter.

3. It is sometimes said, "Gravity always wins." Explain this statement in relationship to stars.

4. What is the difference between a nova and a supernova?

5. Explain how the turnoff point of a cluster of stars plotted on the H–R diagram can reveal the cluster's age.

6. What is the difference between open clusters and globular clusters?

7. Why does a medium-mass star expand into a giant?

Applications

1. What are some of the suggested causes for the asymmetry of planetary nebulae?

2. Explain the Algol paradox.

3. How does the mathematical model of the Chandrasekhar limit compare with observations?

4. Chart the evolutionary process of a medium-mass star like our sun on the H–R diagram from zero-age main sequence (ZAMS) to death.

5. Chart the evolutionary process of a high mass star on the H–R diagram from ZAMS to death.

6. How does the mass transfer between close binary stars affect the evolution of those stars?

Answer Key

Matching I

1. c (p. 196; video lesson; objective 4)

2. i (p. 196; objective 5)

3. g (p. 200; objective 1)

4. b (p. 201; objective 1)

5. k (p. 206; video lesson; objective 3)

6. j (p. 204; video lesson; objective 2)

7. a (p. 204; objective 2)

8. d (p. 204; objective 2)

9. e (p. 205; objective 2)

10. h (p. 207; objective 3)

11. f (p. 207; video lesson; objective 3)

Matching II

1. e (p. 211; objective 4)

2. d (p. 211; objective 4)

3. h (p. 211; objective 4)

4. a (p. 211; objective 4)

5. j (p. 212; objective 4)

6. f (p. 212; objective 4)

7. b (p. 217; video lesson; objective 5)

8. i (p. 217; objective 4)

9. k (p. 217; objective 4)

10. c (p. 217; objective 5)

11. g (p. 219; video lesson; objective 5)

Completion

1. quickly (pp. 203–204; objective 2)

2. younger (pp. 204–205; objective 2)

3. planetary nebula (pp. 206–209; video lesson; objective 3)

4. white dwarf (p. 207; objective 3)

5. Lagrangian point (p. 211; objective 4)

6. type II (p. 217; video lesson; objective 5)

7. ten (p. 201; objective 1)

Self-Test

1. d (pp. 214–216; video lesson; objective 5)

2. a (p. 203; objective 3)

3. c (pp. 206, 208–209; video lesson; objective 3)

4. a (pp. 204–205; objective 2)

5. d (p. 213; video lesson; objective 4)

6. a (p. 217; objective 5)

7. a (p. 212; objective 4)

8. b (p. 211; objective 4)

9. b (p. 205; objective 1)

10. a (p. 201; objective 1)

11. d (p. 201; objective 1)

12. d (p. 207; objective 3)

Short-Answer Questions

1. A type Ia supernova is when a white dwarf gaining mass in a binary star system exceeds the Chandrasekhar limit and collapses. A type Ib supernova occurs when a massive star in a binary system loses its hydrogen-rich outer layers to its companion star. The remains of the massive star could develop an iron core and collapse. A type II supernova occurs when a large-mass star expends its fuel and explodes. (p. 217; video lesson; objectives 4 & 5)

2. The Pauli Exclusion principle states that no two identical electrons can occupy the same energy level. This means that when electrons fill energy levels in pairs, one must spin up and one spin down. For this to occur, the gas must be so dense that the electrons are not free to change their energy and are therefore degenerate. (pp. 199–200; objectives 1 & 3)

3. While a star is on the main sequence, it is in hydrostatic equilibrium. When the star uses all its fuel, it can no longer sustain the pressure to oppose the gravity, gravity wins, and the core will collapse. (pp. 203, 206–207, 214; video lesson; objectives 1, 3, & 5)

4. A nova is an explosion of material that has accumulated on the surface of a white dwarf in a binary system. It is very bright for a short period of time, but does not destroy the star and may be recurring. A supernova destroys the star. (pp. 213–217; objectives 4 & 5)

5. A high-mass star uses its fuel much more quickly than a low-mass star. Therefore the high-mass star turns off the main sequence faster than the smaller, cooler, low-mass stars. A higher turnoff point means a much younger cluster. Stars from older clusters have a much lower turnoff point. (pp. 204–205; objective 2)

6. Open clusters are found in the disk of the galaxy, contain young, hot, bright, blue stars, may contain relatively few stars, and have an abundance of metals. Globular clusters may have up to a million stars; their stars are close together, are found in the halo of the galaxy, appear to have old stars, and are metal-poor. (pp. 204–205; video lesson; objective 1)

7. A flood of energy from the hydrogen-fusion shell pushes toward the surface, heating the outer layers of the star and forcing them to expand

dramatically and this expansion cools the star. (pp. 200–202; video lesson; objective 2)

Applications

1. A disk of gas around a star's equator might form during the slow wind stage and then deflect the fast wind into oppositely directed flows. Another star or planets orbiting the dying star, rapid rotation, or magnetic fields might cause these peculiar shapes. (p. 209; objective 3)

2. The more massive star in the binary expanded to fill its Roche lobe, material transferred to the lower mass star to affect both of their evolutionary sequences. (p. 211; objective 4)

3. The mathematical model suggests that white dwarfs can be no larger than 1.4 solar masses. Observational evidence agrees with this model. No white dwarfs larger than 1.4 solar masses have been found. (pp. 207, 210; objective 4)

4. See the H–R diagrams on pages 199 and 205 of *Horizons*. These stars begin on the zero-age main sequence. As they use their nuclear fuel, they move slightly up and to the right, still on the main sequence. When they run out of hydrogen and become red giants, they will move to the middle right-hand part of the graph. When they turn into white dwarfs, they move to the lower left-hand quadrant. (pp. 199, 205; objectives 2 & 3)

5. See the H–R diagrams on pages 199 and 205 of *Horizons*. These stars also begin on the zero-age main sequence. They move through the main sequence much more quickly and end up as supergiants in the upper right-hand portion of the graph until they explode as a supernova. (pp. 199, 205; objectives 2 & 5)

6. In general, a high-mass star evolves more quickly than a low-mass star. In a close binary with mass transfer, the less massive star may become a giant faster than the more massive star. (p. 211; video lesson; objective 4)

Notes:

Lesson Review

Lesson 10: Stellar Deaths

PLEASE NOTE: Use this matrix to guide your study and achieve the learning objectives of this lesson. It will also help you to view the video, which defines and demonstrates important concepts and principles as they relate to everyday life and actual case studies.

Learning Objective	Textbook	Student Guide
1. Describe the internal processes that cause an aging main-sequence star to expand into a giant.	pp. 198–203	Key Terms: 3, 4; Matching I: 3, 4; Completion: 7; Self-Test: 9, 10, 11; Short-Answer: 2, 3, 6.
2. Explain how studies of star clusters are critical to understanding the evolution of stars.	pp. 202–205	Key Terms: 6, 7, 8, 9; Matching I: 6, 7, 8, 9; Completion: 1, 2; Self-Test: 4; Short-Answer: 5, 7; Applications: 4, 5.
3. Describe the deaths of low- and medium-mass stars, including the formation of planetary nebulae and white dwarfs, and discuss the effects that our sun's death will have on Earth.	pp. 203, 206–210	Key Terms: 5, 10, 11; Matching II: 5, 10, 11; Completion: 3, 4; Self-Test: 2, 3, 12; Short-Answer: 2, 3; Applications: 1, 4.
4. Describe the evolution of close binary stars, including possible effects of mass transfer such as novae and type I supernovae.	pp. 211–214	Key Terms: 1, 12, 13, 14, 15, 16, 17, 19, 20; Matching I: 1; Matching II: 1, 2, 3, 4, 5, 6, 8, 9; Completion: 5; Self-Test: 5, 7, 8; Short-Answer: 1, 4; Applications: 2, 3, 6.
5. Describe the deaths of the most massive stars, including type II supernovae and supernova remnants.	pp. 214–221	Key Terms: 18, 21, 22; Matching I: 2; Matching II: 7, 10, 11; Completion: 6; Self-Test: 1, 6; Short-Answer: 1, 3, 4; Applications: 5.

LESSON
11

Stellar Remnants

Checklist

For the most effective study of this lesson, complete the following activities in this sequence.

Before Viewing the Video

- ❑ Read the Preview, Learning Objectives, and Viewing Notes below.
- ❑ Read Chapter 11, "Neutron Stars and Black Holes," pages 224–247, in the *Horizons* textbook.

What to Watch

- ❑ After reading the textbook chapter, watch the video for Lesson 11, *Stellar Remnants*.

After Viewing the Video

- ❑ Briefly note your answers to questions listed at the end of the Viewing Notes.
- ❑ Review the Summary below.
- ❑ Review all reading assignments for this lesson, especially the Chapter 11 summary on page 246 in *Horizons* and the Viewing Notes in this guide.
- ❑ Write brief answers to the review questions at the end of Chapter 11 in *Horizons*.
- ❑ Complete the Review Exercises below. Check your answers with the Answer Key and review when necessary.
- ❑ Use the Lesson Review matrix found at the end of this lesson to review and assess your knowledge of each Learning Objective.
- ❑ As assigned by your instructor, complete the Applications activities and any additional activities for this lesson.

Preview

In the last two lessons, you discovered how a star lives and how it dies. No matter how a star lives, it collapses into one of three interesting and unusual states: a *white dwarf*, a **neutron star**, or a **black hole**. These are otherwise known as compact objects. Although these objects are difficult to detect, astronomers are on a quest to confirm their presence.

In Lesson 10, you learned how a low- to medium-mass star eventually collapses into a white dwarf and may gradually cool into a *black dwarf*. The core of a more massive star can suddenly collapse triggering a *supernova* explosion while the outer layers of the star shed to produce a *supernova remnant*. Depending on the mass of the collapsed core, it can become either a neutron star or a black hole—compact objects that are smaller, denser, and even more intriguing than a white dwarf.

In this lesson, you'll learn about neutron stars and one type of black hole that is formed from the death of a single massive star. In future lessons, you'll learn about supermassive black holes—as massive as millions of stars—that are concealed in the cores of giant galaxies.

CONCEPTS TO REMEMBER

- Recall from Lesson 10 that an *accretion disk* is the whirling disk of gas that forms around a compact object such as a white dwarf, a neutron star, or a black hole as matter is drawn in. Mass transferred from one star toward the compact object in a binary system must conserve its angular momentum and therefore flows into the rapidly rotating accretion disk (p. 191 in this guide).

- In Lesson 10, you also learned that *angular momentum* is a measure of the tendency of a rotating body to continue rotating. Mathematically, this is the product of mass, velocity, and radius. When mass is transferred from one orbiting object to another—for instance, from a star to a compact object—it must continue rotating because of angular momentum (p. 191 in this guide).

- In Lesson 10, you learned that *degenerate matter* is the extremely high-density matter in which pressure no longer depends on temperature as a result of quantum mechanical effects. Electrons can degenerate and support the mass of a white dwarf; you will learn in Lesson 11 that neutrons can degenerate and support the mass of a neutron star (p. 190 in this guide).

- In Lesson 10, you learned that a *supernova* is the explosion of a star in which its brightness increases by a factor of about a million. The expanding shell of gas resulting from the explosion is called a *supernova remnant* (p. 190 in this guide).

Learning Objectives

After you complete this lesson, you should be able to:

1. Summarize how neutron stars were first predicted and describe their theoretical properties. *HORIZONS* TEXTBOOK PAGES 226–227.

2. Recount the discovery of pulsars and their identification as rapidly spinning neutron stars. *HORIZONS* TEXTBOOK PAGES 227–228, 230–231.

3. Describe the observed properties and evolution of pulsars and binary pulsars, including those with orbiting planets. *HORIZONS* TEXTBOOK PAGES 228–229, 232–237.

4. Describe the theory of the formation of black holes and their predicted properties. *HORIZONS* TEXTBOOK PAGES 237–240.

5. Recount attempts to observe black holes indirectly and describe evidence that supports their existence. *HORIZONS* TEXTBOOK PAGES 240–243.

6. Describe the observed properties of gamma-ray bursts and their possible origins. *HORIZONS* TEXTBOOK PAGES 243–245.

At this point, read Chapter 11, "Neutron Stars and Black Holes," pages 224–247.

The most incomprehensible thing about the world is that it is at all comprehensible.
—Albert Einstein
(1879–1955)

Viewing Notes

The universe is filled with bizarre objects, which include neutron stars, black holes, and gamma-ray bursters. In the video, you'll be offered a glimpse into the evolution of these objects.

The video program contains the following segments:

- ⚙ Neutron Stars
- ⚙ Black Holes
- ⚙ Formation of Black Holes
- ⚙ Gamma-Ray Bursts

The following information will help you better understand the video program:

In the opening segment of the video, you'll learn that a **neutron star** can be formed by the collapsing core of a massive star in a supernova explosion. Similar to how white dwarfs are supported by degenerate electrons, neutron stars are supported by degenerate neutrons—the neutrons become so tightly packed, they can no longer change their energy and so resist any further compression. While stars like our sun will eventually collapse into white dwarfs, stars somewhat more massive than the sun will produce neutron stars. A **pulsar** is a type of neutron star that emits short, precisely timed radio bursts.

Very massive stars are believed to collapse into **black holes**—compact objects whose gravity is so great even light cannot escape from them. In this lesson, you'll learn about these stellar mass black holes and in later lessons you'll be introduced to supermassive black holes.

The video program mentions that black holes can be detected by several different means. The most common way for astronomers to detect a black hole is when a star in a binary system loses mass to its companion black hole. This mass accumulates on the black hole's accretion disk where it is detectable—usually by its X-ray emissions. You learned about this mass transfer in Lesson 10, and the same thing can happen here—a star in a binary system loses mass and it collects on the companion black hole's accretion disk, which is just outside the **event horizon**.

Gamma-ray bursts coming from space were first detected in the 1970s. Astronomers suspect they might be coming from neutron stars or black holes. The objects that produce these bursts are known as gamma-ray bursters. They appear to be associated with the collapse of massive stars.

QUESTIONS TO CONSIDER

- What is the maximum mass a white dwarf can have and what happens when that mass is exceeded?

- What is the maximum mass a neutron star can have and what happens when that mass is exceeded?

- How does Einstein's theory of relativity help to explain the existence of black holes?

- If a person fell into a black hole, how does the perspective of an observer differ from the perspective of the person falling?

- What type of star is believed to create a black hole?

- How can we detect the presence of such a stellar mass black hole?

- How are gamma-ray bursters associated with supernovae? How are they related to black holes?

Watch the video for Lesson 11, *Stellar Remnants.*

Key Terms and Concepts

Page references are keyed to the *Horizons* textbook.

1. **neutron star:** A small, highly dense star composed almost entirely of tightly packed neutrons. (p. 226; video lesson; objective 1)

2. **pulsar:** A source of short, precisely timed radio bursts, believed to be a spinning neutron star. (p. 227; video lesson; objective 1)

3. **pulsar wind:** The breeze of high-energy particles flowing away from a spinning neutron star. (p. 229; objective 3)

4. **gravitational radiation:** Disturbances in a gravitational field traveling at the velocity of light and carrying energy away from an object with a rapidly changing mass distribution. (p. 232; objective 3)

5. **X-ray burster:** An object that produces repeated bursts of X rays. (p. 234; objective 3)

6. **millisecond pulsar:** A pulsar that has a pulse period of only a few thousandths of a second. (p. 234; objective 3)

7. **singularity:** The object of zero radius into which the matter in a black hole is believed to fall. (p. 238; video lesson; objective 4)

8. **black hole:** A mass that has collapsed to such a small volume that its gravity prevents the escape of all radiation. (p. 238; video lesson; objective 4)

9. **event horizon:** The boundary of the region of a black hole from which no radiation may escape. (p. 238; video lesson; objective 5)

10. **Schwarzschild radius (R_s):** The distance from the singularity of a black hole to the event horizon. (p. 238; video lesson; objective 5)

11. **time dilation:** The relationship that states clocks tick slower as they move faster or if they are in strong gravitational fields; derived from the Special and General Theories of Relativity. (p. 239; video lesson; objective 4)

12. **gravitational redshift:** The lengthening of the wavelength of a photon because of its escape from a gravitational field. (p. 239; video lesson; objective 4)

13. **gamma-ray burst:** A sudden, powerful burst of gamma rays. (p. 244; video lesson; objective 6)

14. **magnetar:** A class of neutron star having very strong magnetic fields. (p. 244; objective 6)

15. **hypernova:** Produced when a very massive star collapses into a black hole; another name for a collapsar. (p. 244; video lesson; objective 6)

Summary

A star can die either a spectacular death or a quiet one. Depending on the mass of the stellar remnant, it will form one type of compact object—a white dwarf, a neutron star, or a black hole.

Neutron Stars

In Lesson 10, you learned that a medium-mass star like our sun would collapse into a white dwarf. The maximum mass of a white dwarf—the collapsed remains of the star—is 1.4 solar masses; this is known as the *Chandrasekhar limit*. A somewhat more massive star dies a different death. If the collapsing core is greater than 1.4 solar masses but less than about 3 solar masses the core will collapse into a **neutron star**, while the remainder of the dying star's mass is blown outward as a supernova.

In previous lessons you learned about the *neutron*—an atomic particle with no charge. A neutron behaves similarly to an electron—two identical neutrons cannot occupy the same energy level. Electrons and neutrons in an extremely dense gas can become *degenerate*—their pressure no longer depends on their temperature. If these degenerate neutrons could resist the immense pressure and crushing mass of a star, astronomers in the early part of the twentieth century proposed that neutron stars might exist.

Even today, neutron stars are difficult to detect, but it is believed that they are extremely dense. The density of a single neutron star is about that of an atomic nucleus with all of its empty space squeezed out of it—about 10^{14} g/cm^3. This is difficult to imagine, but on Earth the density would be equivalent to about 50 million cars compressed to the size of a single sugar cube. Theory predicts that a neutron star's radius is only about 10 km. Because of its extreme mass and very small size, a neutron star is believed to spin very rapidly.

When a star's magnetic field is captured into its gases and it collapses, the magnetic field is squeezed into a smaller area. It's theorized that a neutron star has a magnetic force about one trillion times stronger than Earth's. Theory also predicts that the neutron star is very hot and radiates most of its energy in the X-ray portion of the spectrum.

A type of neutron star that emits beams of radiation as it rotates is called a **pulsar**. The first pulsar was discovered in 1967, when astronomers noticed a pattern of regularly spaced pulses on the paper chart from a radio telescope. As astronomers located more pulsars and studied them over the course of months, they discovered that the pulsations were highly precise but slowed down over time. They found that they pulsed anywhere from every 0.033 to 3.75 seconds and the pulse lasted only about 0.001 seconds. Based on their observations, astronomers deduced that the pulsar could not be a normal star because a normal star would be ripped apart at those speeds of rotation.

Although little is known about pulsars, astronomers have developed a model to aid in understanding them. The lighthouse model of a pulsar indicates that a pulsar doesn't really pulse; rather, it emits photons that produce powerful beams of electromagnetic radiation. As it spins, the beams sweep around the sky like a lighthouse. Astronomers tend to detect only those pulsars whose beams sweep over Earth.

It is presumed that when a pulsar forms, it spins very fast and as it ages, it loses energy and it spins more slowly; astronomers can detect this in the frequency of its beams. We can expect that a young pulsar emits more powerful beams of radiation, including those in the visible portion of the spectrum. Although the beams are very powerful, the pulsar emits almost 99.9 percent of its energy in a **pulsar wind**—a breeze of high-energy particles that flows away from it.

Although both pulsars and nebulae are remnants of a supernova, pulsars can exist outside of nebulae and not all nebulae have pulsars in them. An off-center supernova or a supernova that occurs in a binary system can give the pulsar enough velocity to escape the remnant. Not all supernovae leave remnants behind, but those that do scatter their remains into the interstellar medium within about 50,000 years; by contrast, a pulsar can be detected for about 10 million years. A pulsar can also be difficult to detect because it can be hidden in the dense gas of the supernova remnants. Or, if its beams don't sweep directly over Earth, astronomers may not be able to detect it.

Like normal stars, pulsars can be part of a binary system. Astronomers can detect a pulsar in a binary system by observing the change in its pulse frequency. When the pulsar is moving away from Earth in its orbit, we can see the pulse period slightly lengthened and when it moves toward Earth, we can see the pulse period as slightly shortened. This is similar to observing the Doppler shift of a *spectroscopic binary system* that you learned about in Lesson 8.

Through observation of pulsars in a binary system, astronomers can determine a pulsar's radial velocity, orbital period, the shape of its orbit, and its mass. The typical mass of a pulsar is about 1.35 solar masses. You might wonder how a pulsar can be below the Chandrasekhar limit—if the stellar remnant is less than 1.4 solar masses, it should have collapsed into a white dwarf. The reason is not well known, but it's possible that the extra inward momentum of the core implosion assisted the mass in overcoming electron degeneracy.

When a pulsar is in a binary system, its typical companion is a giant star that's losing mass. As the giant star loses mass, the pulsar can gain mass and rotational energy through its *Lagrangian point* (see Lesson 10). Although pulsars typically pulse faster when they are young and pulse slower as they age and lose energy, a pulsar that gains mass from its companion can actually pulse faster as it ages. Because a pulsar is so dense and has a strong gravitational field, it can essentially evaporate the remains of its companion.

Like normal stars, pulsars can have planets. Astronomers discovered planets orbiting a neutron star by observing the variation in its pulsation period much like that caused by the orbital motion of a binary pulsar. They noticed that the Doppler shifts are much smaller than those in a binary star system. According to gravitational theory, when two planets interact, they will modify each other's orbits. They realized that the changes in the pulsation period were caused by the gravitational pull of two planets. Because the pulsar is so hot and its strong gravity would cause a planet to orbit very closely, any normal planet would be vaporized by the pulsar's heat. It's likely that these planets are remnants of the pulsar's stellar companion.

Black Holes

You've already discovered that neutron stars (including pulsars) and white dwarfs are compact objects that are the end stages of a dying star. Another type of compact object is a **black hole**—a mass that has collapsed into such a small volume that its gravity prevents the escape of all radiation. This lesson focuses on black holes that appear to have originated from the death of a very massive single star—one whose collapsed core is greater than 3 solar masses. In later lessons you'll learn about black holes that may be greater than 1 million solar masses.

Black holes are so fascinating and bizarre we might think they only exist in science fiction movies. A black hole is an object in space that has collapsed under its own gravitational force to an extent that its escape velocity is equal to the speed of light. You'll recall from Lesson 4 that the *escape velocity* is the initial velocity an object needs to escape from a celestial body. Einstein's theory of relativity tells us that nothing can travel faster than the speed of light, so an object with an escape velocity faster than that would inhibit anything from escaping—even photons.

When the core of a star contains more than 3 solar masses, it's so massive that no force can stop it from collapsing. Because the degenerate matter can't support its weight, it won't stop collapsing even when it reaches the size of a white dwarf or a neutron star. It will eventually collapse to a zero radius and its density and gravity become infinite—this is called the point of **singularity**.

The boundary between the black hole and the rest of the universe is called the **event horizon** because any event that occurs inside of it cannot be observed. In order to become a black hole, a massive object must be compressed to the radius of the event horizon, otherwise known as its **Schwarzschild radius**. The Schwarzschild radius depends only on the mass of the object: as the mass of an object increases, the radius increases. All objects have a Schwarzschild radius, but only massive ones can collapse under their own weight to form a black hole. An object like the sun has a Schwarzschild radius of about 3 km, but it's not massive enough to become a black hole (see Table 11-1 on p. 239 in the *Horizons* textbook).

Because no *electromagnetic radiation* can escape from a black hole, it cannot be observed. But X rays emitted from the superheated matter flowing toward the event horizon can be detected. The matter falling into the black hole can sometimes form enormous jets that emit radio waves. The gases in these pencil-beam jets seem to follow the magnetic lines that surround the black hole.

The gravitational field of a black hole is very strong, but it looks like any other gravitational field of a massive object. However, when matter gets near the event horizon, it easily gets drawn in. As you might expect, there aren't many objects close enough to a black hole to be affected by its gravitational force. But a black hole in a binary system might receive matter that is transferred from its companion and therefore can be detected.

To begin to understand a black hole, we can look to Einstein's theory that states gravity is the curvature of space-time. Massive objects distort space-time; since a black hole is extremely massive, this distortion of space is severe. A black hole is an object in which space-time is curved so much that it's closed around onto itself. This causes black holes to have very strange properties. For example, it is believed that clocks slow down in the curvature of space-time—this is known as **time dilation**. Another effect is **gravitational redshift**—as light travels out of a gravitational field, it loses energy and its wavelength grows longer.

Some black holes and neutron stars emit powerful bursts of gamma rays. The objects that emit these rays are known as gamma-ray bursters. Gamma rays emitted from these bursters can be detected several times a day all over the sky. They rise to a maximum intensity in a few seconds and then fade away quickly over the course of seconds or minutes. The spectra from the fading glow of the explosion reveal that most gamma-ray bursts occur in distant galaxies, but they can still affect us here on Earth by interfering with radio communications.

Astronomers aren't sure what produces these gamma ray bursts, but two theories exist. One theory proposes that two neutron stars orbiting one another lose energy by gravitational radiation and fall together. The eruption that ensues by the stars colliding could produce gamma rays. The second theory proposes that when a star more massive than 25 solar masses exhausts its nuclear fuel, it collapses into a black hole and leaves behind gas falling in through a hot accretion disk. The resulting energy could cause a supernova explosion and eject high-energy beams that interact with the surrounding gas to produce gamma-ray bursts. These collapsing stars and the ensuing explosions are known as **hypernovae**. Some gamma-ray bursts leave behind a fading glow like that from supernova explosions and it's believed that some bursts are produced by hypernovae.

The study of compact objects such as white dwarfs, neutron stars, and black holes can be quite interesting. In the next few lessons you will explore the galaxies and examine similar objects on a much grander scale.

Review Exercises

Matching I

Match each term with the appropriate definition or description.

1. _____ neutron star	6. _____ millisecond pulsar
2. _____ pulsar	7. _____ singularity
3. _____ pulsar wind	8. _____ black hole
4. _____ gravitational radiation	9. _____ event horizon
5. _____ X-ray burster	

a. A pulsar with a pulse period of only a few thousandths of a second.
b. The breeze of high-energy particles flowing away from a spinning neutron star.
c. The boundary of the region of a black hole from which no radiation may escape.
d. Disturbances in a gravitational field traveling at the velocity of light and carrying energy away from an object with a rapidly changing mass distribution.
e. A small, highly dense star composed almost entirely of tightly packed neutrons.
f. A mass that has collapsed to such a small volume that its gravity prevents the escape of all radiation.
g. An object that produces repeated bursts of X rays.
h. An object of zero radius into which the matter in a black hole is believed to fall.
i. A source of short, precisely timed radio bursts, believed to be spinning neutron stars.

Matching II
Match each term with the appropriate definition or description.

1. _____ Schwarzschild radius		4. _____ gamma-ray burster	
2. _____ time dilation		5. _____ magnetar	
3. _____ gravitational redshift		6. _____ hypernova	

a. Produced when a very massive star collapses into a black hole.
b. A class of neutron star having very strong magnetic fields.
c. A sudden, powerful burst of gamma rays.
d. The relationship that states clocks tick slower as they move faster or if they are in strong gravitational fields.
e. The distance from the singularity of a black hole to the event horizon.
f. The lengthening of the wavelength of a photon as a result of its escape from a gravitational field.

Completion
Fill each blank in the sentences below with the most appropriate term from the list of completion answers that follow. A term may be used once, more than once, or not at all. Check your answers with the Answer Key and review when necessary.

corona	one hundred	ten
fifty	pulsar	three
longer	pulsar wind	white dwarf
one	shorter	X-ray burster

1. A neutron star has an approximate radius of
   _____ km.

2. Roughly 99.9 percent of the energy flowing away from a pulsar is carried as a _____.

3. Most widely accepted calculations suggest that a neutron star cannot be more massive than _____ solar masses.

4. The Crab Nebula is a supernova remnant that contains a
   _____.

5. As light travels out of a gravitational field, it loses energy, and its wavelength grows _____.

Self-Test

Select the best answer.

1. Originally, the signals that were found to be pulsars were thought to be
   a. spinning neutron stars.
   b. spinning white dwarfs.
   c. spinning black holes.
   d. little green men.

2. If you replaced our sun with a one-solar-mass black hole, Earth would
   a. be sucked into the black hole.
   b. be flung into space.
   c. maintain its orbit.
   d. turn into a black hole itself.

3. The Schwarzschild radius of our sun is
   a. 3 km.
   b. 6 km.
   c. 9 km.
   d. 30 km.

4. The maximum mass limit for a _____ is 1.4 solar masses.
   a. red dwarf
   b. white dwarf
   c. neutron star
   d. black hole

5. If a stellar remnant is greater than 1.4 solar masses but less than about 3 solar masses, the resulting object will be a
   a. red dwarf.
   b. white dwarf.
   c. neutron star.
   d. black hole.

6. If a stellar remnant is greater than 3 solar masses, the resulting object will be a
   a. red dwarf.
   b. white dwarf.
   c. neutron star.
   d. black hole.

7. Objects in binary systems consist of a neutron star accumulating material from another star on its surface, where it ignites and produces periodic
   a. type II supernovae.
   b. X-ray bursters.
   c. novae.
   d. helium flashes.

8. Who discovered the peculiar pattern of radio signals that were eventually found to be from a pulsar?
   a. Lev Landau
   b. Fritz Zwicky
   c. Albert Einstein
   d. Jocelyn Bell

9. Black hole candidates are detectable by their accretion disks producing mainly what type of light?
   a. infrared light
   b. ultraviolet light
   c. X rays
   d. gamma rays

10. At the event horizon, the escape velocity is equal to
    a. the speed of sound.
    b. the speed of light.
    c. the speed of a shock wave.
    d. infinity.

11. The first X-ray binary suspected of harboring a black hole is
    a. V404 Cygni.
    b. QZ Vul.
    c. Cygnus X-1.
    d. the Crab Nebula.

12. When were neutron stars first predicted?
    a. 1806
    b. 1932
    c. 1967
    d. 1983

Short-Answer Questions

1. If nothing can escape from a black hole, not even light, how do black holes emit X rays?

2. Briefly explain how a pulsar is like a lighthouse.

3. What would happen if you were to jump feet first into a black hole?

4. How would it appear to an outside observer if someone were to enter a black hole?

5. Why do millisecond pulsars spin so fast?

6. How are X-ray bursters and novae related? How are they different?

7. Why would a neutron star have such a strong magnetic field?

Applications

1. What would be your Schwarzschild radius? Use your mass in kilograms (1 kg = 2.2 pounds).

2. Why does the accretion disk around a compact object get so much hotter than around a main sequence star?

3. What is thought to cause the higher energy gamma-ray bursts, which do not appear to repeat?

4. What connection does the Cold War have with neutron stars and black holes?

5. What objects other than black holes have we studied in previous chapters that have accretion disks and bipolar flows? What is the difference between these objects and black holes?

6. If a pulsar's period is 0.05 seconds, how many times does it spin in one day?

Answer Key

Matching I

1. e (p. 226; video lesson; objective 1)

2. i (p. 227; video lesson; objective 1)

3. b (p. 229; objective 1)

4. d (p. 232; objective 3)

5. g (p. 234; objective 3)

6. a (p. 234; objective 3)

7. h (p. 238; video lesson; objective 4)

8. f (p. 238; video lesson; objective 4)

9. c (p. 238; video lesson; objective 5)

Matching II

1. e (p. 238; video lesson; objective 5)

2. d (p. 239; video lesson; objective 4)

3. f (p. 239; video lesson; objective 4)

4. c (p. 244; video lesson; objective 6)

5. b (p. 244; objective 6)

6. a (p. 244; video lesson; objective 6)

Completion

1. ten (p. 226; video lesson; objective 1)

2. pulsar wind (p. 229; objective 3)

3. three (p. 226; video lesson; objective 1)

4. pulsar (pp. 228, 231; objective 3)

5. longer (p. 239; video lesson; objective 4)

Self-Test

1. d (p. 227; video lesson; objective 2)

2. c (p. 239; video lesson; objective 4)

3. a (p. 239; objective 5)

4. b (p. 226; video lesson; objective 1)

5. c (p. 226; video lesson; objective 1)

6. d (p. 238; video lesson; objective 4)

7. b (p. 234; objective 3)

8. d (p. 227; video lesson; objective 2)

9. c (pp. 240–241; video lesson; objective 5)

10. b (p. 237; objective 4)

11. c (p. 241; objective 5)

12. b (p. 226; video lesson; objective 1)

Short-Answer Questions

1. The X rays are coming from the accretion disk swirling around the edge of the event horizon of the black hole. (p. 241; objective 4)

2. A beam of light sweeps over Earth like a lighthouse beam. (pp. 230–231; video lesson; objective 2)

3. You would probably be ripped apart by the gravitational tidal forces before you reached the event horizon. (pp. 239–240; video lesson; objective 4)

4. The person would appear to go slower and slower and then hover at the edge of the black hole and the image would get dimmer and dimmer. (pp. 239–240; video lesson; objective 4)

5. They are spun up by additional material from a companion star. Additional mass added to the pulsar will increase its angular momentum and make it spin faster. (p. 234; objective 3)

6. Novae are caused by a binary system with a white dwarf. X-ray bursters have mass transfer in a binary system to a neutron star. The difference is in the amount of energy and the type of electromagnetic radiation. (p. 234; objective 3)

7. Whatever magnetic field a star has is captured into the gases. When the star collapses, the magnetic field is carried along and squeezed into a smaller area, which could make the field a billion times stronger. (p. 227; objective 1)

Applications

1. The answer is 1.1×10^{-25} meters. (p. 238; objective 4)

   Assume mass = 75 kg
   $R = 2GM/c^2 = 2(6.67 \times 10^{-11})(75)/(3 \times 10^8)^2$
   This value is more than a billion times smaller than the nucleus of an atom.

2. The accretion disk is made so hot by friction because the compact object has such a strong gravitational field. (p. 243; objective 5)

3. Two theories have been proposed. One suggestion is that they occur when two neutron stars orbiting each other lose energy by gravitational radiation and fall together. The second theory proposes they are produced when a star more massive than 25 solar masses exhausts its nuclear fuels. (pp. 243–244; objective 6)

4. In 1963, a nuclear test ban treaty was signed, and by 1968, the United States was able to put a series of Vela satellites in orbit to watch for nuclear tests through gamma rays. Experts were surprised when they detected about one gamma-ray burst per day. (pp. 243–244; objective 6)

5. There were close binary stars, protostars, and black holes. The difference is the amount of energy produced and the size of the jets. (p. 243; objective 5)

6. The answer is 1.7×10^6 times per day. (p. 227; objective 2)

   0.05 seconds per cycle yields 20 cycles per second (1/0.05)
   (20 spins/second) × (60 seconds/minute) × (60 minutes/hour) × (24 hours/day) = 1,728,000 spins/day.

Lesson Review

Lesson 11: Stellar Remnants

PLEASE NOTE: Use this matrix to guide your study and achieve the learning objectives of this lesson. It will also help you to view the video, which defines and demonstrates important concepts and principles as they relate to everyday life and actual case studies.

Learning Objective	Textbook	Student Guide
1. Summarize how neutron stars were first predicted and describe their theoretical properties.	pp. 226–227	Key Terms: 1, 2, 3; Matching I: 1, 2, 3; Completion: 1, 3; Self-Test: 4, 5, 12; Short-Answer: 7.
2. Recount the discovery of pulsars and their identification as rapidly spinning neutron stars.	pp. 227–228, 230–231	Self-Test: 1, 8; Short-Answer: 2; Applications: 6.
3. Describe the observed properties and evolution of pulsars and binary pulsars, including those with orbiting planets.	pp. 228–229, 232–237	Key Terms: 4, 5, 6; Matching I: 4, 5, 6; Completion: 2, 4; Self-Test: 7; Short-Answer: 5, 6.
4. Describe the theory of the formation of black holes and their predicted properties.	pp. 237–240	Key Terms: 7, 8, 11, 12; Matching I: 7, 8; Matching II: 2, 3; Completion: 5; Self-Test: 2, 6, 10; Short-Answer: 3, 4; Applications: 1.
5. Recount attempts to observe black holes indirectly and describe evidence that supports their existence.	pp. 240–243	Key Terms: 9, 10; Matching I: 9; Matching II: 1; Self-Test: 3, 9, 11; Applications: 2, 5.
6. Describe the observed properties of gamma-ray bursts and their possible origins.	pp. 243–245	Key Terms: 13, 14, 15, 16, 17; Matching II: 4, 5, 6; Applications: 3, 4.

LESSON
12

Our Galaxy: The Milky Way

Checklist

For the most effective study of this lesson, complete the following activities in this sequence.

Before Viewing the Video

- ❏ Read the Preview, Learning Objectives, and Viewing Notes below.
- ❏ Read Chapter 12, "The Milky Way Galaxy," pages 248–275, in the *Horizons* textbook.

What to Watch

- ❏ After reading the textbook chapter, watch the video for Lesson 12, *Our Galaxy: The Milky Way.*

After Viewing the Video

- ❏ Briefly note your answers to questions listed at the end of the Viewing Notes.
- ❏ Review the Summary below.
- ❏ Review all reading assignments for this lesson, especially the Chapter 12 summary on page 274 in *Horizons* and the Viewing Notes in this guide.
- ❏ Write brief answers to the review questions at the end of Chapter 12 in *Horizons.*
- ❏ Complete the Review Exercises below. Check your answers with the Answer Key and review when necessary.
- ❏ Use the Lesson Review matrix found at the end of this lesson to review and assess your knowledge of each Learning Objective.
- ❏ As assigned by your instructor, complete the Applications activities and any additional activities for this lesson.

Preview

Lesson 12 begins with a story of scientific discovery. For thousands of years, civilizations all over the world noticed the path of fuzzy light that arced across the night sky—what has later come to be known as the Milky Way. It took thousands of years before astronomers were able to measure and plot the stars contained in this path of light. The discovery of a certain type of star unlocked the clues astronomers needed to determine the size and shape of our galaxy.

Although we can see stars and planets and we can detect the faint band of light that circles the sky, it's not obvious that we live inside of a galaxy. Nearly every celestial object that we see is in our galaxy, but the casual observer cannot detect how incredibly large and beautiful it is. Astronomers can use radio telescopes to create a map and paint a visual picture of a flat disk with graceful spiral arms—our galaxy, the Milky Way.

CONCEPTS TO REMEMBER

- Recall from Lesson 2 that *apparent visual magnitude (m_v)* is the brightness of a star as seen by human eyes on Earth. It is also known as apparent brightness or apparent magnitude (p. 16 in this guide).

- In Lesson 8, you learned that *absolute visual magnitude (M_v)* is the intrinsic brightness of a star; also the apparent visual magnitude that a star would have if it were 10 parsecs away (p. 146 in this guide).

- In Lesson 8, you also learned that *luminosity (L)* is the total amount of energy a star radiates in one second; also known as maximum intrinsic brightness. Luminosity is measured in energy output per area (watts per meter squared) or more commonly as solar luminosities based on the output of the sun. Astronomers can compare an object's luminosity to its apparent visual magnitude to determine its distance from Earth (p. 146 in this guide).

- In Lesson 10, you learned that a *turnoff point* is the point in the H–R diagram at which a cluster's stars turn off the main sequence and move toward the red giant region. Studying the turnoff point can reveal the approximate age of the cluster (p. 191 in this guide).

Learning Objectives

After you complete this lesson, you should be able to:

1. Describe how William Herschel first mapped the Milky Way Galaxy, and how Harlow Shapley later came to estimate its size and the location of Earth's solar system within it. *HORIZONS* TEXTBOOK PAGES 251–255.

2. Describe the characteristics of the Milky Way Galaxy and its major components, including the nature of the stellar orbits in each region, and explain the importance of 21-centimeter radiation in the work of radio astronomers. *HORIZONS* TEXTBOOK PAGES 255–259.

3. Explain the reason for differing populations of stars within the galaxy and relate their distribution to the process of the formation of the Milky Way. *HORIZONS* TEXTBOOK PAGES 257, 259–264.

4. Describe observations of spiral arms within the Milky Way and other galaxies and explain the two major mechanisms that create spiral structure and their role in star formation. *HORIZONS* TEXTBOOK PAGES 264–269.

5. Describe observations of the nucleus of the Milky Way at nonvisible wavelengths and evidence for the suspected black hole at the center. *HORIZONS* TEXTBOOK PAGES 270–273.

At this point, read Chapter 12, "The Milky Way Galaxy," pages 248–275.

We all travel the milky way together, trees and men ...

—John Muir
American naturalist (1838–1914)

Viewing Notes

Determining the size and shape of our galaxy is not an easy task—a sea of stars and dust obscures our view. In the video program, you will learn the story of how the size and shape of the galaxy was determined—a story filled with observation, speculation, debate, and discovery.

The video program contains the following segments:

- ☼ Counting Stars
- ☼ Human Calculators
- ☼ The Great Debate
- ☼ The Milky Way Galaxy
- ☼ Young Stars & Spiral Arms
- ☼ Old Stars to the Middle
- ☼ The Dark Side

The following information will help you better understand the video program:

Cepheid variable stars can be used to measure distances in the galaxy. By measuring the pulsation period of a Cepheid, we can determine its luminosity by using the **period–luminosity relation**. The longer the pulsation period, the more luminous the Cepheid is. Once we know the luminosity, we can compare it to the apparent brightness to determine its distance from Earth. You'll learn more about the use of Cepheids and other "distance indicators" in Lesson 13.

Astronomers identified a number of Cepheids contained within the concentration of globular clusters between the constellations of Sagittarius and Scorpius. By measuring the distance to the Cepheids and the globular clusters, astronomers were able to estimate the distance of Earth from the galaxy's center.

Gas and dust obscure our view of many distant objects, but astronomers can use radio maps to study the distribution of celestial objects in our galaxy. They can use radio telescopes to detect wavelengths of 21 cm. The 21-centimeter line from atomic hydrogen allows us to probe and detect hydrogen—the main component of gas that fills our galaxy. Radiation at this wavelength easily penetrates the interstellar dust to allow us to compile a complete image of our galaxy.

Stars are born in the spiral arms of our galaxy. It is believed that spiral arms are density waves, or waves of compression. When a large cloud of gas slams into a density wave, the wave is suddenly compressed. Recall from Lesson 9 that star formation will occur where gas clouds are compressed.

QUESTIONS TO CONSIDER

- How did William and Caroline Herschel determine that our galaxy was a flattened disk?

- How did Harlow Shapley determine that our solar system was not near the center of the Milky Way?

- How did Henrietta Leavitt's discovery of the relationship between the pulsation period and luminosity of Cepheids aid in Harlow Shapley's work?

- How can astronomers create a map of our galaxy if clouds of dust and gas obscure most of it?

- In what part of the galaxy are old stars concentrated? Why?

- In what part of our galaxy are young stars concentrated? Why?

Watch the video for Lesson 12, *Our Galaxy: The Milky Way.*

Key Terms and Concepts

Page references are keyed to the *Horizons* textbook.

1. **variable star:** A star whose brightness changes periodically. (p. 251; video lesson; objective 1)

2. **Cepheid variable star:** Variable star with a period of up to 60 days. The period of variation is related to luminosity. (p. 251; video lesson; objective 1)

3. **instability strip:** The region of the H–R diagram in which stars are unstable to pulsation. A star passing through this strip becomes a variable star. (p. 251; objective 1)

4. **RR Lyrae variable star:** Variable star with a period of from 12 to 24 hours. Common in some globular clusters. (p. 251; objective 1)

5. **period–luminosity relation:** The relation between period of pulsation and intrinsic brightness among Cepheid variable stars. (p. 252; video lesson; objective 1)

6. **proper motion:** The rate at which a star moves across the sky. This rate is measured in seconds of arc per year. (p. 254; video lesson; objective 1)

7. **calibration:** The establishment of the relationship between a parameter that is easily determined and a parameter that is more difficult to determine. (p. 254; objective 1)

8. **Shapley–Curtis debate:** A 1920 debate between Harlow Shapley and Heber Curtis on the nature of spiral nebulae. (p. 255; video lesson; objective 1)

9. **disk component:** All material confined to the plane of the galaxy. (p. 256; video lesson; objective 2)

10. **kiloparsec (kpc):** A unit of distance equal to 1000 pc or 3260 ly. (p. 256; objective 2)

11. **spiral arm:** Long spiral pattern of bright stars, star clusters, gas, and dust that extend from the center to the edge of the disk of spiral galaxies. (p. 256; video lesson; objective 2)

12. **spherical component:** The part of the galaxy including all matter in a spherical distribution around the center (the halo and nuclear bulge). (p. 257; video lesson; objective 2)

13. **halo:** The spherical region of a spiral galaxy, containing a thin scattering of stars, globular star clusters, and small amounts of gas. (p. 257; video lesson; objective 2)

14. **nuclear bulge:** The spherical cloud of stars that lies at the center of spiral galaxies. (p. 257; video lesson; objective 2)

15. **rotation curve:** A graph of orbital velocity versus radius in the disk of a galaxy. (p. 258; objective 3)

16. **dark halo:** The low-density extension of the halo of our galaxy believed to be composed of dark matter; also called the galactic corona. (p. 259; video lesson; objective 3)

17. **galactic corona:** The extended, spherical distribution of the low-luminosity matter believed to surround the Milky Way and the other galaxies; also called the dark halo. (p. 259; objective 3)

18. **dark matter:** Nonluminous matter that is detected only by its gravitational influence. (p. 259; video lesson; objective 3)

19. **population I:** Stars rich in atoms heavier than helium. Nearly always relatively young stars found in the disk of the galaxy. (p. 260; objective 3)

20. **population II:** Stars poor in atoms heavier than helium. (p. 260; objective 3)

21. **metals:** In astronomical usage, all atoms heavier than helium. (p. 260; video lesson; objective 3)

22. **spiral tracer:** A celestial object used to map spiral arms. Some examples include O and B associations, young open clusters, clouds of hydrogen ionized by hot stars (emission nebulae), and certain higher-mass variable stars. (p. 264; objective 4)

23. **density wave theory:** Theory proposed to account for spiral arms as compressions of the interstellar medium in the disk of the galaxy. (p. 266; video lesson; objective 4)

24. **flocculent:** Describes a galaxy whose spiral arms have a woolly or fluffy appearance. (p. 268; objective 4)

25. **self-sustaining star formation:** The process by which the birth of stars compresses the surrounding gas clouds and triggers the formation of more stars. (p. 268; objective 4)

26. **Sagittarius A* (Sgr A*):** The powerful radio source located at the core of the Milky Way Galaxy. (p. 272; video lesson; objective 5)

Summary

Almost every celestial object that you can see with the naked eye is a part of our Milky Way Galaxy. Early astronomers noticed a faint band of light that circled the sky, but it wasn't until early in the twentieth century that astronomers realized that we lived in the outskirts of a *galaxy*—a large system of stars, star clusters, gas, dust, and nebulae orbiting a common center of mass.

The Discovery of the Galaxy

In the late 1700s, Sir William Herschel and his sister Caroline attempted to gauge the extent of the "star system" of which our sun is a part. They counted the stars in the sky to find patterns, and in the areas where they found more stars, they concluded the system extended further into space. They noticed empty spots in the sky, but didn't realize that some stars were obscured by dust and gas. That discovery would come much later.

In 1892, Henrietta Leavitt's study of **variable stars** helped astronomers make progress toward estimating the size of our galaxy. Variable stars are those that periodically change—they grow brighter, then dimmer, and then brighter again. They appear to pulsate, but are different from the pulsars that you learned about in Lesson 11. They are unstable stars that have left the main sequence of the H–R diagram and evolved toward the portion of the diagram called the *instability strip*.

A **Cepheid variable star**, one type of variable star, is a giant that has a pulsation period between 1 and 60 days. This pulsation period is much longer than the period of smaller types of variable stars. Leavitt discovered a relationship between the pulsation period and the luminosity of the Cepheid, which was instrumental in estimating the size of our universe. She discovered that the more massive the star, the more luminous it was, and the slower it pulsated. This became known as the **period–luminosity relation** (see Figure 12-4 on p. 253 in the *Horizons* textbook).

Leavitt's discovery of this relationship between luminosity and pulsation period later became instrumental in estimating the size first of our own galaxy and, later, of the universe beyond it. The pulsation period of a Cepheid can be easily determined simply by observing it. Once the pulsation period is identified, we can infer the luminosity based on the period–luminosity relation. Comparing the luminosity to the apparent brightness—which is also easily detected through observation—astronomers can calculate the distance to the Cepheid. They can then use the Cepheid to measure the distance to the celestial object—such as a cluster—which contains it.

In the early twentieth century, Harlow Shapley noticed that *globular star clusters* were scattered all over the sky, but were concentrated between the constellations of Sagittarius and Scorpius. He concluded that the gravitational field of the entire star system must cause this concentration, and he searched for evidence to determine the size and shape of our galaxy.

Since globular clusters are very far away, Shapley was unable to measure their distances using parallax. Instead, he measured the motions of the stars in nearby clusters. All stars move through space and this movement can be detected and measured. These small shifts in the positions of the stars in the sky are called proper motions. Proper motions, on the average, diminish with distance and Shapley found 11 Cepheids with proper motions. Using the Cepheids, he was able to determine how far away the concentration of globular clusters was and was able to estimate the center of the galaxy and its size. Although his estimates about the size of the galaxy proved to be incorrect (because he didn't realize there were two types of Cepheid variables), his conclusion that our solar system was not in the center of the galaxy provided fodder for future discoveries. This illustrates that one person's discovery, even if not completely accurate, can pave the way for future discoveries.

Today, astronomers can explore our galaxy using radio telescopes to detect photons that an optical telescope cannot. A radio telescope can detect photons with a wavelength of 21 cm, which neutral, un-ionized hydrogen radiates. These long-wavelength photons are not scattered or impeded by dust, as is visible light. Observing the galaxy with a radio telescope helps astronomers detect the distribution of gas scattered about our galaxy.

You are probably familiar with the image of our galaxy as seen from afar—it looks like a giant pinwheel with arms spiraling around the center. This luminous part of the galaxy that you would see from a distance is called the **disk component**—the celestial objects, dust, and gas rotate on a plane around the center. Stars like our sun are widely scattered throughout the disk component, but most stars in the spectral class O lie within a narrow disk about 300 ly thick, which gives the disk its bluish glow. The disk component of our galaxy contains **spiral arms**—long curves that contain bright stars, star clusters, gas, and dust. They are where many stars are born because they contain a lot of gas and dust.

The disk does not rotate as a solid body. Stars travel at different orbital velocities, have different orbital periods, and follow a pattern of *differential rotation*. You learned about differential rotation in Lesson 7 when you studied how our sun rotates on its axis. Similarly, the celestial bodies toward the center of the disk orbit at a different rate than the bodies in the outer reaches of the disk.

Most galaxies, including our own, also include a **spherical component**, which contains all the matter scattered in a spherical distribution around the center. The spherical component of our galaxy includes a large halo and a concentrated nuclear bulge. Just as the disk of the galaxy is the incubator where new stars are born, the spherical component is where older stars live.

The **halo** is a spherical cloud of thinly scattered stars and globular clusters. It has little gas and dust and contains only about 2 percent of the stars that the disk contains. Because there is little gas and dust, no new stars are able to form here; it's believed that most of the stars in the halo are old, cool giants, dim lower-main-sequence stars, and old white dwarfs.

Also in the spherical component of the galaxy is the **nuclear bulge**—the dense cloud of stars that surrounds the center of the galaxy. It has a radius of 2,000 pc and is slightly flattened. The nuclear bulge seems to contain little gas and dust, and most of the stars are old, cool stars.

The orbital motions of the stars in the halo are different from the organized motion in the disk. Where the orbits in the disk are like a pinwheel, the orbits in the halo are like a swarm of bees. Each star and globular cluster follows its own randomly tipped elliptical orbit that carries the celestial object into the outer reaches of the galaxy where it slows. As the object continues in its orbit and falls back into the inner part of the galaxy, it speeds up again.

The dimensions of the disk component of our galaxy are uncertain—it's difficult to measure the thickness of the disk and the diameter. Astronomers approximate that the diameter of the disk is about 75,000 ly, or about 25 **kiloparsecs** (25,000 parsecs). Astronomers estimate that Earth is about two-thirds of the way from the center of the galaxy toward the edge. By studying the orbital motion of the celestial objects in the galaxy, astronomers have determined that the combined mass of the disk and spherical components of our galaxy is approximately 100 billion solar masses.

However, significant amounts of matter in our galaxy lie in its outer parts. Evidence suggests that this extra mass could be as great as two trillion solar masses and may extend up to 10 times farther than the edge of the visible disk of our galaxy. This extra mass lies in an extended halo that's sometimes called a **dark halo** or **galactic corona**. The dark halo does contain some low-luminosity stars and white dwarfs, but most of the matter in the dark halo is invisible and may be some form of matter that is yet unknown to astronomers. You will study this so-called **dark matter** in the next three lessons.

The Origin of the Milky Way

As you discovered in Lesson 10, we can approximate the age of a star cluster by analyzing its *turnoff point* on the H–R diagram. Astronomers can study open star clusters and globular clusters to determine the age of our galaxy.

Determining the age of an open cluster can be challenging. The chemical compositions of different clusters may vary and can make their turnoff point vary even if they are the same age. Open clusters are not tightly bound by gravity and the oldest of stars may have simply drifted away making the galaxy seem younger than it actually is.

Finding the age of a globular cluster can also be difficult, but astronomers believe that globular clusters have an average age of about 11 billion years. But like open clusters, globular clusters vary in their chemical composition, which affect the turnoff points of the clusters.

By studying the open clusters, astronomers conclude that the disk of our galaxy is approximately 9 billion years old. By measuring globular clusters of various distances and by studying their distribution, astronomers can approximate that the halo of our galaxy is at least 13 billion years old.

Not only are there variations of clusters within our galaxy, there are also variations in stars. There are two broad categories of stars in the Milky Way: population I stars and population II stars. The stars of the two populations are similar, but the metal content of a star defines its population. Astronomers define **metals** as atoms heavier than helium, including carbon, nitrogen, and oxygen, which is not the same as what non-astronomers call metals.

Population I stars are located in and around the disk of our galaxy and are sometimes called disk population stars. They are relatively young, metal rich, and have circular orbits in the plane of they galaxy. Population I stars range from extreme population I stars to intermediate population I stars. Extreme population I stars are very metal rich, are located primarily in the spiral arms of our galaxy, are about 100 million years in age or younger, and have a nearly circular orbits. Intermediate population I stars are less metal rich, are located throughout the disk, are about 0.2–10 billion years old, and have slightly elliptical orbits.

Population II stars are located in and around the halo of our galaxy and are sometimes called halo population stars. They are relatively old, metal poor, and have randomly tipped elliptical orbits. Population II stars range from extreme population II stars to intermediate population II stars. Intermediate population II stars are less metal rich than population I stars, are found in the nuclear bulge, are approximately 2–10 billion years old, and have moderately elliptical orbits. Extreme population II stars are the most metal-poor stars, are found in the halo and in globular clusters, are about 10–14 billion years old and have highly elliptical orbits.

When comparing population I and population II stars, notice that the youngest stars are the most metal rich. The process that explains this is perhaps the most important process in the history of our galaxy. Stars are mostly comprised of hydrogen and helium; the metals are formed inside the stars as a result of fusion. In Lesson 10 you learned that when massive stars die, their matter—including metals—is interspersed into the interstellar medium from which new stars are formed. The oldest surviving stars that formed about 13 billion years ago were metal poor because there were few stellar remnants from stars that had previously died. Over the years, these stars may have manufactured some metals, but because the stars' interiors are not mixed, the metals remain in their cores and aren't seen in their spectra. As some of these early stars died, they produced supernovae explosions and released their metals into the interstellar medium. Thus, succeeding generations of stars are more metal rich because they were formed partly from the metals released from the deaths of their ancestors.

Understanding how generations of stars formed helps us understand how our galaxy was formed. A spherical gas cloud may have first formed the central bulge of our galaxy, including a very first generation of massive metal-free stars which have long since died. The halo accumulated later from gases that were slightly metal-enriched by their deaths. The disk could have formed still later as the gas rotated and flattened. If our galaxy then merged with other, smaller galaxies, the range of ages in the globular clusters and some other anomalies could be explained. Astronomers are still gathering evidence to understand how our galaxy was formed.

Spiral Arms and Stellar Formation

The spiral arms of the galaxy wind outward from the center of the disk; it's believed that star formation happens in these arms. They contain primarily young star clusters, clouds of dust and gas, and hot, blue stars. Astronomers can search for associations of O and B stars that tend to congregate in the arms to better understand how stars form. Astronomers call these object spiral tracers—young objects that don't live long enough to move away from the spiral arms. Other tracers include young open clusters, emission nebulae, and certain high-mass variable stars.

We can study the spiral arms of our own galaxy by using radio telescopes. Detecting 21-cm radiation coming from cool clouds of hydrogen gas, astronomers can create a radio map that reveals the spiral arms. They can also identify giant molecular clouds, by detecting carbon monoxide in the plane of the galaxy, that reinforce the idea that stars are formed in the spiral arms.

Spiral arms are common in disk-shaped galaxies; the reason the stars, gas, and dust stay together in the arms, rather than being dispersed through space, can be explained by the **density wave theory**. This theory proposes that the spiral arms are waves of compression that move around the galaxy triggering star formation. The waves move very slowly as they rotate around the nucleus of the galaxy and orbiting clouds of gas move up from behind the spiral arms and get caught behind the slow-moving wave.

By studying mathematical models, astronomers believe that the density wave creates the shape of a two-armed spiral pattern with the nuclear bulge at the center. These are called grand-design galaxies. However, not all galaxies have two distinct arms—some have short spiral segments and are called **flocculent**, meaning "woolly."

The **self-sustaining star formation** suggests that star formation can also control the shape of spiral patterns if the cloud of gas can renew itself and continue making new stars. Massive stars explode as supernovae and the expanding gases can compress nearby clouds and trigger star formation. These concentrations of stars are twisted by the galaxy's differential rotation into a cloud of star formation that is shaped like a spiral arm. This can produce branches and spurs common to flocculent galaxies, whereas the spiral density wave generates any underlying two-armed spiral pattern.

The Nucleus

Unlike the brightly lit spiral arms, the nucleus of our galaxy cannot be detected at visual wavelengths. Astronomers can observe the center of our galaxy by analyzing the infrared and radio portions of the spectrum. Astronomers can detect clouds of gas in the center of our galaxy that play a part in star formation. They can also detect supernova remnants that were produced by the deaths of massive stars that were relatively young.

Evidence suggests that the object **Sagittarius A* (Sgr A*)** in the constellation Sagittarius, lies at the center of our galaxy. Although it is only a few AU in diameter, it is a very powerful source of radio and infrared radiation. Coming from the central area of Sgr A*, the infrared radiation appears to be produced by a high concentration of stars and the surrounding gas that they warm.

Evidence suggests that Sgr A* is actually a black hole that is surrounded by a small amount of gas whirling around on its accretion disk. By determining the size and the orbital period of stars circling Sgr A*, astronomers can calculate its mass as 2.6 million solar masses. Only a black hole could contain so much mass in such a small region. Such a supermassive black hole could not be the remains of a single dead star; it may have formed when the galaxy was first formed or may have accumulated over billions of years as matter drifted toward it.

One of the best ways to learn more about our own Milky Way is to study other galaxies. In the next two lessons, you'll discover that not all galaxies are flattened and spiral. They come in a variety of shapes and sizes.

Review Exercises

Matching I

Match each term with the appropriate definition or description.

1. _____ variable star	6. _____ proper motion
2. _____ Cepheid variable star	7. _____ calibration
3. _____ instability strip	8. _____ Shapley–Curtis debate
4. _____ RR Lyrae variable star	9. _____ spherical component
5. _____ period–luminosity relation	

a. The rate at which a star moves across the sky.
b. The establishment of the relationship between a parameter that is easily determined and a parameter that is more difficult to determine.
c. A star whose brightness changes periodically.

d. The relation between period of pulsation and intrinsic brightness among Cepheid variable stars.

e. A 1920 debate between two astronomers on the nature of spiral nebulae.

f. The part of the galaxy including all matter in a spherical distribution around the center (the halo and nuclear bulge).

g. Variable star with a period of up to 60 days; the period of variation is related to luminosity.

h. Variable star with a period of from 12 to 24 hours that is common in some globular clusters.

i. The region of the H–R diagram in which stars are unstable to pulsation. A star passing through this strip becomes a variable star.

Matching II

Match each term with the appropriate definition or description.

1. _____ disk component		5. _____ nuclear bulge	
2. _____ spiral arm		6. _____ rotation curve	
3. _____ kiloparsec (kpc)		7. _____ dark halo	
4. _____ halo		8. _____ dark matter	

a. The spherical region of a spiral galaxy, containing a thin scattering of stars, star clusters, and small amounts of gas.

b. All material confined to the plane of the galaxy.

c. Nonluminous matter that is detected only by its gravitational influence.

d. A unit of distance equal to 1,000 pc or 3,260 ly.

e. The low-density extension of the halo of our galaxy believed to be composed of dark matter.

f. The spherical cloud of stars that lies at the center of spiral galaxies.

g. A graph of orbital velocity versus radius in the disk of a galaxy.

h. Long spiral pattern of bright stars, star clusters, gas, and dust that extends from the center to the edge of the disk of spiral galaxies.

Matching III

Match each term with the appropriate definition or description.

1. _____ galactic corona		6. _____ density wave theory
2. _____ population I		7. _____ flocculent
3. _____ population II		8. _____ self-sustaining star formation
4. _____ metals		
5. _____ spiral tracer		9. _____ Sagittarius A*

a. Describes a galaxy whose spiral arms have a woolly or fluffy appearance.

b. In astronomical usage, all atoms heavier than helium.

c. Stars rich in atoms heavier than helium; nearly always relatively young stars found in the disk of the galaxy.

d. The powerful radio source located at the core of the Milky Way Galaxy.

e. The extended, spherical distribution of the low-luminosity matter believed to surround the Milky Way and the other galaxies.

f. Stars poor in atoms heavier than helium; nearly always relatively old stars found in the halo, globular clusters, or the nuclear bulge.

g. Theory proposed to account for spiral arms as compressions of the interstellar medium in the disk of the galaxy.

h. A celestial object used to map spiral arms.

i. The process by which the birth of stars compresses the surrounding gas clouds and triggers the formation of more stars.

Completion

Fill each blank in the sentences below with the most appropriate term from the list of completion answers that follow. A term may be used once, more than once, or not at all. Check your answers with the Answer Key and review when necessary.

Cepheid	instability	RR Lyrae
circular	9 billion	75
Heber Curtis	nuclear	75,000
disk	older	Harlow Shapley
elliptical	poor	13 billion
flocculent	rich	younger

1. Hot, blue stars tend to be _____ than red stars.

2. Population II stars are metal _____.

3. Some spiral galaxies that have segments that appear fluffy or "woolly" have been termed _____.

4. A star that has a variation in brightness with a period of 1 to 60 days is called a _____ variable star.

5. The section of the H–R diagram where if stars reside they pulsate is called the _____ strip.

6. Stars in the disk of the galaxy tend to move in _____ orbits.

7. The astronomer who calibrated Cepheid variable stars to find the distance to globular clusters was _____.

8. Our galaxy formed about _____ years ago.

9. Our galaxy is _____ light-years in diameter.

10. The dense cloud of stars that surrounds the center of our galaxy is the _____ bulge.

Self-Test

Select the best answer.

1. Of the following stars, the one that has the longest period of variability is
   a. an RR Lyrae star.
   b. the least luminous Cepheid.
   c. the most luminous Cepheid.
   d. pulsars.

2. Compared to open star clusters, globular clusters tend to be
   a. richer in metals and younger.
   b. poorer in metals and younger.
   c. richer in metals and older.
   d. poorer in metals and older.

3. The 21-centimeter line of hydrogen is in the
   a. gamma-ray part of the spectrum.
   b. infrared part of the spectrum.
   c. ultraviolet part of the spectrum.
   d. radio part of the spectrum.

4. The galactic center of the Milky Way lies in the constellation of
   a. Sagittarius.
   b. Cygnus.
   c. Orion.
   d. Leo.

5. The approximate diameter of the disk of the Milky Way Galaxy is
   a. 7.5 light-years.
   b. 75 light-years.
   c. 75,000 light-years.
   d. 75 million light-years.

6. In a typical spiral galaxy, stars in the
   a. disk have highly elliptical orbits.
   b. halo have nearly circular orbits.
   c. galactic plane have highly elliptical orbits.
   d. halo have elliptical orbits.

7. The supermassive black hole at the center of the galaxy is estimated to be
   a. 2.6 solar masses.
   b. 2.6 thousand solar masses.
   c. 2.6 million solar masses.
   d. 2.6 billion solar masses.

8. What spectral classes of stars work best as spiral tracers?
   a. K and M
   b. A and F
   c. R and T
   d. O and B

9. The youngest stars are
   a. Extreme Population I.
   b. Intermediate Population I.
   c. Extreme Population II.
   d. Intermediate Population II.

10. Astronomers use the term "metals" to describe any element heavier than
    a. hydrogen.
    b. helium.
    c. carbon.
    d. iron.

11. A rotation curve measures
    a. luminosity vs. orbital velocity.
    b. orbital velocity vs. orbital radii.
    c. mass vs. luminosity.
    d. proper motion vs. temperature.

12. The 21-centimeter radiation is used to map the galaxy by detecting
    a. neutral (un-ionized) hydrogen.
    b. glowing (ionized) hydrogen.
    c. carbon.
    d. lithium.

13. Which galactic region has the greatest metal content of stars?
    a. halo
    b. nuclear bulge
    c. spiral arms
    d. globular clusters

14. The very center of the Milky Way Galaxy is invisible at what wavelengths?
    a. radio
    b. visible
    c. infrared
    d. X-ray

Short-Answer Questions

1. Explain how Sir William and Caroline Herschel attempted to gauge the true shape of our star system.

2. What makes RR Lyrae variables different from Cepheid variables?

3. What are the differences between population I and population II stars?

4. Why was it difficult to determine the size and shape of our galaxy?

5. Describe the basis for the Shapley–Curtis debate.

6. Describe the different major stages of the history of the Milky Way.

Applications

1. Explain how 21-centimeter radiation is produced from neutral hydrogen.

2. Explain the density wave theory.

3. Explain self-sustaining star formation.

4. What evidence has been found that suggests a black hole at the center of our galaxy?

5. What is the period–luminosity relation and how did this help determine the size of the Milky Way?

6. What part did globular clusters play in the determination of the size and shape of the Milky Way?

Answer Key

Matching I

1. c (p. 251; video lesson; objective 1)

2. g (p. 251; video lesson; objective 1)

3. i (p. 251; objective 1)

4. h (p. 251; objective 1)

5. d (p. 252; video lesson; objective 1)

6. a (p. 254; video lesson; objective 1)

7. b (p. 254; objective 1)

8. e (p. 255; video lesson; objective 1)

9. f (p. 257; video lesson; objective 2)

Matching II

1. b (p. 256; video lesson; objective 2)

2. h (p. 256; video lesson; objective 2)

3. d (p. 256; objective 2)

4. a (p. 257; video lesson; objective 2)

5. f (p. 257; video lesson; objective 2)

6. g (p. 258; bjective 3)

7. e (p. 259; video lesson; objective 3)

8. c (p. 259; video lesson; objective 3)

Matching III

1. e (p. 259; objective 3)
2. c (p. 260; objective 3)
3. f (p. 260; objective 3)
4. b (p. 260; video lesson; objective 3)
5. h (p. 264; objective 4)
6. g (p. 266; video lesson; objective 4)
7. a (p. 268; objective 4)
8. i (p. 268; objective 4)
9. d (p. 272; video lesson; objective 5)

Completion

1. younger (p. 264; objective 3)
2. poor (p. 260; objective 3)
3. flocculent (p. 268; objective 4)
4. Cepheid (p. 251; video lesson; objective 1)
5. instability (p. 251; objective 1)
6. circular (p. 258; objective 3)
7. Harlow Shapley (pp. 253–255; video lesson; objective 1)
8. 13 billion (p. 260; objective 3)
9. 75,000 (p. 256; objective 2)
10. nuclear (p. 257; video lesson; objective 2)

Self-Test

1. c (pp. 252–253; objective 1)
2. d (p. 260; objective 3)
3. d (p. 256; video lesson; objective 4)
4. a (pp. 270–272; video lesson; objective 5)
5. c (p. 256; objective 2)
6. d (p. 258; objective 3)
7. c (p. 271; video lesson; objective 5)
8. d (p. 264; objective 4)
9. a (p. 260; objective 3)
10. b (p. 260; objective 3)
11. b (p. 258; objective 2)

12. a (p. 256; video lesson; objective 2)

13. c (p. 260; objective 3)

14. b (p. 270; video lesson; objective 5)

Short-Answer Questions

1. They counted the stars in 683 different directions in the sky. Where they saw more stars, they assumed the star system extended further into space. The result of their research was an irregular disk shape. (p. 251; objective 1)

2. Cepheid variables have periods from 1 to 60 days and lie in the top part of the instability strip on the H–R diagram. RR Lyrae variable stars have a period of 12–24 hours and lie at the bottom of the instability strip. (pp. 251–252; objective 1)

3. Population I stars are found in the disk of a spiral galaxy, are young, and are metal-rich. Population II stars are found in the halo or the nuclear bulge of a spiral, are old, and are metal-poor. (p. 260; objective 3)

4. We are in the middle of the system so it is difficult to envision what it would look like to an outside observer. Also, dust and gas obscure our views of the core and other portions of the Milky Way Galaxy. (pp. 250–253; video lesson; objective 1)

5. Curtis claimed the faint objects were other galaxies and Shapley argued they were not other galaxies, but were part of our star system. (pp. 254–255; video lesson; objective 1)

6. A spherical cloud of turbulent gas gave birth to the first stars and star clusters. The rotating cloud of gas contracted into a disk while stars and clusters were left behind in the halo. (p. 263; objective 3)

Applications

1. Both the proton and the electron in a neutral hydrogen atom spin and consequently have small magnetic fields. Because they have opposite electrostatic charges, they have opposite magnetic fields when they spin in the same direction. When they spin in opposite directions, their magnetic fields are aligned. Sometimes the electron in the hydrogen atom reverses its spin, which releases a small amount of energy seen at the 21-cm wavelength in the radio region of the spectrum. (pp. 256–257; video lesson; objective 2)

2. The density wave theory proposes that spiral arms are waves of compression, rather like sound waves, that move around the galaxy, triggering star formation. (pp. 266–267; video lesson; objective 4)

3. The self-sustaining star formation is when the intense radiation from a new hot star helps to compress nearby parts of the gas cloud and trigger further star formation. This appears to be happening in the Great Orion Nebula. Star formation might also be sustained by the explosive deaths of massive stars, such as supernovae. (pp. 268–269; objective 4)

4. From the orbits of objects near the very core of the galaxy, there is estimated to be a 2.6 million solar mass object in a very small area of space. There are sources in the radio, X ray, and infrared indicating that the object at the core is very powerful and intense. The only object that fits all these criteria is a supermassive black hole. (pp. 270–271; video lesson; objective 5)

5. The brighter a variable star, the longer its period of variation. This means that if you can measure the cycle of variation, you have a "standard candle" or a known intrinsic brightness. If you have that, you can find the distance to that star. This helped astronomers to map the Milky Way Galaxy. (pp. 251–254; video lesson; objective 1)

6. Harlow Shapley noticed that open clusters were concentrated along the Milky Way, but globular clusters were scattered over the entire sky but strongly concentrated toward Sagittarius and Scorpius. By measuring the distances to globular clusters by identifying the variable stars, he determined by mapping these clusters that the sun was not in the center of the galaxy but in the "suburbs." (pp. 251–254; objective 3)

Lesson Review

Lesson 12: Our Galaxy: The Milky Way

PLEASE NOTE: Use this matrix to guide your study and achieve the learning objectives of this lesson. It will also help you to view the video, which defines and demonstrates important concepts and principles as they relate to everyday life and actual case studies.

Learning Objective	Textbook	Student Guide
1. Describe how William Herschel first mapped the Milky Way Galaxy, and how Harlow Shapley later came to estimate its size and the location of Earth's solar system within it.	pp. 251–255	Key Terms: 1, 2, 3, 4, 5, 6, 7, 8; Matching I: 1, 2, 3, 4, 5, 6, 7, 8; Completion: 4, 5, 7; Self-Test: 1; Short-Answer: 1, 2, 4, 5; Applications: 5.
2. Describe the characteristics of the Milky Way Galaxy and its major components, including the nature of the stellar orbits in each region, and explain the importance of 21-centimeter radiation in the work of radio astronomers.	pp. 255–259	Key Terms: 9, 10, 11, 12, 13, 14; Matching I: 9; Matching II: 1, 2, 3, 4, 5; Completion: 9, 10; Self-Test: 5, 11, 12; Applications: 1.
3. Explain the reason for differing populations of stars within the galaxy and relate their distribution to the process of the formation of the Milky Way.	pp. 257, 259–264	Key Terms: 15, 16, 17, 18, 19, 20, 21; Matching II: 6, 7, 8; Matching III: 1, 2, 3, 4; Completion: 1, 2, 6, 8; Self-Test: 2, 6, 9, 10, 13; Short-Answer: 3, 6; Applications: 6.
4. Describe observations of spiral arms within the Milky Way and other galaxies and explain the two major mechanisms that create spiral structure and their role in star formation.	pp. 264–269	Key Terms: 22, 23, 24, 25; Matching III: 5, 6, 7, 8; Completion: 3; Self-Test: 3, 8; Applications: 2, 3.

Learning Objective	Textbook	Student Guide
5. Describe observations of the nucleus of the Milky Way at nonvisible wavelengths and evidence for the suspected black hole at the center.	pp. 270–273	Key Terms: 26; Matching III: 9; Self-Test: 4, 7, 14; Applications: 4.

Notes:

LESSON
13

Galaxies

Checklist

For the most effective study of this lesson, complete the following activities in this sequence.

Before Viewing the Video

- ❑ Read the Preview, Learning Objectives, and Viewing Notes below.
- ❑ Read Chapter 13, "Galaxies," pages 276–299, in the *Horizons* textbook.

What to Watch

- ❑ After reading the textbook chapter, watch the video for Lesson 13, *Galaxies*.

After Viewing the Video

- ❑ Briefly note your answers to questions listed at the end of the Viewing Notes.
- ❑ Review the Summary below.
- ❑ Review all reading assignments for this lesson, especially the Chapter 13 summary on page 298 in *Horizons* and the Viewing Notes in this guide.
- ❑ Write brief answers to the review questions at the end of Chapter 13 in *Horizons*.
- ❑ Complete the Review Exercises below. Check your answers with the Answer Key and review when necessary.
- ❑ Use the Lesson Review matrix found at the end of this lesson to review and assess your knowledge of each Learning Objective.
- ❑ As assigned by your instructor, complete the Applications activities and any additional activities for this lesson.

Preview

Until the early twentieth century, astronomers believed that we lived in a star system with nothing outside except for some distant nebulae. Today we know that there are billions of galaxies in the universe.

When we think about what a galaxy looks like, we might envision our own Milky Way. The long spiral arms gracefully swirl around a dense central bulge. But, galaxies come in many shapes and sizes. Some are a chaotic mix of gas, dust, and stars with no visible center.

You may think that galaxies are isolated, but over billions of years they have interacted with one another and evolved: they collide, blend, shred, create star nurseries, and interact in other ways. In fact, our own Milky Way Galaxy is in the process of swallowing the Sagittarius Galaxy and will likely merge with the Andromeda Galaxy someday.

In this lesson you will discover the different types of galaxies and how they've evolved and interacted with others. In Lesson 14, you will study violently active galaxies that output tremendous amounts of energy.

CONCEPTS TO REMEMBER

- Recall from Lesson 2 that *apparent visual magnitude (m$_v$)* is the brightness of a star as seen by human eyes on Earth. It is also known as apparent brightness or apparent magnitude (p. 16 in this guide).

- In Lesson 8, you learned that *absolute visual magnitude (M$_v$)* is the intrinsic brightness of a star. It is the apparent visual magnitude that a star would have if it were 10 parsecs away (p. 146 in this guide).

- In Lesson 8, you also learned that *luminosity (L)* is the the total amount of energy a star radiates in one second. It is also known as maximum intrinsic brightness. Luminosity is measured in Joules per second or, more commonly, solar luminosities based on the output of the sun. Astronomers can compare an object's luminosity to its apparent visual magnitude to determine its distance from Earth (p. 146 in this guide).

Learning Objectives

After you complete this lesson, you should be able to:

1. Identify the major galaxy classifications and describe their characteristics. *HORIZONS* TEXTBOOK PAGES 279–282.

2. Discuss the observable properties of galaxies and the importance of each. *HORIZONS* TEXTBOOK PAGES 283–291.

3. State the Hubble law and explain the result of Hubble's comparison of the distances and redshifts of distant galaxies. *HORIZONS* TEXTBOOK PAGES 285–287.

4. Describe the factors involved in galaxy formation, including the role of interactions between galaxies and within clusters of galaxies. *HORIZONS* TEXTBOOK PAGES 291–297.

At this point, read Chapter 13, "Galaxies," pages 276–299.

Viewing Notes

At one time, it was believed that our galaxy—the Milky Way—was the only galaxy in the universe. In this video, you'll explore a variety of galaxies that are quite different from our own.

The video program contains the following segments:

- ❂ Stellar Beacons
- ❂ Distance
- ❂ Classification of Galaxies
- ❂ Interaction & Evolution
- ❂ The Galactic Census

The following information will help you better understand the video program:

By measuring the pulsation period of a *Cepheid variable star*—a giant star that pulsates within a period of 1–60 days—astronomers can determine their luminosity by using the *period–luminosity relation*. The longer the pulsation period, the more luminous the Cepheid is. Once they know the luminosity, they can compare it to the apparent brightness to determine the Cepheid's distance from Earth. By comparing Cepheids in nearby galaxies to Cepheids of known distances—distance indicators—they can determine the distance to the galaxies.

Astronomers can also use planetary nebulae and type Ia supernovae as distance indicators to measure distances to more remote galaxies.

The video program discusses that astronomers realized the universe is expanding when they measured the redshifts in the spectra of all galaxies. The amount of the redshift is greater the farther away the galaxy is, which indicates that the farther away a galaxy is, the faster it is moving away from us. This redshift is similar to the Doppler effect that you learned about in Lesson 6.

QUESTIONS TO CONSIDER

- How did Hubble determine there were other galaxies in the universe?

- What are the characteristics of the different types of galaxies?

We find them smaller and fainter, in constantly increasing numbers, and we know that we are reaching into space, farther and farther, until, with the faintest nebulae that can be detected with the greatest telescopes, we arrive at the frontier of the known universe.

—Edwin Hubble
U.S. astronomer
(1889–1953)

- How can astronomers use distance indicators, such as Cepheids, planetary nebulae, and type Ia supernovae, to measure the distance to galaxies?

- How are galaxies distributed around the universe? How does this distribution affect the evolution of galaxies?

Watch the video for Lesson 13, *Galaxies.*

Key Terms and Concepts

Page references are keyed to the *Horizons* textbook.

1. **elliptical galaxy:** A galaxy that is round or elliptical in outline and contains little gas and dust, no disk or spiral arms, and few hot, bright stars. (p. 280; video lesson; objective 1)

2. **spiral galaxy:** A galaxy with an obvious disk component containing gas; dust; hot, bright stars, and spiral arms. (p. 280; video lesson; objective 1)

3. **barred spiral galaxy:** A spiral galaxy with an elongated nucleus resembling a bar from which the arms originate. (p. 280; video lesson; objective 1)

4. **irregular galaxy:** A galaxy with a chaotic appearance, large clouds of gas and dust, and both population I and II stars, but without spiral arms. (p. 281; video lesson; objective 1)

5. **Large Magellanic Cloud:** An irregular galaxy that is a satellite of our Milky Way Galaxy; the larger of two prominent galaxies visible in the southern sky. (p. 281; objective 1)

6. **Small Magellanic Cloud:** An irregular galaxy that is a satellite of our Milky Way Galaxy; the smaller of two prominent galaxies visible in the southern sky. (p. 281; objective 1)

7. **megaparsec (Mpc):** A unit of distance equal to 1,000,000 parsecs or 3.26 million light-years. (p. 283; video lesson; objective 2)

8. **distance indicator:** Object whose luminosity or diameter is known; used to find the distance to a star cluster or galaxy. (p. 283; video lesson; objective 2)

9. **standard candle:** A distance indicator whose brightness is known. (p. 283; objective 2)

10. **look-back time:** The amount by which we look into the past when we look at a distant galaxy. A time equal to the distance to the galaxy in light-years. (p. 285; video lesson; objective 2)

11. **Hubble law:** The linear relation between the distances to galaxies and their velocity of recession. (p. 285; video lesson; objective 3)

12. **Hubble constant (H):** A measure of the rate of expansion of the universe. The average value of velocity of recession divided by distance; presently believed to be about 70 km/s/Mpc. (p. 285; objective 3)

13. **rotation curve method:** A method of determining a galaxy's mass by observing the orbital velocity and radius of stars in the galaxy. (p. 288; objective 2)

14. **cluster method:** The method of determining the masses of galaxies based on the motions of galaxies in a cluster. (pp. 288–289; objective 2)

15. **velocity dispersion method:** A method of finding a galaxy's mass by observing the range of velocities within the galaxy. (p. 289; objective 2)

16. **gravitational lensing:** The process by which the gravitational field of a massive object focuses the light from a distant object to produce multiple images of the distant object or to make the distant object look brighter. (p. 289; objective 2)

17. **rich galaxy cluster:** A cluster containing more than 1,000 galaxies, mostly elliptical, scattered over a volume about 3 Mpc in diameter. (p. 291; objective 4)

18. **poor galaxy cluster:** An irregularly shaped cluster that contains fewer than 1,000 galaxies, many spiral, and no giant ellipticals. (p. 291; objective 4)

19. **Local Group:** The small cluster of a few dozen galaxies that contains our Milky Way Galaxy. (p. 291; video lesson; objective 4)

20. **tidal tail:** A long streamer of stars, gas, and dust torn from a galaxy during its close interaction with another passing galaxy. (p. 294; objective 4)

21. **galactic cannibalism:** The theory that large galaxies absorb smaller galaxies. (p. 294; objective 4)

22. **ring galaxy:** A galaxy that resembles a ring around a bright nucleus. It is believed to be the result of a head-on collision of two galaxies. (p. 295; video lesson; objective 4)

23. **starburst galaxy:** A galaxy undergoing a rapid burst of star formation. (pp. 293, 296; video lesson; objective 4)

Summary

It wasn't until the 1920s that astronomers began to realize that there were other galaxies in the universe. When Edwin Hubble photographed distant spiral nebulae, he discovered that they contained individual stars. With his discovery, the universe suddenly became a much larger place.

The Family of Galaxies

Galaxies come in all shapes and sizes; astronomers try to create order by classifying them. Using a classification system is very common in science—it reveals the relationships between different kinds of objects like galaxies. When grouping galaxies, the most commonly used classification system is based on shape—a system developed by Edwin Hubble. Hubble grouped galaxies into three classifications: elliptical, spiral, and irregular galaxies. The active galaxy, which you will learn about in the next lesson, is a recent discovery that sometimes defies the rules of classification.

An **elliptical galaxy** has a round or elliptical outline. It contains little gas or dust, and has neither a disk nor spiral arms. This type of galaxy typically contains very few hot and bright stars; because of the older, cooler stars contained within, the galaxy tends to have a reddish tint. Elliptical galaxies are classified with a numerical index ranging from 1 to 7 preceded by the letter E for elliptical. E0s are round, while E7s are highly elliptical.

A **spiral galaxy** contains an obvious disk component surrounding a spherical center with spiral arms embedded within it. A spiral galaxy contains a lot of gas and dust, tending to be blue because of an abundance of young, hot, bright stars. Spiral galaxies are classified a, b, c, preceded by the letter S for spiral. Sa galaxies have large nuclei and tightly wound arms, less dust and gas, and fewer hot, bright stars. Sc galaxies have small nuclei and more open arms, lots of gas and dust, and many hot, bright stars. Sb galaxies are somewhere in between Sa and Sc. S0 galaxies have an obvious disk and nuclear bulge but no visible gas and dust; they have no spiral arms and no hot, bright stars.

A type of spiral galaxy is the **barred spiral galaxy**, which contains a nucleus shaped like a bar. Roughly two-thirds of all spiral galaxies are this type, including our own Milky Way Galaxy. They are classified as SBa, SBb, and SBc (the SB stands for "spiral bar"). They have similar characteristics to galaxies Sa, Sb, and Sc, respectively.

An **irregular galaxy** has a chaotic, disorganized appearance. Irregular galaxies are classified simply as Irr. They have large clouds of dust and gas and both hot, bright, blue stars and older, cool, red stars.

To summarize, irregular galaxies contain both young and old stars scattered randomly within them. Spiral galaxies contain mostly young stars in their disks and older stars in their centers. Elliptical galaxies contain mostly old stars throughout.

When astronomers learned more about the evolution of galaxies, they began to assign more classifications to them. When two galaxies collide, they can form a different type of galaxy: these include ring galaxies and starburst galaxies. **Ring galaxies** resemble a ring around a bright nucleus; they are believed to be the result of a head-on collision of two galaxies. They tend to have nearby companions, which supports the theory that there was a recent interaction between galaxies. **Starburst galaxies** are undergoing a rapid burst of star formation. They are very luminous in the infrared because a collision triggered the burst that is heating the dust, which reradiates the energy in the infrared.

At least 100 billion galaxies are visible with today's telescopes and still others are too faint to detect. Of those galaxies that are cataloged, about 70 percent of them are spiral galaxies. But, this might not represent the true proportion of galaxies in our universe. Spiral galaxies contain hot, bright stars that are easier to detect than galaxies with cool dim stars.

Measuring the Properties of Galaxies

As you learned in Lesson 8, once you know the distance to a star, you can calculate its luminosity, diameter, and mass. Astronomers can apply a similar technique in their study of galaxies. It is difficult to determine the distance to galaxies so far away, so astronomers use **distance indicators**—objects whose luminosity or diameter is known—to measure how far away a distant star cluster or galaxy is. Astronomers use the term "**standard candle**" to refer to a source of a known amount of light. Cepheid variable stars and type Ia supernovae are used as standard candles to measure distance. Globular clusters and planetary nebulae can also be used to estimate distances.

Astronomers most often use *Cepheid variables* as distance indicators because their pulse period is related to their luminosity. You discovered in Lesson 12 that that the luminosity of a Cepheid is directly related to its period of variation: a massive Cepheid is more luminous and has a longer pulsation period than a less massive Cepheid. Once we know a Cepheid's pulsation period, we can determine its luminosity or absolute magnitude by plotting it on the period–luminosity diagram (see Figure 12-4 on p. 253 in the *Horizons* textbook). We can then compare the luminosity or absolute magnitude to the apparent magnitude to determine the distance to the Cepheid. Comparing the apparent magnitude of the Cepheid to that of a galaxy can help astronomers learn the distance to galaxies that are relatively close.

When studying more remote galaxies, astronomers cannot use Cepheids because they are too faint at great distances. One way for them to measure distances is by comparing *globular clusters* in the galaxies. Astronomers have found that the brightest globular clusters have absolute magnitudes of about −10. If they find a globular cluster in a distant galaxy, they can assume that it has an absolute magnitude of about −10. By comparing this to its apparent magnitude, they can approximate its distance.

Astronomers can also use *planetary nebulae* as distance indicators. Recall from Lesson 10 that a planetary nebula is an expanding shell of gas ejected from a star during the latter stages of its evolution. The central stars in the planetary nebula are small and hot, and radiate most of their energy in the ultraviolet. The planetary nebula absorbs the ultraviolet radiation from the central stars and reradiates it as visible emission lines. Astronomers can compare the planetary nebulae in distant galaxies to those of known distances.

Type Ia supernovae can also be used as distance indicators; because they are so bright they can be seen from a great distance. Type Ia supernovae—those that are caused by the collapse of a white dwarf—all reach about the same absolute visual magnitude. Again, astronomers can compare the absolute visual magnitude to the apparent visual magnitude to determine the distance of the supernova. The drawback of using supernovae as distance indicators is that they are rare and may never occur in a galaxy that is studied over several lifetimes.

Some galaxies are so far away that astronomers can't use distance indicators to estimate their distances from Earth. Astronomers can use the Hubble law to estimate these distances. As you've learned in previous lessons, celestial objects that are moving away or toward us display a redshift or blueshift, respectively. By viewing the spectra of galaxies, astronomers have inferred that galaxies have large redshifts—meaning they have large radial velocities and are receding from Earth at a very fast pace.

In 1929, Edwin Hubble plotted the apparent velocity of recession versus distance for a number of galaxies; this relationship is now known as the **Hubble law**. Hubble discovered that the farther away a galaxy is from us, the faster it's moving. The Hubble law is important to astronomers because it provides evidence that the universe is expanding. It also is used to estimate the distance to remote galaxies.

When we look into the vast expanse of the universe, we are actually looking back into time. Light travels very fast—186,000 miles per second. When we observe a galaxy that is millions of light-years away, we see it as it was millions of years ago when its light began its journey toward Earth. For example, the Andromeda Galaxy is 2.3 million light-years away, so the light reaching us today began its journey 2.3 million years ago. Since the light takes

2.3 million years to reach us, when we observe the galaxy, we look back into time 2.3 million years. This is known as the **look-back time**—the amount by which we look into the past when we look at a distant galaxy is equal to its distance in light-years.

The Evolution of Galaxies

In previous lessons, you learned about star clusters and the evolution of stars. Galaxies can be found in clusters as well, and their gravitational interactions can alter them. Galaxy clusters are receding from each other as the universe expands. Galaxies can also collide with one another and merge.

The clustering of galaxies can help you understand their evolution. Galaxy clusters can be sorted into rich galaxy clusters and poor galaxy clusters. **Rich galaxy clusters** contain over a thousand galaxies, mostly elliptical and they often contain one or more giant elliptical galaxies at the center. By contrast, **poor galaxy clusters** are irregularly shaped and contain fewer than 1,000 galaxies. They tend to have small groupings of galaxies within a cluster rather than having giant galaxies at their centers. They contain a larger percentage of spirals. These observations have led astronomers to believe that the secrets to galactic evolution lie in the interactions between galaxies.

Stars almost never interact with one another because there are great distances between them—the average distance between stars is about 10^7 times their diameters. But galaxies are much closer together—their average separation is only about 20 times their diameters. Two galaxies can interact with one another and trigger the formation of spiral arms or trigger star formation. Large galaxies can absorb smaller galaxies in a process that's called **galactic cannibalism**.

Evidence suggests that galaxies didn't form in isolation—many have interacted or merged with others in the past. Some galaxies show evidence of recent interactions—by collision, merger, or cannibalism. Astronomers believe that a merger of at least two or three galaxies forms most elliptical galaxies. During a collision, a galaxy can use up all its gas and dust in a burst of star formation thereby creating a starburst galaxy. Others may be stripped of their gas and dust when they pass through a cluster of galaxies. Our own Milky Way Galaxy is believed to have quietly cannibalized some of its neighbors and is in the process of merging with the much smaller galaxy Sagittarius.

Others galaxies don't seem like they've suffered from any recent interactions. For example, many spiral galaxies have thin, delicate disks. If one had interacted with another large galaxy, tidal forces would have destroyed the delicate disk.

Astronomers can study the farthest galaxies to understand their history. Blue dwarf galaxies can only be found at distant look-back times, and they no longer exist in the present universe. They are small and irregularly shaped; their blue color indicates that stars are rapidly forming within them. At the limits of the most sophisticated telescopes, astronomers can detect faint galaxies that appear red because of their redshift—these may have been the first galaxies to appear after the beginning of our universe. Other galaxies that were formed long ago were more compact and irregular than more recently formed galaxies. About one-third to one-half of distant galaxies are in close pairs, while only 7 percent of nearby galaxies are in pairs. This supports the hypothesis that galaxies have evolved by merger.

The galaxies that you studied in this lesson may have interacted with others in the past, but are now relatively inactive. In the next lesson, you'll learn about galaxies that suffer from tremendous explosions and activity in their center— these are called active galaxies.

Review Exercises

Matching I

Match each term with the appropriate definition or description.

1. _____ elliptical galaxy	7. _____ megaparsec
2. _____ spiral galaxy	8. _____ distance indicator
3. _____ barred spiral galaxy	9. _____ look-back time
4. _____ irregular galaxy	10. _____ Hubble law
5. _____ Large Magellanic Cloud	11. _____ Hubble constant
6. _____ Small Magellanic Cloud	

a. An irregular galaxy that is a satellite of our Milky Way Galaxy; the larger of two prominent galaxies visible in the southern sky.

b. A unit of distance equal to 1,000,000 parsecs or 3.26 million light-years.

c. A galaxy with an obvious disk component containing gas; dust; hot, bright stars, and spiral arms.

d. The amount by which we look into the past when we look at a distant galaxy; a time equal to the distance to the galaxy in light-years.

e. An irregular galaxy that is a satellite of our Milky Way Galaxy; the smaller of two prominent galaxies visible in the southern sky.

f. A measure of the rate of expansion of the universe; the average value of velocity of recession divided by distance, which is presently believed to be about 70 km/s/Mpc.

g. A galaxy that is round or elliptical in outline and contains little gas and dust, no disk or spiral arms, and few hot, bright stars.

h. An object whose luminosity or diameter is known; used to find the distance to a star cluster or galaxy.

i. A galaxy with a chaotic appearance, large clouds of gas and dust, and both population I and II stars, but without spiral arms.

j. A spiral galaxy with an elongated nucleus resembling a bar from which the arms originate.

k. The linear relation between the distances to galaxies and their velocity of recession.

Matching II

Match each term with the appropriate definition or description.

1. _____ rotation curve method	6. _____ poor galaxy cluster
2. _____ cluster method	7. _____ Local Group
3. _____ velocity dispersion method	8. _____ tidal tail
4. _____ gravitational lensing	9. _____ galactic cannibalism
5. _____ rich galaxy cluster	10. _____ ring galaxy
	11. _____ starburst galaxy

a. The method of determining the masses of galaxies based on the motions of galaxies in a cluster.

b. A cluster containing more than 1,000 galaxies, mostly elliptical, scattered over a volume about 3 Mpc in diameter.

c. A method of determining a galaxy's mass by observing the orbital velocity and radius of stars in the galaxy.

d. The process by which the gravitational field of a massive object focuses the light from a distant object to produce multiple images of the distant object or to make the distant object look brighter.

e. A method of finding a galaxy's mass by observing the range of velocities within the galaxy.

f. The theory that large galaxies absorb smaller galaxies.

g. An irregularly shaped cluster that contains fewer than 1,000 galaxies, many spiral, and no giant ellipticals.

h. A galaxy undergoing a rapid burst of star formation.

i. Small cluster of a few dozen galaxies that contains our Milky Way Galaxy.

j. A galaxy that resembles a ring around a bright nucleus.

k. A long streamer of stars, gas and dust torn from a galaxy during its close interaction with another passing galaxy.

Completion

Fill each blank in the sentences below with the most appropriate term from the list of completion answers that follow. A term may be used once, more than once, or not at all. Check your answers with the Answer Key and review when necessary.

Andromeda	rich
bluer	ring
cannibalism	rotation curve
cluster	slower
faster	Sagittarius
poor	starburst
redder	

1. The Hubble law states that the further a galaxy is away, the _____ it appears to be moving from us.

2. Our Local Group of galaxies is a _____ galaxy cluster.

3. The Milky Way is currently colliding with the _____ Galaxy.

4. The engulfing of one galaxy by another is called galactic _____.

5. A galaxy that interacts with another galaxy causing many hot bright blue stars to be formed is called a _____ galaxy.

6. In a high-speed interaction, when a smaller galaxy passes through another galaxy almost perpendicular to the disk, disrupting the spiral arms but leaving the nuclear bulge and a circle of stars and gas, a _____ is produced.

7. The method used to determine the mass of galaxies using the radial velocities of many galaxies in a group is called the _____ method.

8. The halo of a spiral galaxy is dimmer and _____ than the disk.

Self-Test

Select the best answer.

1. Which scientist used the 100-inch telescope on Mount Wilson in 1923 to measure the distance to Andromeda?
   a. Albert Einstein
   b. Edwin Hubble
   c. Jocelyn Bell
   d. Stephen Hawking

2. Which is NOT a good distance indicator for galaxies?
   a. globular clusters
   b. type Ia supernovae
   c. white dwarfs
   d. Cepheid variables

3. Which major galaxy classification contains no visible gas or dust and lacks hot bright stars?
   a. spiral
   b. irregular
   c. elliptical
   d. barred spiral

4. The Milky Way Galaxy is best described as a _____ galaxy.
   a. spiral
   b. irregular
   c. elliptical
   d. barred spiral

5. The Large and Small Magellenic Clouds are best described as _____ galaxies.
   a. spiral
   b. irregular
   c. elliptical
   d. barred spiral

6. Which major galaxy classification has the greatest variation in size and mass from dwarf to giant?
   a. spiral
   b. irregular
   c. elliptical
   d. barred spiral

7. The most spherical elliptical galaxies would be classified as
   a. E0.
   b. E3.
   c. E5.
   d. E7.

8. Barred spiral galaxies make up what portion of all spiral galaxies?
   a. one-tenth
   b. one-half
   c. two-thirds
   d. nine-tenths

9. The Hubble law compares what two variables?
   a. distance and size
   b. apparent velocity of recession and distance
   c. luminosity and mass
   d. luminosity and period

10. The most precise measurements of the Hubble constant yield a value of
    a. 20 km/s/Mpc.
    b. 50 km/s/Mpc.
    c. 70 km/s/Mpc.
    d. 90 km/s/Mpc.

11. What is the chance that stars from two interacting galaxies will collide?
    a. almost certain
    b. likely
    c. about 50/50
    d. very unlikely

12. Observations of galaxies and clusters of galaxies reveal that our
    universe is _____ percent dark matter.
    a. 5 to 10
    b. 20 to 25
    c. 50 to 55
    d. 80 to 90

Short-Answer Questions

1. What is the difference between rich and poor galaxy clusters?

2. How many galaxies are visible with today's telescopes?

3. How can we tell there are supermassive black holes at the centers of most large galaxies?

4. Explain how a galaxy gets tidal tails.

5. Explain how Cepheid variables are used to determine distances to nearby galaxies.

6. Explain why "standard candles" are so vitally important in astronomy.

7. What is the difference between a spiral galaxy and an elliptical galaxy?

Applications

1. If a galaxy is 20 Mpc away, what is the apparent radial velocity of recession?

2. If the radial velocity of a galaxy is measured to be 1,500 km/s, how far away is it?

3. Describe the cluster method for measuring a galaxy's mass.

4. Describe the velocity dispersion method for measuring a galaxy's mass.

5. Describe the rotation curve method for measuring a galaxy's mass.

6. How are the methods described in Questions 4 and 5 above the same and how are they different?

Answer Key

Matching I

1. g (p. 280; video lesson; objective 1)
2. c (p. 280; video lesson; objective 1)
3. j (p. 280; video lesson; objective 1)
4. i (p. 281; video lesson; objective 1)
5. a (p. 281; objective 1)
6. e (p. 281; objective 1)
7. b (p. 283; video lesson; objective 2)
8. h (p. 283; video lesson; objective 2)
9. d (p. 285; video lesson; objective 2)
10. k (p. 285; video lesson; objective 3)
11. f (p. 285; objective 3)

Matching II

1. c (p. 288; objective 2)
2. a (p. 288; objective 2)
3. e (p. 289; objective 2)
4. d (p. 289; objective 2)
5. b (p. 291; objective 4)
6. g (p. 291; objective 4)
7. i (p. 291; video lesson; objective 4)
8. k (p. 294; objective 4)
9. f (p. 294; objective 4)
10. j (p. 295; video lesson; objective 4)
11. h (p. 293; video lesson; objective 4)

Completion

1. faster (pp. 285–287; video lesson; objective 3)
2. poor (p. 291; video lesson; objective 4)
3. Sagittarius (p. 293; video lesson; objective 4)
4. cannibalism (p. 294; objective 4)
5. starburst (pp. 293, 296; video lesson; objective 4)
6. ring (p. 295; video lesson; objective 4)

7. cluster (pp. 288–289; objective 2)

8. redder (pp. 279–281; objective 1)

Self-Test

1. b (video lesson; objective 2)

2. c (p. 283; video lesson; objective 2)

3. c (pp. 280–281; video lesson; objective 1)

4. d (p. 280; video lesson; objective 1)

5. b (p. 281; video lesson; objective 1)

6. c (p. 280; video lesson; objective 1)

7. a (p. 280; video lesson; objective 1)

8. c (p. 280; video lesson; objective 1)

9. b (p. 285; video lesson; video lesson; objective 3)

10. c (p. 286; objective 3)

11. d (p. 292; video lesson; objective 4)

12. d (p. 291; objective 2)

Short-Answer Questions

1. Rich galaxy clusters contain more than 1,000 galaxies, mostly elliptical, scattered through a spherical volume about 3 Mpc in diameter. Such a cluster is very crowded with the galaxies more concentrated toward the center. Poor galaxy clusters contain fewer than 1,000 galaxies and are irregularly shaped and tend to have small groupings within the clusters. (p. 291; objective 4)

2. 100 billion (p. 282; objective 1)

3. Measurements show that the stars near the centers of most galaxies are orbiting very rapidly. To hold stars in such small, short period orbits, the centers of galaxies must contain masses of a million to a few billion solar masses. Yet no object is visible so most astronomers believe that the nuclei of galaxies contain supermassive black holes. (p. 289; objective 2)

4. When galaxies pass near each other, the tidal gravity can distort a galaxy and produce long streamers called tidal tails. (p. 294; objective 4)

5. If you know the period of the star's variation, you can use the period-luminosity diagram to determine its absolute magnitude. If you know the apparent and absolute magnitudes, you can determine distance. (p. 283; video lesson; objective 2)

6. A standard candle is a source of a known amount of light. They help to calibrate other distance measurements. Without standard candles, we could not accurately find distance. (p. 283; objective 2)

7. Elliptical galaxies are round or elliptical, contain no visible gas and dust and lack hot, bright stars. Spiral galaxies contain a disk and spiral arms. They contain gas and dust and hot, bright O and B stars. Star formation is occurring in spirals, but not on a large scale in ellipticals. (p. 280; video lesson; objective 1)

Applications

1. The answer is 1,400 km/s. (p. 287; objective 3)

   $V = Hd$
   $H = 70$ km/s/Mpc
   $V = 70$ km/s/Mpc (20 Mpc) = 1,400 km/s

2. The answer is 21.4 Mpc. (p. 287; objective 3)
   $V = Hd$
   $V/H = d$
   1,500 km/s / (70 km/s/Mpc) = 21.4 Mpc

3. In the cluster method of finding mass, you measure the radial velocities of the galaxies in a cluster and can calculate the mass from the gravity needed to have that orbital motion. (pp. 288–289; objective 2)

4. In the velocity dispersion method of finding mass, a spectrum is taken. The broader the spectral lines, the faster the stars and gas are moving, one side of the line is redshifted, the other side is blueshifted. The faster the dust and gas move, the more massive the galaxy. (p. 289; objective 2)

5. In the rotation curve method of finding mass, spectra are taken of different parts of the galaxy and a rotation curve is produced. By looking at the velocities of the different parts orbiting the center of the galaxy, we can determine mass. (p. 288; objective 2)

6. Both techniques use the Doppler effect to determine velocity and gravity to determine mass from this velocity. The difference is that the rotation curve method can only be used on the nearest of galaxies because you must be able to measure different parts of the galaxy. The velocity dispersion method will work for distant galaxies. (pp. 288–289; objective 2)

Lesson Review

Lesson 13: Galaxies

PLEASE NOTE: Use this matrix to guide your study and achieve the learning objectives of this lesson. It will also help you to view the video, which defines and demonstrates important concepts and principles as they relate to everyday life and actual case studies.

Learning Objective	Textbook	Student Guide
1. Identify the major galaxy classifications and describe their characteristics.	pp. 279–282	Key Terms: 1, 2, 3, 4, 5, 6; Matching I: 1, 2, 3, 4, 5, 6; Completion: 8; Self-Test: 3, 4, 5, 6, 7, 8; Short-Answer: 2, 7.
2. Discuss the observable properties of galaxies and the importance of each.	pp. 283–291	Key Terms: 7, 8, 9, 10, 13, 14, 15, 16; Matching I: 7, 8, 9; Matching II: 1, 2, 3, 4; Completion: 7; Self-Test: 1, 2, 12; Short-Answer: 3, 5, 6; Applications: 3, 4, 5, 6.
3. State the Hubble law and explain the result of Hubble's comparison of the distances and red shifts of distant galaxies.	pp. 285–287	Key Terms: 11, 12; Matching I: 10, 11; Completion: 1; Self-Test: 9, 10; Applications: 1, 2.
4. Describe the factors involved in galaxy formation, including the role of interactions between galaxies and within clusters of galaxies.	pp. 291–297	Key Terms: 17, 18, 19, 20, 21, 22, 23; Matching II: 5, 6, 7, 8, 9, 10, 11; Completion: 2, 3, 4, 5, 6; Self-Test: 11; Short-Answer: 1, 4.

Notes:

LESSON
14

Active Galaxies

Checklist

For the most effective study of this lesson, complete the following activities in this sequence.

Before Viewing the Video

❑ Read the Preview, Learning Objectives, and Viewing Notes below.

❑ Read Chapter 14, "Galaxies with Active Nuclei," pages 300–319, in the *Horizons* textbook.

What to Watch

❑ After reading the textbook chapter, watch the video for Lesson 14, *Active Galaxies*.

After Viewing the Video

❑ Briefly note your answers to questions listed at the end of the Viewing Notes.

❑ Review the Summary below.

❑ Review all reading assignments for this lesson, especially the Chapter 14 summary on page 318 in *Horizons* and the Viewing Notes in this guide.

❑ Write brief answers to the review questions at the end of Chapter 14 in *Horizons*.

❑ Complete the Review Exercises below. Check your answers with the Answer Key and review when necessary.

❑ Use the Lesson Review matrix found at the end of this lesson to review and assess your knowledge of each Learning Objective.

❑ As assigned by your instructor, complete the Applications activities and any additional activities for this lesson.

Preview

In previous lessons you've learned about the tremendous amount of energy in the universe. The sun emits enough energy to sustain life on our planet; the Milky Way Galaxy has as much energy as ten billion suns. Supernovae inject massive amounts of energy and matter into the interstellar medium. But, this energy output pales by comparison to the energy emitted by some of the mysterious galaxies in our universe—these are called active galaxies.

An **active galaxy** emits vast amounts of energy in wavelengths not emitted from a typical galaxy like the Milky Way. Astronomers have concluded that only one thing could explain the tremendous amount of energy emitted by an active galaxy, and it will be revealed in this lesson.

CONCEPTS TO REMEMBER

- Recall from Lesson 13 that *look-back time* is the amount by which we look into the past when we look at a distant galaxy. If we look to a great enough distance, we can get a glimpse of what the universe was like when it was very young (p. 253 in this guide).

- In Lesson 6, you learned that *redshift* is a shift in an object's spectrum toward longer wavelengths because of velocity of recession. If we measure the amount of redshift in a distant object's spectrum, we can determine how fast it's moving away from Earth (p. 105 in this guide).

Learning Objectives

After you complete this lesson, you should be able to:

1. Describe observations of active galaxies and the physical processes proposed to explain their energy output and other characteristics. *HORIZONS* TEXTBOOK PAGES 300–311.

2. Describe the discovery and characteristics of quasars and discuss how they relate to the topic of active galaxies. *HORIZONS* TEXTBOOK PAGES 311–313.

3. Explain how quasar distances are estimated. *HORIZONS* TEXTBOOK PAGES 313–314.

4. Describe the current explanation of quasars and their energy sources and the implications of this explanation for the formation of galaxies and early history of the universe. *HORIZONS* TEXTBOOK PAGES 314–317.

At this point, read Chapter 14, "Galaxies with Active Nuclei," pages 300–319.

Viewing Notes

The Milky Way Galaxy has the luminosity of ten billion suns, but this energy output is insignificant compared to that of an active galaxy. The nucleus of an active galaxy can be so bright that it can overwhelm the rest of the galaxy, making it appear as a single point of light.

The greater the sphere of our knowledge, the larger the surface of its contact with the infinity of our ignorance.

—Arthur Berry
A Short History of Astronomy
(1898)

The video program contains the following segments:

◎ Seyfert Galaxies

◎ Radio Galaxies

◎ Quasars

◎ Unified Theory

◎ The Young Universe & Quasars

The following information will help you better understand the video program:

The discovery of the Seyfert galaxy in the 1940s revealed some unique spectral signatures in active galaxies. Their spectra indicated that they contain highly excited ionized gases swirling around a small, high-energy core; the source of this high-energy output was a mystery. In the 1950s, astronomers discovered the sources that emitted vast amounts of radio waves. Astronomers didn't know what they were at first and deemed them radio galaxies—galaxies that emit strong radio waves. This discovery added to the intrigue of active galaxies.

In the 1960s the discovery of the quasar—a quasi-stellar radio source—provided the final piece of the active galaxy puzzle. Quasars appeared to have properties similar to those of radio galaxies, but displayed extremely high redshifts in their spectra. The quasar appeared to be energy emitted from a point source, but it turned out to be an extremely energetic center of a galaxy that was very far away. Because they are so far away and the nucleus is so bright, we cannot see the rest of the galaxy surrounding the point source.

When astronomers pieced together the puzzle of Seyfert galaxies, radio galaxies, and quasars, they came to a conclusion: only a supermassive black hole could be the source of these extreme amounts of energy.

QUESTIONS TO CONSIDER

- How does the spectrum of a Seyfert galaxy differ from the spectrum of a typical galaxy, like the Milky Way?

- How does the spectrum of a radio galaxy differ from the spectrum of a typical galaxy?

- How does a supermassive black hole explain the properties of quasars and active galaxies?

- What evidence has led astronomers to conclude that quasars are extremely distant objects?

- How does the study of quasars give us a glimpse into the time when galaxies were first forming?

Watch the video for Lesson 14, *Active Galaxies*.

Key Terms and Concepts

Page references are keyed to the *Horizons* textbook.

1. **radio galaxy:** A galaxy that is a strong source of radio signals. (p. 300; video lesson; objective 1)

2. **active galaxy:** A galaxy whose center emits large amounts of excess energy, often in the form of radio emission. (p. 302; video lesson; objective 1)

3. **active galactic nuclei (AGN):** The centers of active galaxies that are emitting large amounts of excess energy. (p. 302; video lesson; objective 1)

4. **Seyfert galaxy:** An otherwise normal spiral galaxy with an unusually bright, small core that fluctuates in brightness. (p. 302; video lesson; objective 1)

5. **double-lobed radio source:** A galaxy that emits radio energy from two regions (lobes) located on opposite sides of the galaxy. (p. 304; video lesson; objective 1)

6. **double-exhaust model:** The theory that double radio lobes are produced by pairs of jets emitted in opposite directions from the centers of active galaxies. (p. 304; video lesson; objective 1)

7. **hot spot:** In radio astronomy, a bright spot in a radio lobe. (p. 306; objective 1)

8. **unified model:** An attempt to explain the different types of active galactic nuclei using a single model viewed from different locations. (p. 308; video lesson; objective 1)

9. **BL Lac object:** Thought to be highly luminous cores of distant active elliptical galaxies. (p. 309; video lesson; objective 1)

10. **blazar:** Another name for a BL Lac object. (p. 309; video lesson; objective 1)

11. **quasar:** Small, powerful sources of energy believed to be the active cores of very distant galaxies. (p. 311; video lesson; objective 2)

Summary

Some galaxies in the universe have nuclei that are small and extremely powerful. These nuclei are known as **active galactic nuclei (AGN)**, which produce the vast amount of energy that powers active galaxies. The energy emitted by **active galaxies** is unlike that emitted by "normal" galaxies—astronomers can detect a tremendous amount of ultraviolet and X rays. Some galaxies even emit gamma rays—the highest energy form of light. Others emit vast amounts of radio waves—the lowest energy form of light. The one thing they have in common is the vast amount of energy they emit.

Active Galactic Nuclei

The light from a normal galaxy comes from its stars; its spectrum is a composite of all of the stars observed within it. But **Seyfert galaxies**, the spiral galaxies that Carl Seyfert classified, have small, highly luminous nuclei that produce unique spectra. They display both normal absorption lines and broad emission lines, indicating the presence of hot, low-density, gas within a small volume of space.

There are two kinds of Seyfert galaxies: Type 1 and Type 2. Both types have emission spectral lines in their nuclei, which is evidence of a highly excited gas. The difference between the two arises in the shape of the emission lines. Type 1 Seyfert galaxies emit strong X rays and ultraviolet light and the emission lines are very broad, suggesting that the gases in them are travelling very fast. The emission lines of Type 2 Seyfert galaxies are narrower, suggesting that the gas in these galaxies is moving more slowly.

The brightness in a Seyfert galaxy can change rapidly—within a few minutes. This indicates that its diameter is no more than a few light-minutes across—not much larger than Earth's orbit around the sun. These extremely small galaxies produce a tremendous amount of energy—as much as 100 times the energy produced in the entire Milky Way Galaxy.

About 25 percent of Seyfert galaxies have irregular shapes, suggesting that they've had gravitational interactions with other galaxies. Something may have triggered these galaxies to interact—perhaps a collision with another galaxy or an interaction with a companion. Some Seyfert galaxies expel matter in jets, similar to the jets emitted from the poles of the accretion disk of a stellar mass black hole (see Lesson 11).

Another amazing discovery was that of the **double-lobed radio source**—a galaxy that emits radio energy from two regions (lobes) located at opposite sides of the galaxy. Gravitational forces between the galaxies can cause the jets to twist, while interactions and mergers between galaxies can trigger an eruption in the jets. Double-lobed radio sources in active galaxies are similar in shape to double lobes that we see in the planetary nebula, the jets from T-Tauri stars, and stellar black holes.

The small size and high speed of the motion in the core of a galaxy surrounded by a double lobe indicates that a galaxy spewing these jets must contain a great amount of mass in a very small volume. Similarly, the rapidly fluctuating nuclei in the Seyfert galaxies indicate that they are small but must contain a large amount of mass. Double-lobed galaxies and Seyfert galaxies seem to have similar energy sources—which wasn't understood until the discovery of the quasar.

Quasars

In the 1960s, astronomers discovered faint points of light that had unusual emission spectra. They called these objects **quasars**, or quasi-stellar objects. These objects sometimes emitted radio signals like those from radio galaxies but they determined that these objects were not normal radio galaxies.

By studying the spectra of the hazy objects near quasars and the spectra of the faint nebulae surrounding a quasar, astronomers have concluded that quasars are the active cores of very distant galaxies. Some quasars even have jets and radio lobes.

The spectra from quasars contain bright emission lines, especially the Balmer hydrogen lines. But their spectra were so unusual that astronomers determined that the patterns must have been caused by extreme redshifts. The Balmer lines can be shifted completely out of the visible spectrum and ultraviolet lines could be shifted into the visible spectrum.

Recall from Lesson 13 that these huge redshifts are caused by large velocities of recession—the object is receding from Earth at a very high velocity. The apparent velocity of recession is proportional to the object's distance from Earth, so a quasar must be very far away. Some quasars have redshifts larger than any known galaxy, yet they can be easily photographed; this indicates that these quasars must be ultraluminous—from 10 to 1000 times the luminosity of a large galaxy. Astronomers discovered that these quasars fluctuate in brightness in as little as a few days or a few weeks, indicating that they are very small objects—no larger than a few light-days or light-weeks in diameter.

Knowing the distance to a quasar is an important part in understanding them. When estimating distances to nearby stars and galaxies, astronomers use the Doppler formula (see Lesson 6). But, some quasars have redshifts greater than 1 and the largest known are over 6. Using the Doppler formula, this would indicate that these quasars are receding faster than the speed of light. Since Einstein's Theory of Relativity states that nothing can travel faster than the speed of light, we can conclude that the issue lies with the Doppler formula. For relatively close galaxies and stars, astronomers can measure the redshifts and convert them into apparent velocities of recession using the traditional Doppler formula. But when estimating distances to objects that are extremely far away, astronomers have to use relativistic equations—the traditional Doppler formula doesn't work. Astronomers still use the term "redshift" when using this relativistic formula, as it is akin to the Doppler effect.

Galaxies all formed about the same time, but because quasars are very distant, the look-back time is large. The extremely large redshifts indicate that quasars are very far away—some are more than 10 billion ly away. Therefore, when we look at a quasar, we are looking back to a time when the universe was very young. When the universe was young, galaxies were forming out of clouds of gas and dust. Some of the matter formed supermassive black holes at the centers of these clouds that would eventually become the nuclear bulges of galaxies.

Most quasars have a redshift of about 2 indicating that we are seeing them at the time galaxies were actively forming. Quasars with higher redshifts are more rare; this would be looking back to a time when the universe was so young that few galaxies were forming and therefore fewer quasars existed.

A Unified Model

Many astronomers believe that the different types of active galaxies are really all the same phenomenon simply seen from different viewing angles. In trying to make sense of it all, astronomers have developed a **unified model**. This model can help astronomers think about how galaxies with active nuclei are formed.

Astronomers have concluded that only one thing could explain the tremendous amount of energy emitted by an active galaxy—a supermassive black hole. The model suggests that it's possible that these active galaxies are similar to one another and each has a supermassive black hole in the center. We detect different properties because we are looking at the galaxies from different angles.

To illustrate this model, we can imagine throwing thousands of compact discs into the air. If we took a picture of them, we would see some of them edge-on and they would look like very thin lines. We'd see others face on and we'd be looking directly into the hole of the CD. We might see others tipped at various angles toward or away, to the left or to the right. If we didn't know that all of the objects tossed into the air were compact discs, then they might seem like very different objects—we are looking at them from different angles and we therefore detect different properties.

Like the CDs tossed into the air, galaxies, too, are at random angles. The differences between active galaxies may be explained by our viewing angles rather than a difference in the fundamental properties of the galaxies. If we are looking at an accretion disk that is slightly tipped, we are viewing the black hole nearly along the jets. The gas is very hot and moving rapidly and the high velocity broadens the spectral lines. This could explain Type 1 Seyfert galaxies—the emission lines are very broad, suggesting that the gases in them are travelling very fast.

If we look at a galaxy almost directly through the accretion disk—edge-on—then the disk blocks the light from the central black hole and the jet. The light we see comes from slower-moving gas farther from the black hole above and below the central disk, producing narrower spectral lines with smaller Doppler shifts. This could explain the Type 2 Seyfert galaxies—the emission lines are narrower, suggesting that the observable gas in these galaxies is moving more slowly.

If we looked directly at the face of the accretion disk, down the center of the black hole, we would detect a **BL Lac object** (or **blazar**). Once thought of as variable stars, they may actually be the cores of active galaxies in which the jets are facing directly toward Earth.

This unified model could also help to explain quasars. If a quasar were indeed the active nucleus of a galaxy, it would behave like a supermassive black hole in the active nucleus of an active galaxy. The orientation of the accretion disk surrounding the nucleus may explain the different kinds of quasars that astronomers have detected. If we are looking at the disk face on—with the jets coming toward us—we may detect one type of quasar. If we are looking at the disk edge-on, we may detect a different type of quasar. This evidence supports that quasars are the ultraluminous, active cores of galaxies that are very far away.

Most galaxies, including our own Milky Way Galaxy, are believed to have a supermassive black hole in the center. But, since only a small percentage of galaxies have active nuclei, this leads astronomers to believe that the black holes in "normal" galaxies are dormant—they are not consuming large amounts of matter and therefore don't produce jets or release large amounts of nonstellar energy.

Review Exercises

Matching

Match each term with the appropriate definition or description.

1. _____ radio galaxy		6. _____ double-exhaust model	
2. _____ active galaxy		7. _____ hot spot	
3. _____ active galactic nuclei (AGN)		8. _____ unified model	
		9. _____ BL Lac object	
4. _____ Seyfert galaxy		10. _____ blazar	
5. _____ double-lobed radio source		11. _____ quasar	

    a. In radio astronomy, a bright spot in a radio lobe.

    b. Another name for a BL Lac object.

c. A galaxy that is a strong source of radio signals; the strength of the signal depends on the orientation of the active galaxy.

d. An otherwise normal spiral galaxy with an unusually bright, small core that fluctuates in brightness.

e. Thought to be highly luminous cores of distant active elliptical galaxies.

f. The centers of active galaxies that are emitting large amounts of excess energy.

g. An attempt to explain the different types of active galactic nuclei using a single model viewed from different locations.

h. The theory that double radio lobes are produced by pairs of jets emitted in opposite directions from the centers of active galaxies.

i. A galaxy that emits radio energy from two regions (lobes) located on opposite sides of the galaxy.

j. Small, powerful sources of energy believed to be the active cores of very distant galaxies.

k. A galaxy whose center emits large amounts of excess energy, often in the form of radio emission.

Completion

Fill each blank in the sentences below with the most appropriate term from the list of completion answers that follow. A term may be used once, more than once, or not at all. Check your answers with the Answer Key and review when necessary.

emission spectra   Seyfert
Hawking   short
high   small
long   synchrotron
low   unified
photometry

1. The spectral lines of quasars tend to have very _____ redshifts.

2. Active galaxies vary their energy and brightness over very _____ intervals of time.

3. The theory that explains all active galaxies with a single mechanism is called the _____ model.

4. Rapid fluctuations of quasars show that the objects must be very _____.

5. Quasars were first identified as star-like objects that had peculiar _____.

6. The twisted magnetic field around a supermassive black hole confines the jets in a narrow beam and causes _____ radiation.

Self-Test

Select the best answer.

1. Of the following objects, the one that must be a spiral galaxy is a
   a. radio galaxy.
   b. BL Lac object.
   c. Seyfert.
   d. quasar.

2. The speed of a very highly redshifted quasar as determined by the relativistic Doppler formula is
   a. greater than that determined by the classical formula.
   b. less than that determined by the classical formula.
   c. exactly the same as that determined by the classical formula.
   d. impossible to determine because of how far quasars are from Earth.

3. Of the following objects, the one that produces the least energy is a
   a. supernova.
   b. Seyfert galaxy.
   c. BL Lac object.
   d. quasar.

4. A BL Lac object is the same as a
   a. Type 1 Seyfert.
   b. Type 2 Seyfert.
   c. hypernova.
   d. blazar.

5. Because of the large look-back times of quasars, they appear as they were when the universe was only _____ percent of its present age.
   a. 10
   b. 25
   c. 50
   d. 75

6. Most of the lines in the spectrum of the quasar 3C 273 are caused by
   a. calcium.
   b. carbon.
   c. helium.
   d. hydrogen.

7. Quasar distances are estimated by
   a. observations of Cepheids within them.
   b. observations of supernovae within them.
   c. parallax.
   d. their redshifts.

8. Eruptions of supermassive black holes in quasars may have been caused by
   a. novae.
   b. collisions or mergers with other galaxies.
   c. depleting all the material from the accretion disk.
   d. supernovae.

9. The objects that are detected at the points where the gas of a radio jet hits the extragalactic medium are called
   a. quasars.
   b. Herbig–Haro objects.
   c. Dyson spheres.
   d. hot spots.

10. The evidence leads modern astronomers to conclude that at the cores of active galaxies are supermassive black holes with masses as high as
   a. ten solar masses.
   b. a thousand solar masses.
   c. a million solar masses.
   d. a billion solar masses.

Short-Answer Questions

1. What makes a Seyfert galaxy different than a BL Lac object?

2. Is a quasar with a redshift of $z = 2$ going faster than the speed of light? Why or why not?

3. What are the characteristics of a quasar?

4. What is the difference between a Type 1 Seyfert galaxy and a Type 2 Seyfert galaxy?

5. What do the spectra of Seyfert galaxies tell astronomers?

6. How do quasars relate to active galaxies?

Applications

1. Explain what the gravitational lensing effect on quasars tells us about these objects.

2. What is the similarity between active galaxies, stellar black holes, and protostars?

3. Explain the unified model of active galactic nuclei and how each is viewed.

4. How were quasars discovered?

5. Why do we not see quasars near our galaxy?

Answer Key

Matching

1. c (p. 300; video lesson; objective 1)

2. k (p. 302; video lesson; objective 1)

3. f (p. 302; video lesson; objective 1)

4. d (p. 302; video lesson; objective 1)

5. i (p. 304; video lesson; objective 1)

6. h (p. 304; video lesson; objective 1)

7. a (p. 306; objective 1)

8. g (p. 308; video lesson; objective 1)

9. e (p. 309; video lesson; objective 1)

10. b (p. 309; video lesson; objective 1)

11. j (p. 311; video lesson; objective 2)

Completion

1. high (or large) (p. 313; video lesson; objective 3)

2. short (p. 313; video lesson; objective 4)

3. unified (p. 308; video lesson; objective 4)

4. small (p. 314; video lesson; objective 3)

5. emission spectra (pp. 311–313; objective 4)

6. synchrotron (p. 307; objective 4)

Self-Test

1. c (p. 302; video lesson; objective 1)

2. b (pp. 313–314; objective 3)

3. a (pp. 214–217, 302, 309, 311; objective 2)

4. d (p. 309; video lesson; objective 1)

5. a (p. 314; objective 4)

6. d (pp. 311–313; objective 2)

7. d (pp. 311–313; video lesson; objective 3)

8. b (p. 310; video lesson; objective 4)

9. d (p. 306; objective 1)

10. d (p. 303; objective 1)

Short-Answer Questions

1. A Seyfert galaxy is a spiral galaxy with an active core. A BL Lac object is an elliptical galaxy with an active core. (pp. 302, 309; video lesson; objective 1)

2. No, the classical redshift equation breaks down at high speeds. Then you have to use the relativistic equations. (pp. 311–313; objective 3)

3. A quasar is a very luminous object that is very small, very active and very distant. They are thought to be active galaxies and because of their great distance and great look-back time, we are seeing them as they appeared when the galaxies were just beginning to form. (pp. 311–316; video lesson; objective 2)

4. Type 1 Seyfert galaxies are very luminous at X-ray and UV wavelengths and have the typical broad emission lines with sharp, narrow cores. Type 2 Seyfert galaxies have much weaker X-ray emission and have emission lines that are narrower than Type 1 Seyfert galaxies, but still broader than regular galaxies. (pp. 302–304; video lesson; objective 1)

5. The spectra of Seyfert galaxies contain broad emission lines of highly ionized atoms. This suggests that the object is very hot and energetic and the gases are moving very quickly. (pp. 302–303; video lesson; objective 1)

6. A quasar is thought to be an active galaxy very far away and therefore we are seeing these objects as they appeared in the relatively early universe. (p. 314; objective 2)

Applications

1. The gravitational lensing effect tells us that these objects are very far away if normal galaxies are the lensing objects and that they must be very luminous. (p. 314; objective 2)

2. They all have the same geometry, dense central nucleus with an accretion disk and jets. (pp. 177, 238, 302–304; objective 1)

3. The Seyfert galaxies are looking at the accretion disk from different vantage points, the more face-on to the disk, the stronger the radiation. Blazars are looking at the accretion disk face-on or "looking down the throat of the dragon." (pp. 308–309; video lesson; objective 1)

4. Quasars were star-like objects with very unusual spectra. It was discovered that these strange emission lines were actually hydrogen, but the lines astronomers were used to seeing were shifted to another part of the spectrum. This large redshift indicated that these objects were moving very fast and were therefore very far away. (pp. 311–313; video lesson; objective 2)

5. When we see quasars very far away, we are looking very far back in time. We are viewing galaxies in early formation. We do not see quasars around us because all of the active objects have evolved. (pp. 315–317; video lesson; objective 4)

Lesson Review

Lesson 14: Active Galaxies

PLEASE NOTE: Use this matrix to guide your study and achieve the learning objectives of this lesson. It will also help you to view the video, which defines and demonstrates important concepts and principles as they relate to everyday life and actual case studies.

Learning Objective	Textbook	Student Guide
1. Describe observations of active galaxies and the physical processes proposed to explain their energy output and other characteristics.	pp. 300–311	Key Terms: 1, 2, 3, 4, 5, 6, 7, 8, 9, 10; Matching: 1, 2, 3, 4, 5, 6, 7, 8, 9, 10; Self-Test: 1, 4, 9, 10; Short-Answer: 1, 4, 5; Applications: 2, 3.
2. Describe the discovery and characteristics of quasars and discuss how they relate to the topic of active galaxies.	pp. 311–313	Key Terms: 11; Matching: 11; Self-Test: 3, 6; Short-Answer: 3, 6; Applications: 1, 4.
3. Explain how quasar distances are estimated.	pp. 313–314	Completion: 1, 4; Self-Test: 2, 7; Short-Answer: 2.
4. Describe the current explanation of quasars and their energy sources and the implications of this explanation for the formation of galaxies and early history of the universe.	pp. 314–317	Completion: 2, 3, 5, 6; Self-Test: 5, 8; Applications: 5.

Notes:

LESSON
15

Cosmology

Checklist

For the most effective study of this lesson, complete the following activities in this sequence.

Before Viewing the Video

❑ Read the Preview, Learning Objectives, and Viewing Notes below.

❑ Read Chapter 15, "Cosmology in the 21st Century," pages 320–345, in the *Horizons* textbook.

What to Watch

❑ After reading the textbook chapter, watch the video for Lesson 15, *Cosmology.*

After Viewing the Video

❑ Briefly note your answers to questions listed at the end of the Viewing Notes.

❑ Review the Summary below.

❑ Review all reading assignments for this lesson, especially the Chapter 15 summary on pages 343–344 in *Horizons* and the Viewing Notes in this guide.

❑ Write brief answers to the review questions at the end of Chapter 15 in *Horizons.*

❑ Complete the Review Exercises below. Check your answers with the Answer Key and review when necessary.

❑ Use the Lesson Review matrix found at the end of this lesson to review and assess your knowledge of each Learning Objective.

❑ As assigned by your instructor, complete the Applications activities and any additional activities for this lesson.

Preview

The branch of astronomy that is concerned with the universe as a whole is called **cosmology**. Cosmologists pursue answers to questions regarding the origin, the evolution and the fate of the universe—they look at the big picture.

It was once thought that the universe was eternal—there was no beginning and no end. But when a relationship between galactic redshifts and distances was discovered, astronomers realized that the universe was expanding. If the universe were in fact expanding, it would stand to reason that there was a moment when it began.

In this lesson, you will explore a number of theories and concepts that cosmologists study. You will begin to understand that our infinite universe has a beginning point in time and it continues to expand at an ever-increasing rate. You will also discover how cosmologists determine the shape of the universe—this is important because its shape will ultimately determine its fate.

CONCEPTS TO REMEMBER

- Recall from Lesson 13 that *look-back time* is the amount by which we look into the past when we look at a distant galaxy. This time is equal to the distance to the object in light-years (p. 253 in this guide).

- In Lesson 12, you learned that *dark matter* is nonluminous matter that is detected only by its gravitational influence (p. 230 in this guide).

Learning Objectives

After you complete this lesson, you should be able to:

1. Explain Olbers's paradox and how it relates to the darkness of the night sky. HORIZONS TEXTBOOK PAGES 323–324.

2. Describe the observational evidence that there was a big bang, such as expansion of the universe and cosmic background radiation. HORIZONS TEXTBOOK PAGES 323, 325–329.

3. Explain how curved space-time resolves the edge-center problem and suggests possible futures for the universe. HORIZONS TEXTBOOK PAGES 322–323, 332–333, 339–343.

4. Recount currently accepted theories about the early history of the universe from the big bang until the beginning of the formation of galaxies. HORIZONS TEXTBOOK PAGES 329–331.

5. Identify and define isotropy, homogeneity, and the resulting cosmological principle. HORIZONS TEXTBOOK PAGE 332.

6. Describe observations that establish the existence of dark matter and its role in the origin of galaxies and large-scale structure. *HORIZONS* TEXTBOOK PAGES 333–335.

7. Describe the flatness and horizon problems, and explain how they are resolved by an inflationary universe. *HORIZONS* TEXTBOOK PAGES 335–336.

8. Describe observations that imply an accelerating universe and explain their impact on cosmology. *HORIZONS* TEXTBOOK PAGES 337–339.

At this point, read Chapter 15, "Cosmology in the 21st Century," pages 320–345.

The Universe, as has been observed before, is an unsettlingly big place, a fact which for the sake of a quiet life most people tend to ignore.

—Douglas Adams
The Restaurant at the End of the Universe (1980)

Viewing Notes

The video program addresses the quest to determine the origin, the shape, and the fate of our universe.

The video program contains the following segments:

- ☺ Paradox & the Poet
- ☺ The Expanding Universe
- ☺ The Big Bang & the Cooling Universe
- ☺ Building the Universe
- ☺ Matter & Energy

The following information will help you better understand the video program:

The video opens with a discussion of Olbers's paradox. Although it is not really a paradox—a seemingly contradictory statement—it states that if the universe were indeed infinite and eternal, your eyes should fall upon the light of a star no matter where you looked in the night sky. Our observations tell us that this is not true; Olbers suggested that the night sky was dark because clouds of matter in space absorbed the light. Astronomers have since concluded that Olbers's conclusion was mostly incorrect.

It was the poet Edgar Allan Poe who established the idea of an **observable universe**—the portion of the universe that we can observe from our location in space and time. The night sky was dark because the observable universe was not infinite—not that the stars were obscured by gas and dust as Olbers had suggested.

Edwin Hubble's observations of an expanding universe supported Poe's idea. Hubble noticed that the spectra of galaxies had redshifts. The size of the redshift was proportional to the distance to the galaxy, which indicated that the universe is expanding uniformly.

Astronomers proposed the big bang theory to explain how space, time, and matter became separate and unique entities—there was a huge explosion of energy, which then transformed partly into matter. It wasn't until the discovery of **cosmic background radiation**—the leftover energy that first escaped from the matter—that the big bang theory was more widely accepted.

In order to understand the age, the composition, and the fate of the universe, cosmologists search for evidence of its geometry, or its curvature. In a **closed universe**—the shape of a beach ball—there is enough matter to slow the expansion to a halt and possibly reverse back on itself. A **flat universe**— shaped like a piece of paper—must be infinite and eternal. An **open universe**— the shape of a saddle—will also expand forever. To determine the shape of the universe, astronomers have studied the cosmic background radiation and determined that the universe is open and tending toward flat.

QUESTIONS TO CONSIDER

- Why does the night sky have a dark background with scattered starlight?

- What evidence supports the idea that the universe is expanding?

- What is the origin of cosmic background radiation? How does it support the theory of the big bang?

- What reasons do we have to conclude that the universe contains a substantial amount of dark matter?

- How does the shape of the universe determine its fate?

Watch the video for Lesson 15, *Cosmology.*

Key Terms and Concepts

Page references are keyed to the *Horizons* textbook.

1. **cosmology:** The study of the nature, origin, and evolution of the universe. (p. 320; video lesson; objective 1)

2. **Olbers's paradox:** The conflict between observation and the theory about why the night sky should or should not be dark. (p. 323; video lesson; objective 1)

3. **observable universe:** The part of the universe that we can see from our location in space and in time. (p. 323; video lesson; objective 1)

4. **big bang:** The high-density, high-temperature state from which the expanding universe of galaxies began. (p. 326; video lesson; objective 2)

5. **Hubble time:** The age of the universe equivalent to 1 divided by the Hubble constant. (p. 326; objective 2)

6. **cosmic microwave background radiation:** Radiation from the hot clouds of the big bang explosion. (p. 329; video lesson; objective 2)

7. **antimatter:** Matter composed of anti-particles, which upon colliding with a matching particle of normal matter annihilate and convert the mass of both particles into energy. (p. 329; objective 4)

8. **recombination:** The stage within 300,000 years of the big bang, when the gas became transparent to radiation. (p. 330; objective 4)

9. **dark age:** The period of time after the glow of the big bang faded into the infrared and before the birth of the first stars, during which the universe expanded in darkness. (p. 331; video lesson; objective 4)

10. **reionization:** The stage in the early history of the universe when ultraviolet photons from the first stars ionized the gas filling space. (p. 331; objective 4)

11. **isotropy:** The assumption that in its general properties the universe looks the same in every direction. (p. 332; objective 5)

12. **homogeneity:** The assumption that, on the large scale, matter is uniformly spread through the universe. (p. 332; objective 5)

13. **cosmological principle:** The assumption that any observer in the universe sees the same general features of the universe. (p. 332; objective 5)

14. **critical density:** The average density of the universe needed to make its curvature flat. (p. 333; objective 5)

15. **closed universe:** A model of the universe in which the average density is great enough to stop the expansion and make the universe contract. (p. 333; video lesson; objective 3)

16. **flat universe:** A model of the universe in which space-time is not curved. (p. 333; video lesson; objective 3)

17. **open universe:** A model of the universe in which the average density is less than the critical density needed to halt the expansion. (p. 333; video lesson; objective 3)

18. **nonbaryonic matter:** Proposed dark matter made up of particles other than protons and neutrons. (p. 335; objective 6)

19. **hot dark matter:** Dark matter made up of particles such as neutrinos traveling at or nearly at the speed of light. (p. 335; objective 6)

20. **cold dark matter:** Mass in the universe, as yet undetected except for its gravitational influence, which is made up of slow-moving particles. (p. 335; objective 6)

21. **flatness problem:** In cosmology, the peculiar circumstance that the early universe must have contained almost exactly the right amount of matter to make space-time flat. (p. 335; objective 7)

22. **horizon problem:** In cosmology, the circumstance that the primordial background radiation seems much more isotropic than can be explained by the standard big bang theory. (p. 336; objective 7)

23. **inflationary universe:** A version of the big bang theory that includes a rapid expansion when the universe was very young; derived from grand unified theories. (p. 336; objective 7)

24. **grand unified theories (GUTs):** Theories that attempt to describe in a similar way the electromagnetic, weak, and strong forces of nature. (p. 336; video lesson; objective 7)

25. **cosmological constant:** A constant in Einstein's equations of space and time that represents a force of repulsion. (p. 337; video lesson; objective 8)

26. **quintessence:** Postulated energy that fills empty space and drives the acceleration of the universe; also known as dark energy. (p. 337; objective 8)

27. **dark energy:** The energy believed to fill empty spaces and drive the acceleration of the expanding universe; also known as quintessence. (p. 338; video lesson; objective 8)

28. **big rip:** The fate of the universe if dark energy increases with time, and galaxies, stars, and even atoms are eventually ripped apart by the accelerating expansion of the universe. (p. 339; objective 8)

29. **large-scale structure:** The distribution of clusters and superclusters of galaxies in filaments and walls enclosing voids. (p. 339; video lesson; objective 4)

30. **supercluster:** A cluster of galaxy clusters. (p. 339; objective 4)

Summary

When you look up into the night sky, you can see a dark background with scattered stars. Astronomers once thought that the universe was infinite and the sky was dark because clouds of gas and dust were absorbing starlight. Astronomers now know that this is not the case.

Introduction to the Universe

If the universe were infinite, then you would assume that anywhere you looked, your eyes would fall upon the light of a star, much like if you looked in any direction in a forest, you'd see a tree. If you assume that the universe is infinite and uniformly filled with stars, you might conclude that the night sky should glow as brightly as the surface of the average star. But, through your observations, you know that this is not the case. This conflict between theory and observation is called **Olbers's paradox**.

Heinrich Olbers, and many scientists before him, assumed that the universe was infinite. He concluded that the reason that the sky wasn't as bright as the surface of a star was that clouds of dust and gas absorbed the radiation from distant stars before it could reach Earth. Today's astronomers know that these clouds—the interstellar medium—would be heated from the distant stars and glow as brightly as the stars themselves. Clearly, this is not the case. Our night sky is not as bright as the surface of a star. If it were, we wouldn't be able to live on Earth. There was a flaw in Olbers's assumption—he assumed that the universe was infinite not only in space but also in time.

There are several possible resolutions of Olbers's paradox. The first is that the universe is not infinite, that it ends somewhere. The second is if you go farther from Earth, or some other point in space, the universe changes its character in some way. A third way to solve Olbers's paradox is to say that the universe is infinite in extent—but it's *not* infinite in time—it has no edge but it has a certain age. Because it has a certain age, you can only see out to a certain distance, which is equal to the age of the universe in light-years. Astronomers have accepted this third solution, rendering Olbers's paradox not a paradox at all.

As you learned in Lesson 13, astronomers today believe that the night sky is dark because the universe was created sometime in the past—about 14 billion years ago. The more distant stars are so far away that their light hasn't yet reached us. If you looked deep enough into space, you would look back to a time before stars began to shine. That is, the *look-back time* would be greater than the age of the universe.

The universe is everything that exists, but the poet Edgar Allan Poe established the idea of the **observable universe**—everything that we can detect. Poe also suggested that, at one point in time, all the material of the universe was compressed to an individual point. These were radical ideas at the time, but years later evidence supported his ideas. In the early 1900s, Einstein had reinforced this concept of a static universe in his theory of relativity, which discredited Poe's ideas. It wasn't until 1929 that Edwin Hubble found evidence that contradicted the idea of a static universe.

Hubble and other astronomers noticed that the spectra of all distant galaxies had redshifts. No matter where they looked, the galaxies had redshifts, which implied that galaxies are receding from each other and the universe is expanding. The size of the redshift was proportional to the distance to the galaxy, which indicated that the universe is expanding uniformly. Galaxies that were nearby had small redshifts, while galaxies that were very far away had larger redshifts. This was in line with Poe's idea that at one point in time, all the material of the universe was compressed to an individual point.

The discovery of these galactic redshifts led to the **big bang** theory, which states that the expanding universe began from a moment of extreme conditions. The big bang was not an explosion at a single point; rather, it filled the entire volume of the universe. When the theory was first proposed it was referred to as "the day without a yesterday," and it was not widely accepted.

Over the years, several observations supported the big bang theory. When scientists thought about the big bang, they concluded that if the universe began with a great expansion outward, then it would still be expanding today and we should be able to detect the radiation generated from it. When this **cosmic microwave background radiation** was detected in the 1960s, there was finally enough evidence to support the big bang theory. This set the stage for a major paradigm shift in the way astronomers modeled the universe—the idea of an infinitely old and static universe was discredited. The universe had a beginning and it was expanding.

Thirty years later, another discovery supported the big bang theory. If the theory were correct, the radiation coming from the big bang would look like *black body radiation* coming from a very cool source. It wasn't until 1990 that satellite measurements were able to confirm this assumption. The gas clouds that emitted the cosmic background radiation had a temperature of about 3000 K, but the apparent temperature of this radiation was about 2.7 K—about 1,100 times cooler than it really was when the universe began. Scientists concluded that the radiation coming from almost the time the universe began—14 billion light-years away—would have a tremendous redshift. Because of the expansion of the universe, it redshifted the light and made it appear cooler. The radiation appears to be so cool because of the tremendous redshift.

The big bang theory was also able to explain the abundance of hydrogen and helium in the universe. Within the first few minutes after the universe began, all of the ordinary matter of the universe was created—essentially it condensed out of a high-energy cosmic soup. As the universe cooled in the first three minutes, electrons, protons, and neutrons were created out of this soup. As it expanded and cooled further, protons and neutrons eventually combined to form deuterium—heavy hydrogen—and further reactions converted deuterium into helium. This explains the vast amounts of these elements that are present in the universe today.

The Shape of Space and Time

The model of an expanding universe is based on the fact that when we look in a certain direction, we see galaxies, no matter where we look, and they appear to be moving away from us.

Any observer in the galaxy sees the same features of the universe as any other observer any place else—the universe is the same in all directions. There are local variations of course, one observer might live in a spiral galaxy in a small cluster of galaxies and another might live in an elliptical galaxy in a large cluster. But, the general features of the universe are the same no matter where the observer is. This is the **cosmological principle**—there are no special places in the universe and any observer in any galaxy sees the same general features of the universe, but observers living at different times will see stars and galaxies at different stages in their evolution.

The concept of an expanding universe where everything is the same—there is no center or an edge—is difficult to wrap your mind around. How can something expand if it has no edge and no center? We can look to Einstein's theory of general relativity to help us understand.

Einstein's theory of general relativity predicts that the presence of mass can curve space-time; we experience this curvature as gravity. If we knew how much mass is in the universe, then we would know how it is curved. There are three possible ways our universe might be curved: positive curvature, zero curvature, or negative curvature. A positively curved universe is known as a **closed universe** because it is finite and has no edge. A **flat universe**—one with zero curvature—must be infinite, otherwise, it would have a center or an edge. A negatively curved universe is known as an **open universe**; it, too, must be infinite (see p. 342 in the *Horizons* textbook).

Since the presence of mass can curve space-time, the general curvature of the universe is determined by its density. **Critical density** is the average density needed to make the universe flat, or have zero curvature. If the average density of the universe is equal to the critical density, the universe would be completely flat. If the average density of the universe is less than the critical density, the universe is negatively curved, or open. If the average density of the universe is more than the critical density, the universe is positively curved, or closed.

You might think that astronomers could estimate the mass of the universe and therefore be able to determine its curvature and age. But, there are vast amounts of *dark matter* in the universe, which is difficult to detect and even more difficult to measure. Additionally, energy has a mass equivalent, so all of the energy in the universe would have to be measured as well. You can see how it would be difficult to obtain such measurements.

Dark matter is not like "normal matter"—the type of matter that makes stars, Earth, and humans. The protons and neutrons in normal matter belong to a family of subatomic particles called *baryons*. Dark matter may be made up primarily of nonbaryonic particles and is therefore described as **nonbaryonic matter**. Because dark matter doesn't interact with normal matter, it was not affected by the extreme radiation that existed during the early years of our universe.

Hot dark matter is made of neutrinos and other particles that are moving at or near the speed of light. Because the particles are moving so fast, they cannot clump together and cannot have played a part in the formation of stars and galaxies. **Cold dark matter** is made up of slower moving particles and can clump together more easily. The abundance of these different types of dark matter makes it difficult to detect the curvature of the universe.

21st-Century Cosmology

Although the big bang theory was able to explain a number of mysteries in the universe, it wasn't able to explain two important features of the universe: its curvature and its uniformity of temperature, respectively deemed the **flatness problem** and the **horizon problem**.

The flatness problem asks, "*Why* is the universe so flat?" Recall that the general curvature of the universe is determined by its density. The density of the universe is within a factor of 10 of the critical density needed to make the universe flat, so it seems nearly flat. Given a range of possibilities between 0 and infinity, these densities are relatively close. When we take into consideration the amount of dark matter that must be in the universe, the difference between the density of the universe and the critical density is closer to being within a factor of 3. The big bang theory could not explain why the curvature of the universe is nearly flat.

The horizon problem asks, "How did every part of the universe get to be the same temperature when recombination occurred?" **Recombination** occurred as the young universe cooled—the protons were able to capture the free electrons to form neutral hydrogen. The free electrons were no longer scattered through the universe and the photons no longer interacted with the neutral gas. Those photons—originally with the temperature of 3000 K and seen today as 2.7 K—make up the cosmic microwave background radiation. The cosmic background radiation created during the beginning of the universe is the same no matter where we look, and the big bang theory could not explain this observation either.

To solve the flatness problem and the horizon problem, a new variation of the big bang was created. The **inflationary universe** explains these two problems—it predicts that there was an extremely sudden expansion when the universe was very young. In Lesson 7, you learned about the four forces in nature—gravity, electromagnetic force, strong force, and weak force. According to the theory, when the universe was very young, the forces were indistinguishable—they all had the same effect. As the universe expanded and began to cool, the forces separated and began to behave in different ways. This triggered inflation in the size of the universe, which suddenly expanded by a factor between 10^{20} and 10^{30}. Even with this rapid inflation, the observable universe was still no larger than the volume of an atom.

This rapid expansion can explain the flatness problem—it would have forced whatever curvature the universe had toward zero. Because the observable universe was no larger than an atom, its temperature was equalized before it expanded further. This is why the background radiation that we detect today is the same temperature no matter where we look; this solves the horizon problem.

When we think about the expansion of the universe, common sense would tell us that it would only expand so far before gravity would kick in and begin to slow the expansion. We would expect that the expansion in a low-density, open universe would be slowed slightly by the force of gravity. The expansion in a high-density, closed universe would slow down dramatically and the universe would eventually contract. If we could measure the rate of the slowing, then we would know the true curvature of the universe.

However, astronomers have discovered that the expansion of the universe is not slowing down—it's actually accelerating. Astronomers can determine this by using *distance indicators*; recall from Lesson 13, that astronomers use Cepheid variable stars and type Ia supernovae as distance indicators to determine the distance to celestial objects. By comparing distant galaxies to these distance indicators astronomers were able to conclude that about 6 billion years ago, the expansion of the universe shifted gears from deceleration to acceleration.

Something must be fueling the expansion; it must be strong enough to win the battle with gravity but cannot be detected. Astronomers refer to this energy as **dark energy**. It represents a repulsive force that is a property of empty space and thus actually increases with distance. When the universe was young, galaxies were close together and gravity was able to overpower the dark energy and slow the expansion. When the expansion reached a certain point, about 6 billion years ago, the galaxies were so far apart, that gravity could no longer keep them together to slow the expansion of the universe—acceleration began.

You've already learned that dark matter and baryonic matter make up about a third of the critical density to make the universe flat. If these make up only a third, you might wonder how astronomers can say that the universe is flat. Recall that energy has a mass equivalent—the dark energy in the universe makes up the remainder of the critical mass needed to make the universe flat.

The largest structures in the universe pose another problem for astronomers. Galaxies are clustered together and those clusters are gathered around **superclusters**. Astronomers have created maps of the universe that reveal that galaxies are clustered in filaments that surround great voids of empty space. These filaments make up the largest structures in space. The problem this poses lies in trying to explain how, at the time of recombination, the uniform gases coalesced so quickly to form galaxies. The answer lies in the presence of dark matter—it must have provided the gravity needed to pull the baryonic matter together.

The clumping of the matter in the universe could have been triggered by subatomic fluctuations in space. There would have been very subtle variations in the gravitational fields. These variations can be detected in the background radiation—some spots on the sky look hotter and brighter than other spots. If the universe were flat, then these irregularities would be about 1 degree in diameter. Careful measurements reveal that they are in fact approximately this size, which supports the theory that the universe is nearly flat.

In this lesson, you've studied some of the most perplexing questions in cosmology—does the universe have an edge in space or in time? How can the universe expand if it has no edge? What is the shape of the universe? As cosmologists search for answers to these questions, they come closer to learning the fate of our universe.

Review Exercises

Matching I

Match each term with the appropriate definition or description.

1. _____ cosmology	6. _____ cosmic microwave background radiation
2. _____ Olbers's paradox	
3. _____ observable universe	7. _____ antimatter
4. _____ big bang	8. _____ recombination
5. _____ Hubble time	9. _____ dark age
	10. _____ reionization

a. The age of the universe equivalent to 1 divided by the Hubble constant.

b. The stage within 300,000 years of the big bang, when the gas became transparent to radiation.

c. The part of the universe that we can see from our location in space and in time.

d. Matter composed of anti-particles, which upon colliding with a matching particle of normal matter annihilate and convert the mass of both particles into energy.

e. The stage in the early history of the universe when ultraviolet photons from the first stars ionized the gas filling space.

f. The high-density, high-temperature state from which the expanding universe of galaxies began.

g. Radiation from the hot clouds of the big bang explosion.

h. The period of time after the glow of the big bang faded into the infrared and before the birth of the first stars, during which the universe expanded in darkness.

i. The study of the nature, origin, and evolution of the universe.

j. The conflict between observation and the theory about why the night sky should or should not be dark.

Matching II

Match each term with the appropriate definition or description.

1. _____ isotropy	6. _____ flat universe
2. _____ homogeneity	7. _____ open universe
3. _____ cosmological principle	8. _____ nonbaryonic matter
4. _____ critical density	9. _____ hot dark matter
5. _____ closed universe	10. _____ cold dark matter

a. The assumption that, on the large scale, matter is uniformly spread through the universe.

b. The assumption that in its general properties the universe looks the same in every direction.

c. Proposed dark matter made up of particles other than protons and neutrons.

d. The average density of the universe needed to make its curvature flat.

e. The assumption that any observer in the universe sees the same general features of the universe.

f. A model of the universe in which space-time is not curved.

g. A model of the universe in which the average density is great enough to stop the expansion and make the universe contract.

h. Dark matter made up of particles such as neutrinos traveling at or nearly at the speed of light.

i. A model of the universe in which the average density is less than the critical density needed to halt the expansion.

j. Mass in the universe, as yet undetected except for its gravitational influence, which is made up of slow-moving particles.

Matching III

Match each term with the appropriate definition or description.

1. _____ flatness problem		6. _____ quintessence	
2. _____ horizon problem		7. _____ dark energy	
3. _____ inflationary universe		8. _____ big rip	
4. _____ grand unified theories (GUTs)		9. _____ large-scale structure	
5. _____ cosmological constant		10. _____ supercluster	

a. In cosmology, the peculiar circumstance that the early universe must have contained almost exactly the right amount of matter to make space-time flat.

b. A constant in Einstein's equations of space and time that represents a force of repulsion.

c. A cluster of galaxy clusters.

d. The energy believed to fill empty space and drive the acceleration of the expanding universe; also known as quintessence.

e. The postulated energy that fills empty space and drives the acceleration of the universe; also known as dark energy.

f. The distribution of clusters and superclusters of galaxies in filaments and walls enclosing voids.

g. A version of the big bang theory that includes a rapid expansion when the universe was very young; derived from grand unified theories.

h. In cosmology, the circumstance that the primordial background radiation seems much more isotropic than can be explained by the standard big bang theory.

i. Theories that attempt to describe in a similar way the electromagnetic, weak, and strong forces of nature.

j. The fate of the universe if dark energy increases with time and galaxies, stars, and even atoms are eventually ripped apart by the accelerating expansion of the universe.

Completion

Fill each blank in the sentences below with the most appropriate term from the list of completion answers that follow. A term may be used once, more than once, or not at all. Check your answers with the Answer Key and review when necessary. If a question requires two or more answers in succession, they may be in any order, otherwise unless indicated.

accelerating   decelerating   lengthens
annihilate   8.1 K   light energy
astrology   flatness   recombination
bang   homogeneity   rip
closed   horizon   shortens
cosmological   inflationary   30,000 K
cosmology   isotropy   2.7 K
dark energy

1. The universe is expanding and _____ its expansion.

2. The property of being uniform in every direction is _____.

3. The property of being uniform at every distance is _____.

4. The cosmic background radiation correlates with an apparent black body temperature of _____.

5. When a particle of matter meets its antimatter counterpart, the two particles _____ each other.

6. Einstein added a component to his general relativity equations to balance gravity called the _____ constant.

7. The energy that drives the acceleration of the universe but does not contribute to the formation of starlight or the cosmic microwave background radiation is _____.

8. The study of the universe as a whole is called the study of

   _____.

9. As the falling temperature of the universe after the big bang reached 3000 K, protons were able to capture and hold free electrons to form neutral hydrogen, a process called _____.

10. The stretching of space-time not only moves the galaxies away from each other, but it _____ the wavelength of photons.

11. The scenario of the universe accelerating faster and faster until it pulls galaxy, stars and eventually, atoms apart is called the big _____.

12. A theory that describes the sudden expansion when the universe was very young is called the _____ universe.

13. The improbability of the universe being balanced near the boundary between an open and closed universe is called the _____ problem.

Self-Test

Select the best answer.

1. The universe has
   a. an edge and a center.
   b. an edge but no center.
   c. a center but no edge.
   d. no edge and no center.

2. The main reason the night sky is dark is that the universe is
   a. static.
   b. infinite in size.
   c. finite in age.
   d. expanding.

3. Isotropy is the assumption that
   a. matter is uniformly spread throughout space.
   b. the universe looks the same in every direction.
   c. the physical laws we know on Earth apply everywhere in the universe.
   d. the universe is expanding.

4. According to the Hubble law, if Galaxy A is 10 times more distant than Galaxy B, then Galaxy A is
   a. receding 10 times faster than Galaxy B.
   b. approaching 10 times faster than Galaxy B.
   c. receding 100 times faster than Galaxy B.
   d. approaching 100 times faster than Galaxy B.

5. The primordial background radiation is observable primarily at
   a. radio wavelengths.
   b. ultraviolet wavelengths.
   c. X-ray wavelengths.
   d. gamma ray wavelengths.

6. Of the following groupings of galaxies, the one that is largest in size is the
   a. Local Supercluster.
   b. Local Group.
   c. filaments and voids.
   d. Coma Cluster.

7. The force that is NOT unified with the others in a Grand Unified Theory is the
   a. strong force.
   b. electromagnetic force.
   c. weak force.
   d. gravitational force.

8. The type of universe in which the average density is predicted to be greatest is the
   a. open universe.
   b. flat universe.
   c. closed universe.
   d. negatively curved universe.

9. What did Arno Penzias and Robert Wilson initially think caused the background radiation in their microwave antenna?
   a. the thermal "echo" of the big bang
   b. pigeon droppings
   c. interference from Earth-based radios
   d. microwave ovens

10. The type of universe that will keep expanding forever is
    a. an open universe.
    b. a closed universe.
    c. an oscillating universe.
    d. all of the above types.

11. A nuclear particle theory predicts the existence of massive particles that interact weakly with normal matter. These are all classified as
    a. WIMPs.
    b. MACHOs.
    c. neutrinos.
    d. plutinos.

Short-Answer Questions

1. What did Einstein call his "greatest blunder"?

2. What is the flatness problem?

3. Explain why Olbers's paradox is not a paradox.

4. What evidence demonstrates that the universe is expanding?

5. Why was the idea that the universe is eternal and unchanging abandoned?

6. What is quintessence?

7. If most galaxies are moving away from us, are we at the center of the universe? Why or why not?

8. What are the possible geometric models and their respective corresponding fate for the universe?

Applications

1. Calculate the age of the universe if the Hubble constant were 65 km/s/Mpc.

2. What would be the Hubble constant if the universe were 18 billion years old?

3. How are isotropy, homogeneity, and the cosmological principle related?

4. Explain the evidence that supports the big bang theory.

5. Explain why it is suggested that we live in an inflationary universe.

6. What does the size difference of the irregularities in the background radiation tell us about the universe?

Answer Key

Matching I

1. i (p. 320; video lesson; objective 1)
2. j (p. 323; video lesson; objective 1)
3. c (p. 323; video lesson; objective 1)
4. f (p. 326; video lesson; objective 2)
5. a (p. 326; objective 2)
6. g (p. 329; video lesson; objective 2)
7. d (p. 329; objective 4)
8. b (p. 330; objective 4)
9. h (p. 331; video lesson; objective 4)
10. e (p. 331; objective 4)

Matching II

1. b (p. 332; objective 5)
2. a (p. 332; objective 5)
3. e (p. 332; objective 5)
4. d (p. 333; objective 3)
5. g (p. 333; video lesson; objective 3)
6. f (p. 333; video lesson; objective 3)
7. i (p. 333; video lesson; objective 3)
8. c (p. 335; objective 6)
9. h (p. 335; objective 6)
10. j (p. 335; objective 6)

Matching III

1. a (p. 335; objective 7)
2. h (p. 336; objective 7)
3. g (p. 336; objective 7)
4. i (p. 336; video lesson; objective 7)

5. b (p. 337; video lesson; objective 8)

6. e (p. 337; objective 8)

7. d (p. 338; video lesson; objective 8)

8. j (p. 339; objective 8)

9. f (p. 339; video lesson; objective 4)

10. c (p. 339; objective 4)

Completion

1. accelerating (p. 337; video lesson; objective 8)

2. isotropy (p. 332; objective 5)

3. homogeneity (p. 332; objective 5)

4. 2.7 K (p. 329; video lesson; objective 2)

5. annihilate (p. 329; objective 4)

6. cosmological (p. 337; video lesson; objective 8)

7. dark energy (p. 338; video lesson; objective 8)

8. cosmology (p. 320; video lesson; objective 1)

9. recombination (pp. 330–331; objective 4)

10. lengthens (p. 332; objective 2)

11. rip (p. 339; objective 8)

12. inflationary (p. 336; objective 7)

13. flatness (p. 335; objective 7)

Self-Test

1. d (p. 322; video lesson; objective 3)

2. c (p. 323; video lesson; objective 1)

3. b (p. 332; objective 5)

4. a (pp. 326–327; objective 2)

5. a (pp. 328–329; video lesson; objective 2)

6. c (p. 339; video lesson; objective 6)

7. d (p. 336; objective 7)

8. c (p. 333; video lesson; objective 3)

9. b (pp. 327–328; video lesson; objective 4)

10. a (p. 333; video lesson; objective 3)

11. b (p. 335; objective 6)

Short-Answer Questions

1. Einstein said that introducing a cosmological constant into his equations describing space and time was his biggest blunder. It turns out that his "mistake" may have been correct for identifying an accelerating expanding universe. (p. 337; video lesson; objective 8)

2. The universe seems to be balanced near the boundary between an open and a closed universe. That is, it seems nearly flat. Given the vast range of possibilities, it seems peculiar that the density of the universe is within a factor of 10 of the critical density. (pp. 335–336; objective 7)

3. It is not a paradox if you understand the nature of the universe. Olbers's paradox suggests that if the universe were infinite in size and infinite in age, every direction you look, your eye should fall on a bright object. Therefore, the sky should be bright at night. The sky is dark at night because the universe is finite in size, finite in age, or both. (p. 323; video lesson; objective 1)

4. Hubble showed that the farther galaxies are from us, the faster they are apparently moving away from us. This suggests the universe is expanding. (pp. 323–327; video lesson; objective 2)

5. This theory was rejected after the detection of the cosmic background radiation. This is the leftover blackbody radiation from the big bang. This suggested that the universe was expanding, not eternal and unchanging. (pp. 323–329; objective 2)

6. The theoretical energy in empty space which causes the acceleration of the expansion of the universe outward against the attractive force of gravity. (pp. 337–338; objective 8)

7. No, we are not the center of the universe. Because space itself is expanding, every galaxy on the large scale is moving away from every other galaxy, like the raisins in the rising raisin bread dough analogy. (p. 325; objective 2)

8. The universe can be closed, in which it would collapse back in on itself; flat, in which it would stop its expansion; or open, in which it would keep expanding forever. (p. 333; objective 3)

Applications

1. The answer is 15.4 billion years. (p. 326; objective 2)

   $t_o = 1/H \times 10^{12}$ years
   $t_o = 1/(15.4) \times 10^{12}$ years

2. The answer is 55.6 km/s/Mpc. (p. 326; objective 2)

   $t_o = 1/H \times 10^{12}$ years
   $18 = 1/H \times 10^{12}$ years

3. Isotropy is when the universe looks the same in any direction. Homogeneity is when the universe looks the same at any distance. The cosmological principle says that any observer sees the same general features of the universe. (p. 332; objective 5)

4. The main evidence for the big bang theory is the expansion of the universe shown by the expansion of the space between the galaxies. Also, the cosmic background radiation is the thermal "echo" of the big bang. (pp. 323–329; video lesson; objective 2)

5. Using the grand unified theories and quantum mechanics, the inflationary universe would solve the flatness problem and the horizon problem. (pp. 335–337; objective 7)

6. Careful measurements of the size of the irregularities in the cosmic background radiation show that the observations fit the theory very well for a flat universe. (pp. 341–343; objective 6)

Notes:

Lesson Review

Lesson 15: Cosmology

PLEASE NOTE: Use this matrix to guide your study and achieve the learning objectives of this lesson. It will also help you to view the video, which defines and demonstrates important concepts and principles as they relate to everyday life and actual case studies.

Learning Objective	Textbook	Student Guide
1. Explain Olbers's paradox and how it relates to the darkness of the night sky.	pp. 323–324	Key Terms: 1, 2, 3; Matching I: 1, 2, 3; Completion: 8; Self-Test: 2; Short-Answer: 3.
2. Describe the observational evidence that there was a big bang, such as expansion of the universe and cosmic background radiation.	pp. 323, 325–329	Key Terms: 4, 5, 6; Matching I: 4, 5, 6; Completion: 4, 10; Self-Test: 4, 5; Short-Answer: 4, 5, 7; Applications: 1, 2, 4.
3. Explain how curved space-time resolves the edge-center problem and suggests possible futures for the universe.	pp. 322–323, 332–333, 339–343	Key Terms: 15, 16, 17; Matching II: 4, 5, 6, 7; Self-Test: 1, 8, 10; Short-Answer: 8.
4. Recount currently accepted theories about the early history of the universe from the big bang until the beginning of the formation of galaxies.	pp. 329–331	Key Terms: 7, 8, 9, 10, 29, 30; Matching I: 7, 8, 9, 10; Matching III: 9, 10; Completion: 5, 9; Self-Test: 9.
5. Identify and define isotropy, homogeneity, and the resulting cosmological principle.	p. 332	Key Terms: 11, 12, 13, 14; Matching II: 1, 2, 3; Completion: 2, 3; Self-Test: 3; Applications: 3.

Learning Objective	Textbook	Student Guide
6. Describe observations that establish the existence of dark matter and its role in the origin of galaxies and large-scale structure.	pp. 333–335	Key Terms: 18, 19, 20; Matching II: 8, 9, 10; Self-Test: 6, 11; Applications: 6.
7. Describe the flatness and horizon problems, and explain how they are resolved by an inflationary universe.	pp. 335–336	Key Terms: 21, 22, 23; Matching III: 1, 2, 3, 4; Completion: 12, 13; Self-Test: 7; Short-Answer: 2; Applications: 5.
8. Describe observations that imply an accelerating universe and explain their impact on cosmology.	pp. 337–339	Key Terms: 25, 26, 27, 28; Matching III: 5, 6, 7, 8, 11; Completion: 1, 6, 7, 11; Short-Answer: 1, 6.

LESSON 16

Solar Systems

Checklist

For the most effective study of this lesson, complete the following activities in this sequence.

Before Viewing the Video

❑ Read the Preview, Learning Objectives, and Viewing Notes below.

❑ Read Chapter 16, "The Origin of the Solar System," pages 346–369, in the *Horizons* textbook.

What to Watch

❑ After reading the textbook chapter, watch the video for Lesson 16, *Solar Systems*.

After Viewing the Video

❑ Briefly note your answers to questions listed at the end of the Viewing Notes.

❑ Review the Summary below.

❑ Review all reading assignments for this lesson, especially the Chapter 16 summary on page 368 in *Horizons* and the Viewing Notes in this guide.

❑ Write brief answers to the review questions at the end of Chapter 16 in *Horizons*.

❑ Complete the Review Exercises below. Check your answers with the Answer Key and review when necessary.

❑ Use the Lesson Review matrix found at the end of this lesson to review and assess your knowledge of each Learning Objective.

❑ As assigned by your instructor, complete the Applications activities and any additional activities for this lesson.

Preview

In the last few lessons, you learned about the origin of the universe and how galaxies were formed. Now it's time to look inside the Milky Way Galaxy at our own solar system. In this lesson, you will study how the interstellar medium came together to give birth to the planets that orbit our sun.

You may have wondered how Earth became a planet that can sustain life or how Saturn's beautiful rings were formed. The celestial objects in our own solar system are captivating, and because they are so close we can study them in detail. You might be surprised to learn that there are probably more planets in the universe than stars. Earth-like planets are quite rare, but you'll soon discover there are other types of planets in our solar system and throughout the universe. By understanding how our solar system was formed, we can better understand how other solar systems are formed and how there can be an abundance of planets.

This lesson begins with the exploration of the theories that explain the origin of our solar system. You will learn about the characteristics of our solar system and be introduced to the story of planet building. In the next three lessons, you'll study the objects in our solar system much more closely.

CONCEPTS TO REMEMBER

- Recall from Lesson 10 that *angular momentum* is a measure of the tendency of a rotating body to continue rotating. This helps explain why all the planets in our solar system revolve around the sun in a counterclockwise direction and most rotate on their axes in the same direction (p. 191 in this guide).

- In Lesson 6, you learned that the *Doppler effect* is the change in the wavelength of radiation caused by the relative radial motion of source and observer. Astronomers can study the Doppler shift in a star's spectrum to detect motion (p. 105 in this guide).

- In Lesson 9, you learned that *interstellar medium* is the gas and dust distributed between the stars. This medium can coalesce to form solar systems and give new life to the universe (p. 168 in this guide).

- In Lesson 9, you also learned that a *protostar* is a collapsing cloud of gas and dust that is destined to become a star (p. 169 in this guide).

Learning Objectives

After you complete this lesson, you should be able to:

1. Describe the solar nebula theory, noting the observed evidence from our own solar system that supports it. *HORIZONS* TEXTBOOK PAGES 348–362.

2. Describe the observations made and conclusions drawn in the search for extrasolar planets. *HORIZONS* TEXTBOOK PAGES 362–367.

3. List the characteristic properties of the solar system, and discuss how astronomers have estimated its age. *HORIZONS* TEXTBOOK PAGES 351, 354–356.

4. Describe the characteristics of the terrestrial and Jovian planets and compare the processes that resulted in the formation of each. *HORIZONS* TEXTBOOK PAGES 350, 352–353.

5. Identify the minor members of the solar system ("space debris") and describe their characteristics, typical locations, and origins. *HORIZONS* TEXTBOOK PAGES 350–351.

6. Outline and compare theories on the formation of planetesimals and the growth of protoplanets. *HORIZONS* TEXTBOOK PAGES 357–360.

At this point, read Chapter 16, "The Origin of the Solar System," pages 346–369.

What we know is not much. What we do not know is immense.
—Pierre-Simon Laplace
French astronomer and mathematician (1749–1827)

Viewing Notes

In order to understand how our solar system was formed, astronomers can compare and contrast the planets and speculate what conditions existed to form them. Astronomers then seek observational evidence to support their theories.

The video program contains the following segments:

⚙ Our Solar System

⚙ Formation of the Solar System

⚙ Terrestrial & Jovian Planets

⚙ The Final Stage

The following information will help you better understand the video program:

According to the **solar nebula theory** of planetary formation, the planet-building process begins with a cloud of dust and gas that's swirling around. As matter falls toward the center of this cloud, the size of the cloud decreases, and begins to rotate faster. As the velocity of rotation increases, the cloud of gas and dust begins to flatten into a disk with a large central bulge where the protostar is forming. As the protostar forms, the gas and dust within the forming disk may eventually clump together to form planets through the processes of **condensation** and **accretion**. When the star becomes hot enough, it blows away its dust and gas cocoon and may leave behind planets. After the nebula is cleared, the planets can no longer accumulate great amounts of mass.

The video program introduces the term **snowline**. This is the demarcation of the inner solar system where the terrestrial planets reside from the outer solar system where the gaseous Jovian planets reside. Beyond this boundary, water is stable and could condense and form ice particles. In the outer reaches of our solar system, beyond Neptune, astronomers have discovered thousands of icy bodies that orbit the sun in a region known as the **Kuiper belt**. Pluto is one of the biggest objects in the belt. You will learn more about the Kuiper belt in Lessons 18 and 19.

To learn more about the formation of our solar system, astronomers can study other solar systems and **extrasolar planets**, which orbit stars elsewhere in the universe. These are sometimes called exosolar planets. Astronomers can detect them by observing the Doppler shift of the star's spectrum—as the planet orbits the star, it tugs on the star and causes it to move slightly. Astronomers can also use the method of photometry to detect extrasolar planets—as a planet orbits in front of a star, the amount of light from the star that makes it to Earth is diminished. Because they are the easiest to detect, only massive, Jovian-like worlds unexpectedly close to their stars have been discovered so far.

QUESTIONS TO CONSIDER

- Why is most of the motion in our solar system going in the same direction?

- How does the solar nebula theory explain how our solar system was formed?

- What are the primary differences between the terrestrial planets and the Jovian planets?

- How is the solar nebula cleared away?

- How can studying extrasolar planets help us to understand our own solar system?

Watch the video for Lesson 16, *Solar Systems.*

Key Terms and Concepts

Page references are keyed to the *Horizons* textbook.

1. **solar nebula theory (or hypothesis):** The theory that the planets formed from the same cloud of gas and dust that formed the sun. (p. 348; video lesson; objective 1)

2. **extrasolar planet:** A planet orbiting a star other than the sun; also known as exosolar planet. (p. 364; video lesson; objective 2)

3. **asteroid:** A small, rocky world. Most asteroids lie between Mars and Jupiter in the asteroid belt. (p. 350; video lesson; objective 5)

4. **comet:** One of the small, icy bodies that orbit the sun and produce tails of gas and dust when they approach the sun. (p. 351; video lesson; objective 5)

5. **terrestrial planet:** An earthlike planet; small, dense, and rocky. (p. 352; video lesson; objective 4)

6. **Jovian planet:** Jupiter-like planet with a large diameter and low density. (p. 352; video lesson; objective 4)

7. **Galilean satellites:** The four largest satellites of Jupiter, named after their discoverer, Galileo. (p. 353; objective 4)

8. **meteor:** A small bit of matter heated by friction to incandescent vapor as it falls into Earth's atmosphere. Also refers to the flash of light seen in the sky (i.e., a "shooting star"). (p. 351; objective 5)

9. **meteorite:** A piece of space debris that produces a meteor and survives its passage through the atmosphere and strikes the ground. (p. 351; objective 5)

10. **meteoroid:** A meteor in space before it enters Earth's atmosphere. The small piece of space debris. (p. 351; objective 5)

11. **half-life:** The time required for half of the atoms in a radioactive sample to decay. (p. 351; objective 3)

12. **gravitational collapse:** The process by which a forming body, such as a planet, gravitationally captures gas from its surroundings. (p. 358; objectives 4 & 6)

13. **uncompressed density:** The density that a planet would have if its gravity did not compress it. (p. 357; objectives 4 & 6)

14. **condensation sequence:** The sequence in which different materials condense from the solar nebula as we move outward from the sun. (p. 357; objectives 4 & 6)

15. **planetesimal:** One of the small bodies that formed from the solar nebula and eventually grew into protoplanets. (p. 357; video lesson; objective 6)

16. **condensation:** The growth of a particle, atom by atom, by addition of material from surrounding gas. (p. 357; video lesson; objectives 4 & 6)

17. **accretion:** The sticking together of solid particles to produce a larger particle. (p. 357; video lesson; objectives 4 & 6)

18. **protoplanet:** Massive object resulting from the coalescence of planetesimals in the solar nebula and destined to become a planet. (p. 358; video lesson; objective 6)

19. **differentiation:** The separation of planetary material according to density. (p. 358; video lesson; objectives 4 & 6)

20. **outgassing:** The release of gases from a planet's interior. (p. 359; objective 4)

21. **heat of formation:** In planetology, the heat released by infalling matter during the formation of a planetary body. (p. 359; objectives 4 & 6)

22. **radiation pressure:** The force exerted on the surface of a body by its absorption of light. Small particles floating in the solar system can be blown outward by the pressure of the sunlight. (p. 361; video lesson; objective 1)

23. **heavy bombardment:** The intense cratering during the first 0.5 billion years in the history of the solar system. (p. 361; video lesson; objective 1)

24. **ice line (or snowline):** The demarcation point between the inner and outer regions of our solar system where temperatures are low enough for water to condense out of the solar nebula and form ice. These ice particles stick to rocky planetesimals, allowing for the rapid growth into larger planets. (p. 357; objectives 4 & 6)

25. **Kuiper belt:** The collection of icy planetesimals believed to orbit in a region from just beyond Neptune out to 100 AU or more. (p. 350; video lesson; objective 5)

26. **Oort cloud:** The hypothetical source of comets. A swarm of icy bodies believed to lie in a spherical shell extending to 100,000 AU from the sun. (video lesson; objective 5)

27. **debris disks:** Cold dust disks that form around some stars. (p. 363; objective 2)

Summary

As you look into the night sky, you see mostly stars, so you might be surprised to learn that there are probably more planets in the universe than there are stars. By studying the origin of our own solar system, you will understand how there can be so many other planets in the universe.

The Great Chain of Origins

The solar nebula theory is the most widely accepted theory that explains how our solar system and other solar systems like it were formed. It proposes that the planets formed from the same cloud of gas and dust that formed the sun 5 billion years ago. In Lesson 9, you learned about the *interstellar medium*—the gas and dust that is distributed between the stars. It consists of a complex tangle of cool, dense clouds that are twisted by currents of hot, low-density gas. One of these large clouds can become unstable and something triggers its collapse—perhaps a spiral arm's shock wave or a supernova explosion. Gravity pulls the dust and gas together toward the center of the cloud causing it to take a spherical shape.

Because of the *conservation of angular momentum,* as the size of the cloud decreases, it begins to rotate faster. Conservation of angular momentum causes the collapsing cloud to spin more rapidly. This increase in rotation causes the matter to flatten into a disk. As the cloud continues to collapse, the central portion of the disk becomes hot and a protostar begins to form. When temperatures are sufficiently high, hydrogen within the protostar can begin to fuse and a star is born.

The whirling dust and gas in the disk surrounding the protostar begin to condense in various locations. Gravitational and other forces pull the gas and dust together to form planets—they are formed along with the star they orbit. When the star becomes hot enough, it blows away its dust and gas cocoon and may leave behind planets. The angular momentum of the original solar nebula comes to reside in the orbital motion of the planets and the spinning of the star.

If this solar nebula theory were correct, this would lead us to believe that most stars in the universe have planets that formed in the disks that surrounded them. To confirm this theory, astronomers look for evidence to support it. Astronomers have noticed that disks around young stars are common and they have discovered two types of disks: low density and high density. Low-density disks are produced by dust that came from many collisions among comets, asteroids, and other celestial objects; they provide evidence that planetary systems have already been formed. High-density disks are more compact and may be where planets are currently forming.

Planets orbiting other stars are called **extrasolar planets** (sometimes called exosolar planets), and we can study them to learn how our own solar system evolved. These planets are quite faint compared to their stars and are therefore almost impossible to be seen directly. Astronomers can use two common techniques to determine if a star has planets revolving around it: first, they can determine if the star wobbles as the planets orbit it. When planets orbit around a common center of mass, the gravitational forces tug on the star and the star moves slightly in its orbit. This can be detected by the Doppler shifts in the star's spectrum. Second, they can detect any change in the brightness of the star. As the planets cross in front of the star relative to Earth, the decrease in brightness can be detected.

Astronomers have found mostly high-mass planets with small orbits because they are the easiest to detect. These types of planets have a greater tug on their stars, making the Doppler shift in the star's spectrum greater. Stars with shorter orbital periods are easier to detect than ones with longer periods because they can be studied in a shorter period of time.

A Survey of the Solar System

There is a common motion in our disk-shaped solar system. This motion of the planets around the sun, the direction that each planet in our solar system spins, and the degree to which the equators are tipped all seem to be related to the disk shape of our solar system.

Most of the planets spin in the same counterclockwise direction. Nearly all of the moons in the solar system revolve around their planets in this same counterclockwise direction.

The planets in our solar system have orbits that lie nearly in the same plane; all the planets' orbits lie within 17.2° of Earth's orbital plane. The degree to which the planets' equators are tipped is also related to the disk-shape of our solar system. The equator of the sun is inclined 7.25° to Earth's orbit and most of the other planets' equators are tipped less than 30° relative to the plane of Earth's orbit. Uranus and Pluto are the exceptions. Uranus and Pluto rotate on their sides with their equators nearly perpendicular to their orbits around the sun. This may have been caused by a catastrophic event, such as an off-center collision with a large **planetesimal**—an object destined to become a planet.

Most of the planets rotate in the same counterclockwise direction, except for Venus. There are two theories that can explain why Venus spins in a clockwise direction. An evolutionary theory suggests that the sun produced tides—or gravitational forces—in the thick atmosphere of Venus and eventually reversed its rotation. The second theory is a catastrophic one—its rotation was altered by an off-center impact.

For the purposes of organization and comparison, astronomers have divided the planets into two groups. The inner planets are called terrestrial (Earthlike) and the outer planets consist of Pluto and the four Jovian (Jupiterlike) planets. Pluto, the farthest planet from the sun, is neither terrestrial nor Jovian and may not be considered a planet at all. It is one of the largest objects in the **Kuiper**

belt—a region beyond Neptune where roughly a thousand icy bodies orbit the sun. You will study this further in Lessons 18 and 19.

Terrestrial planets are similar to Earth—they are small, dense, rocky planets with little or no atmosphere. They are the four planets closest to the sun: Mercury, Venus, Earth, and Mars. Of the terrestrial planets, Earth is the largest. You will learn more about these planets in Lesson 17.

Jovian planets are those that are similar to Jupiter—they are low-density planets with thick atmospheres and liquid interiors. They are the next four planets farther from the sun: Jupiter, Saturn, Uranus, and Neptune. The Jovian planets are much more massive than Earth; for example, Jupiter is more than 300 Earth masses. All of these planets have ring systems: Saturn's rings are made of ice particles, while the less-noticeable rings around Jupiter, Uranus, and Neptune are made of rocky particles. Large systems of satellites revolve around the Jovian planets; the Galilean satellites—four large moons that were discovered by Galileo in 1610—orbit Jupiter. You will learn more about the Jovian planets in Lesson 18.

The delineation between the rocky terrestrial planets and massive Jovian planets offers evidence to support the solar nebula theory. The sun and the planets can also give us some clues about the origin of our solar system, but they have evolved since they were first formed. Astronomers can look to other celestial objects to gain more information to support their theory. They can look to space debris—asteroids, comets, and meteoroids—to help them unravel the history of our solar system.

Asteroids are small, rocky, irregularly shaped objects that are sometimes called minor planets. Most can be found in the asteroid belt between Mars and Jupiter. Modern astronomers believe that asteroids in this belt are the debris of a planet that failed to form about 2.8 AU from the sun. More than 2,000 asteroids have been identified, 200 of which are larger than 100 km (60 mi) in diameter. Although tens of thousand of asteroids are bigger than 10 km (6 mi), the majority of them are small.

Comets are small objects that orbit the sun and produce tails of gas and dust when they approach the sun. The nucleus of a comet is believed to be made of rock, dust, and ice—the remains of the solar nebula. They are sometimes likened to a dirty snowball, but since they are primarily rock and dust, they are more like an icy mud ball. The nucleus is only a few tens of kilometers in diameter and remains frozen and inactive when it is far from the sun. As the comet approaches the sun, the heat from the sun vaporizes the comet's nucleus and the solar wind pushes the gas and dust away forming the long tail. The tail of the comet can be longer than 1 AU and generally points away from the sun. It appears that most comets originate in one of two places: the Kuiper belt just beyond Neptune or the **Oort cloud**—a spherical cloud of icy bodies believed to extend from 10,000 to 100,000 AU from the sun.

Meteors are commonly called "shooting stars" but they are not stars at all. They are the streaks of light that are produced from small bits of rock and metal falling into Earth's atmosphere. As they fall, they become vaporized at about 80 km (50 mi) from Earth's surface due to friction with the air. While this rocky object is still in space, it is called a **meteoroid**, and if any bit of it makes its way to Earth without being vaporized, it becomes known as a **meteorite**. Meteoroids that weigh less than one gram produce nearly all of the meteors that you see in the sky. Meteoroids must be much larger to make their way through Earth's atmosphere, but by the time they reach the surface, the meteorites are typically smaller than pebbles. Impacts are common on all of the planets, and craters can be seen on the planets that have solid surfaces—evidence that they were once bombarded with enormous meteorites.

Astronomers can gauge the age of the solar system by analyzing the composition of the celestial objects that they have thus far been able to obtain: from Earth, the moon, Mars, and meteorites. When a rock solidifies, it is made up of known percentages of chemical elements. Those elements will decay into other elements, known as daughter elements. This radioactive decay can give us clues as to how old these objects are. The time it takes for half of the radioactive element's atoms to decay into the daughter element is known as its **half-life**.

For example, an isotope of uranium (^{238}U) will decay into an isotope of lead (^{206}Pb). The half-life of uranium—the time it takes half of its atoms to decay into lead—is 4.5 billion years. Because astronomers know the percentages of elements a rock contains when it solidifies, they know how much uranium a rock should have contained when it was first formed. They can measure the present abundance of uranium and lead and determine the rock's age in a process known as uranium-lead dating. Astronomers can use other radioactive elements in this process of radioactive dating; they can use potassium-argon dating and rubidium-strontium dating to find the age of mineral samples.

By using this method of radioactive dating, astronomers have been able to determine that the age of the oldest rock found on Earth is about 4.3 billion years old. They have found rocks on the moon that are 4.48 billion years old, while minerals from Mars have been identified as being 4.6 billion years old. It is widely accepted that the solar system is about 4.56 billion years old.

The Story of Planet Building

Evidence suggests that the solar nebula from which our solar system was formed was a fragment of an interstellar gas cloud. By studying the solar spectrum, astronomers can deduce that the solar nebula was comprised of mostly hydrogen, a quarter of its mass was helium, and about 2 percent was made of the heavier elements. We have to wonder how the gas and dust in the solar nebula was able to form planets.

The gas and dust swirled around the solar nebula as the solar nebula condensed. These specks of matter gently collided with one another and grew from their microscopic size by condensation and accretion. In the process of **condensation**, the solid particle gains one atom at a time from the surrounding gas. In the process known as **accretion**, solid particles stick together to form bigger particles. These particles continued to join other particles, and when they reached about 1 km in size, they could be considered planetesimals. As they grew, they underwent a series of changes and continued to grow at increasing rates.

As the planetesimals orbited around in the solar nebula, they gently collided with other planetesimals and fused together. As they grew larger, their gravitational fields increased and were able to sweep up more dust and debris at increasing rates. Some of these planetesimals continued to gain mass until they formed **protoplanets**—massive objects that would eventually become planets.

Planets can be formed in one of two ways; the Jovian planets formed differently than the terrestrial planets. The Jovian planets began aggregating pieces of rock and ice. Once the planets gained enough mass, they could capture gas directly from the solar nebula in a process called **gravitational collapse**. These planets became hydrogen-rich planets with very low density. By contrast, the terrestrial planets have very low masses and didn't have enough gravity to capture gas from the solar nebula. They contain little hydrogen and helium and are composed of the heavier elements from the solar nebula.

The difference between the Jovian and the terrestrial planets is the result of the different way the gases condensed into solids in the different parts of the solar nebula. The kinds of matter that condensed in a particular region of the solar nebula depended on the temperature of the gas there. The **condensation sequence** illustrates the sequence in which the different materials condense from the gas as you move away from the sun (see Table 16–3 on p. 357 in the *Horizons* textbook). Additionally, the solar nebula did not remain the same temperature over time. It may have grown progressively cooler—as it cooled, lower-density materials were able to condense in regions where only metals once solidified.

The region closest to the sun would have been the hottest (about 1500 K) and compounds with high melting points like pure metals and metal oxides would have condensed to eventually form the planet Mercury. A little farther away from the sun, the nebula would have been between 680 K and 1200 K. In this region, silicates, feldspars, and troilite could have condensed to eventually form Venus, Earth, and Mars. The demarcation point between the inner and outer regions of our solar system is known as the **ice line (or snowline)**. In the cold, outer regions of the solar nebula, the sun's radiation vaporized fewer materials and they were able to condense. In addition to higher-density materials, low-density materials such as water, ammonia, and methane were also able to condense and eventually form the Jovian planets.

Astronomers aren't certain how the planets in our solar system were formed, but we can study the formation of Earth to deduce how the terrestrial planets were formed. One model of planet building suggests that all of the terrestrial planetesimals had the same chemical composition—mainly rock and metal. They accumulated material to form protoplanets that were the same chemical composition throughout. Evidence suggests that the solar nebula contained short-lived radioactive elements that were captured by the young planets. As these elements decayed, they produced enough heat to cause the interior to **differentiate**—the materials separated according to their densities. Iron, nickel, and other heavy metals settled in the core and the lighter silicates floated to the surface to form Earth's crust. The planet never gained enough mass to grow by gravitational collapse, but the gravitational forces would have captured some of the gases in the solar nebula to form a primitive atmosphere. This primitive atmosphere was eventually driven off and a new atmosphere was formed from the process called **outgassing**—the atmosphere was created from the planet's interior.

This simplified model of planet building assumes that the solar nebula didn't have a drastic change in temperature during the formation of the planets—but what if it had? A modified version of the simple model proposes that if the nebula cooled during planet formation, the planetesimals would have accumulated material that varied in chemical composition. In the hot nebula, the first materials to condense would have been the heavy metals. By the process of accretion, the protoplanets would have collected these metals to form their cores. As the nebula cooled, silicates formed; and the protoplanets captured them to form their mantles.

The simple model of planet building also assumes that the planets grew slowly. The modified version of the simple model suggests that the planets may have grown so rapidly that the heat released by infalling particles—the **heat of formation**—didn't have time to escape. The planet would have differentiated as it formed—not many years later as the simple model suggests. Jupiter grew hot enough to glow with a luminosity of about 1 percent of the present-day sun, but it was never hot enough to generate its own energy by nuclear fusion. Even today Jupiter and Saturn radiate more heat than they absorb from the sun— evidence that they are still cooling.

By studying the formation of Earth, we can learn how the terrestrial planets were formed. Evidence suggests that the Jovian planets were formed in much of the same way, but they grew much faster. The outer solar nebula contained solid bits of metal and silicates as well as a lot of ice. The Jovian planets quickly accumulated these materials and as they drew in gases from the solar nebula, they were able to grow by gravitational collapse. If this model were correct, the Jovian planets would have taken about 10 million years to form—by the time the solar nebula was blown away from the sun. Most of the Jovian planet's mass was accumulated before the sun reached high enough temperatures to dissipate and blow away the remaining gases of the solar nebula. The terrestrial planets, composed of metals and silicates that have a higher melting point, continued to grow as the solar nebula cooled.

Two observations have made astronomers revisit their theories regarding the formation of the Jovian planets. First, astronomers have found that Jovian-like planets are common in other solar systems. The second observation doesn't seem to coincide with the first: gas and dust disks around newborn stars don't last very long—perhaps only 100,000 to 7 million years. They may be evaporated by the intense radiation from hot stars within the nebula or stripped away by the gravitational forces of other stars. This doesn't seem long enough to form a Jovian planet, but since Jovian planets are common, there must be another explanation.

Based on mathematical models of the solar nebula, astronomers have revised their model of Jovian planet formation. The solar nebula may have become unstable and the massive planets could have quickly formed from the gas in the nebula rather than first forming a dense core by accretion. This revised model may help explain how Uranus and Neptune were formed: they may have formed closer to the sun and drifted outwards by gravitational interactions with other planets.

The solar nebula eventually cleared away and the planets could no longer gain vast amounts of mass. Four internal processes eventually destroyed the nebula. The first process was that of **radiation pressure**—the radiation from the luminous sun pushed low-mass specks of dust and atoms of gas out of the solar system. The second was a series of surges in the solar wind—the flow of ionized hydrogen and other atoms—which pushed gas and dust out of the solar system. The third process was the planets' sweeping up of the space debris with their gravitational fields. At this time, there may have been steady stream of meteorites crashing into the planets and moon in a period of **heavy bombardment**. The fourth process was the ejection of material from the solar system. Small objects such as planetesimals may have come close to the high-mass Jovian planets. They gained energy from the planets' gravitational fields and were thrown out of the solar system. These four processes, in addition to the intense radiation and gravitation from nearby stars, caused the solar nebula to dissipate.

This concludes our survey of our solar system and introduction to how it was formed. In the next two lessons, you will have the opportunity to examine the terrestrial and the outer planets much more closely. In Lesson 19, you will study comets, meteors, and asteroids. In this lesson, you've discovered the abundance of planets in the universe, and in Lesson 20, you'll examine the possibility of life on other planets.

Review Exercises

Matching I

Match each term with the appropriate definition or description.

1. _____ solar nebula theory	8. _____ comet
2. _____ gravitational collapse	9. _____ uncompressed density
3. _____ terrestrial planet	10. _____ meteor
4. _____ Jovian planet	11. _____ meteoroid
5. _____ extrasolar planet	12. _____ meteorite
6. _____ Galilean satellites	13. _____ debris disk
7. _____ asteroids	

a. A planet with a large diameter and low density.
b. An object that survives its passage through the atmosphere and strikes the ground.
c. A rock in space before it enters Earth's atmosphere.
d. The theory that the planets and the sun formed from the same cloud of gas and dust.
e. A small, dense, rocky planet.
f. The density the planets would have if gravity did not compress them.
g. A small, icy body that orbits the sun.
h. The process that can only occur when a planet has grown to about 15 Earth masses, thus giving it enough gravity to begin capturing gas directly from the solar nebula.
i. The four largest satellites of Jupiter.
j. A streak of light in the sky.
k. Minor planets, most of which orbit the sun between Mars and Jupiter.
l. A planet orbiting a distant star (not the sun).
m. A cold dust disk.

Matching II

Match each term with the appropriate definition or description.

1. _____ condensation sequence	8. _____ heat of formation
2. _____ ice line	9. _____ radiation pressure
3. _____ planetesimal	10. _____ outgassing
4. _____ differentiation	11. _____ heavy bombardment
5. _____ accretion	12. _____ Kuiper belt object
6. _____ protoplanet	13. _____ Oort cloud
7. _____ half-life	

    a. The energy released by infalling of matter during formation of a planetary body.

    b. The sticking together of solid particles to produce a larger particle.

    c. The order in which different materials condense from a gas in relationship to distance from the sun.

    d. The time it takes for half the atoms of a radioactive element in a sample of material to decay.

    e. Small bodies that formed from the solar nebula and eventually grew into protoplanets.

    f. The force exerted on the surface of a body by its absorption of light.

    g. Separation of planetary material according to density.

    h. The creation of a planet's atmosphere from the planet's interior.

    i. The process that cratered all of the solid worlds as the last debris in the solar nebula was swept up.

    j. A massive object resulting from the coalescence of planetesimals in the solar nebula.

    k. The demarcation point between the inner and outer regions of our solar system where temperatures are low enough for water to condense out of the solar nebula and form ice particles.

    l. An icy planetesimal believed to orbit in a region from just beyond Neptune.

    m. A spherical shell of icy bodies some 100,000 AU from the sun that is believed to be the source of some comets.

Completion

Fill each blank in the sentences below with the most appropriate term from the list of completion answers that follow. A term may be used once, more than once, or not at all. Check your answers with the Answer Key and review when necessary.

asteroids	ice	planetesimals
comets	Jovian	protoplanets
condensation	moons	solar nebula
dirty snowball	nitrogen	terrestrial
dust	oxygen	water

1. The adding of material an atom at a time is the process of
   _____.

2. The _____ model is used to describe the physical structure of _____.

3. The pressure of sunlight was one of the forces that cleared the
   _____.

4. The _____ planets have lower average densities than the _____ planets.

5. Planets begin to form when _____ join to make larger objects called _____.

Self-Test

Select the best answer.

1. Evidence of cratering on the moon and the other terrestrial planets suggests that Earth
   a. never experienced a similar kind of cratering.
   b. experienced a similar kind of cratering earlier in its history.
   c. will experience a similar kind of cratering in the future.
   d. will never experience a similar kind of cratering.

2. Because of Earth's low mass, our planet's atmosphere
   a. contains mostly gases that are similar to those on the Jovian planets.
   b. contains the same percentages of gases that exist throughout the solar system.
   c. contains very little hydrogen and helium.
   d. is composed primarily of carbon dioxide.

3. The orbits of the planets
    a. all lie in approximately the same plane.
    b. lie perpendicular to Earth's orbit.
    c. are perfectly circular.
    d. are all highly inclined to the ecliptic.

4. Each of the planets in our solar system orbits the sun
    a. in circular orbits.
    b. in the direction opposite to which they rotate, clockwise as seen
       from the north (above).
    c. in the same direction, counterclockwise as seen from above (north).
    d. opposite in direction to the orbital motion of Earth.

5. The orbits of the planets lie mainly in the same plane because the planets
    a. formed from a large thin disk of material surrounding the new sun.
    b. were captured by an already-formed sun.
    c. were pulled out of the sun by the gravitational influences of a star
       passing nearby.
    d. formed at a variety of locations but were pulled into a single plane
       by the gravitational influence of the sun.

6. The solar system has relatively little gas and dust between the planets.
   The lack of this material is probably the result of the
    a. "vacuuming up" of the material by the gravitational attraction of the
       planets.
    b. condensation properties of the Jovian planets.
    c. radiation pressure from the sun.
    d. accretion processes of the terrestrial planets.

7. The solar nebula theory describes how the planets formed from
    a. the capture of planetesimals in orbit around a nearby star.
    b. the condensation of a cloud of pure hydrogen and helium.
    c. the accretion of high-density matter in the vicinity of the newly
       forming sun.
    d. a disk-shaped cloud of gas and dust around the newly forming sun.

8. The terrestrial planets nearest the sun are
    a. large and composed of ice and hydrogen and helium gas.
    b. small, solid, and rocky.
    c. irregular in shape and composed mostly of ice and rock.
    d. spherical in shape with solid crusts and gaseous interiors.

9. The atmospheres of the Jovian planets are made primarily of
   a. hydrogen and helium.
   b. methane and ammonia.
   c. liquid hydrogen.
   d. icy rock in solid form.

10. Most asteroids revolve around the sun between the orbits of
    a. Jupiter and Saturn.
    b. Earth and Mars.
    c. Saturn and Uranus.
    d. Mars and Jupiter.

11. Objects made mostly of ice that form a large gaseous "tail" that is directed away from the sun at all times are called
    a. asteroids.
    b. meteors.
    c. meteorites.
    d. comets.

12. Formation of planetesimals by condensation refers to the process of
    a. meteorites colliding and sticking together.
    b. asteroids colliding and breaking apart.
    c. matter building one atom at a time.
    d. asteroid collisions and adhesion.

13. Solid particles sticking together to form larger particles describes the process of
    a. accretion.
    b. adhesion.
    c. condensation.
    d. adherence.

14. The separation of low-density and high-density materials in a planet during the early stages of its formation is referred to as
    a. accretion.
    b. differentiation.
    c. homogeneous condensation.
    d. heterogeneous condensation.

15. The interiors of Jupiter and Saturn consist mostly of
    a. icy rock in solid form.
    b. solid hydrogen and helium.
    c. liquid hydrogen.
    d. methane and ammonia.

16. Low-density and cold dust disks have been found around many stars. These disks are believed to be regions where planets have formed or are in the process of forming. The disks are most readily detected by
    a. infrared telescopes.
    b. X-ray sensitive satellite detectors.
    c. optical telescopes.
    d. orbiting ultraviolet telescopes.

17. A planet that has formed around the star 51 Pegasi has been detected by
    a. direct observation in an optical telescope.
    b. infrared sensors on the Hubble Telescope.
    c. Doppler shifts in the star's spectrum as the planet orbits, causing the star to wobble back and forth.
    d. no such planet has been observed.

18. The source of comets is
    a. the Kuiper belt.
    b. the asteroid belt.
    c. the Oort cloud.
    d. both a and c.

19. Our solar system may have a cold dust disk similar to that found around the star Beta Pictoris. Astronomers believe that this disk is located in the _____ of objects.
    a. asteroid belt
    b. Oort cloud
    c. Kuiper belt
    d. Jovian class

Short-Answer Questions

1. The sun and most of the planets revolve and rotate in the same direction. Furthermore, the planetary orbits (except for Pluto) lie in nearly the same plane. How do these observations support the solar nebula hypothesis for the formation of the solar system?

2. Explain how a radioactive isotope can be used to determine the age of a sample of rock or other substance.

3. What is the difference between asteroids and comets? Refer to both their orbital and physical characteristics.

4. What is differentiation and what role does it play in the growth of protoplanets?

5. Briefly describe the evidence that suggests the existence of extrasolar/exosolar planets.

6. What is the difference between the process of condensation and the process of accretion?

Applications

1. If the half-life of a radioactive isotope is 3.5 minutes, how many grams of a 200-gram sample would be left after 24.5 minutes?

2. The half-life of radioactive carbon (^{14}C) is 5,370 years. If a sample of organic material contains 100 grams of ^{14}C and it is estimated that the

organism originally contained 1,600 grams, approximately how old is the sample?

3. Using data from Table A-6 on 479 and Table A-10 on page 482 in the *Horizons* textbook, how long does it take for a beam of light from the sun to cover the distance to Pluto?

4. Build a scale model of the solar system, using data from Table A-10 on page 482 of the *Horizons* textbook. Set the diameter of Earth (12,656 km) to equal 1 inch. Draw and cut out of a piece of construction paper a circle of diameter 1 inch. Calculate the diameter of the planet Jupiter on this scale. Cut out a circle equal to that diameter from a piece of construction paper. You may be surprised at the comparison of Earth and Jupiter.

5. Again, using data from Table A-10, "Properties of the Planets" on page 482 of the *Horizons* textbook, build a scale model of the solar system. Set the distance between the Earth and the sun (149.6×10^6 km) to equal 1 foot. Cut a piece of adding machine tape to equal the distance between Earth and the sun. Calculate the distance to the planet Jupiter on this scale. Cut a piece of adding machine tape equal to the distance between Jupiter and the sun. You may be surprised at the comparison of the Earth-sun distance and the Jupiter-sun distance.

6. Just for fun, cut a circle out of construction paper that represents the sun on the same scale that you used for question 4 above.

Answer Key

Matching I

1. d (p. 348; video lesson; objective 1)
2. h (p. 358; objectives 4 & 6)
3. e (p. 352; video lesson; objective 4)
4. a (p. 352; video lesson; objective 4)
5. l (p. 364; video lesson; objective 2)
6. i (p. 353; objective 4)
7. k (p. 350; video lesson; objective 5)
8. g (p. 351; video lesson; objective 5)
9. f (p. 357; objectives 4 & 6)
10. j (p. 351; objective 5)
11. c (p. 351; objective 5)
12. b (p. 351; objective 5)
13. m (p. 363; objective 2)

Matching II

1. c (p. 357; objectives 4 & 6)
2. k (p. 357; objectives 4 & 6)
3. e (p. 357; video lesson; objective 6)
4. g (p. 358; video lesson; objectives 4 & 6)
5. b (p. 357; video lesson; objectives 4 & 6)
6. j (p. 358; video lesson; objective 6)
7. d (p. 351; objective 3)
8. a (p. 359; objectives 4 & 6)
9. f (p. 361; video lesson; objective 1)
10. h (p. 359; objective 4)
11. i (p. 361; video lesson; objective 1)
12. l (p. 350; video lesson; objective 5)
13. m (video lesson; objective 5)

Completion

1. condensation (p. 357; video lesson; objectives 4 & 6)
2. dirty snowball, comets (p. 351; video lesson; objective 5)
3. solar nebula (p. 348; video lesson; objective 1)

4. Jovian, terrestrial (p. 352; video lesson; objective 4)

5. planetesimals, protoplanets (pp. 357–358; objective 6)

Self-Test

1. b (p. 361; objective 4)

2. c (p. 359; objective 4)

3. a (p. 350; objective 1)

4. c (p. 350; objective 1)

5. a (p. 350; objectives 1 & 3)

6. c (p. 361; video lesson; objective 1)

7. d (p. 348; video lesson; objective 1)

8. b (pp. 352–353; video lesson; objective 4)

9. a (pp. 359–360; objective 4)

10. d (p. 350; video lesson; objective 5)

11. d (p. 351; objective 5)

12. c (p. 357; objectives 4 & 6)

13. a (p. 357; video lesson; objectives 4 & 6)

14. b (p. 358; video lesson; objectives 4 & 6)

15. c (p. 353; objective 4)

16. a (p. 363; objective 2)

17. c (pp. 364, 366; objective 2)

18. d (p. 350; video lesson; objective 5)

19. c (p. 363; objective 5)

Short-Answer Questions

1. Once the region of the solar nebula that was to eventually become the sun and solar system began to contract, it began to spin a little more rapidly. As gravity worked to pull in more matter into the forming sun, the sun and disk spun even faster. (Very much like an ice skater who draws in her arms and spins faster.) The entire nebula was spinning in the same direction and beginning to flatten. Within the nebula there were smaller regions of higher density that also began to contract, eventually becoming the planets. The rotations of these smaller regions (future planets) were in the same direction. Since the entire nebula was spinning in the same direction, these regions orbited around the sun in the same direction. Eventually, the nebula formed a flattened disk constraining the orbits of the newly forming planets into a common plane. These planets were all moving and spinning in the same

direction, counterclockwise when viewed from the north (above the north pole of Earth). (pp. 348–350; objective 1)

2. Radioactive isotopes change to totally different isotopes as a result of nuclear processes (events that occur within the nucleus of the atom) in a time called the half-life. This time specifies how long it takes for half of the number of atoms in a sample to change to a different isotope. For example, uranium has a half-life of 4.5 billion years. If we had a sample of pure uranium, half of it would decay into an isotope of lead in 4.5 billion years. Scientists can compare the amount of uranium in a sample with the daughter elements the decay produces and determine how long the decay process has continued. Knowing the amount of isotope that began the process and knowing how much remains, scientists can determine the age of the sample. (pp. 351, 354–355; objective 3)

3. Asteroids are rocky objects, most of which circle the sun between the orbits of Mars and Jupiter. Most asteroids are irregular in shape and have multiple craters. Comets are made up of about 50 percent dirt and dust and about 50 percent ice. Their orbits are highly elongated bringing some closer to the sun than Mercury and farther from the sun than the orbit of Uranus. They are relatively fragile objects, some having been observed to break apart as they approached the sun. Viewed in a telescope, comets appear as a fuzzy ball of light with a "tail" of gas and dust. Asteroids, on the other hand, appear as tiny stars revealing themselves only after several nights of observation having moved against the background stars. (pp. 350–351; video lesson; objective 5)

4. Differentiation is relevant to the formation of the terrestrial planets. Differentiation is the separation of material according to density. As the interior of the newly forming planet heats and melts, material of high density (iron and nickel) sinks towards the center of the planet forming a core. At the same time, material of lower density (silicates) rises to the surface, cools, and forms a crust. (pp. 358–359; objectives 4 & 6)

5. The strongest evidence for the existence of an extrasolar/exosolar planet is the observation of the wobble in the motion of the star that such a planet produces as it orbits. This wobble, detected by Doppler shift measurements, is the result of the gravitational tug of war between the planet and its star. Further evidence is found in the presence of gas and dust disks around many stars. Although this observation alone does not provide information about the size of the planet, it does indicate that a solar system may have formed or is in the process of forming. (pp. 364–367; objective 2)

6. Condensation is the growth of particles one atom at a time. It is limited to the smallest particles, dust grains that accumulate matter as the result of atoms within the solar nebula colliding with and sticking to the grains. This increases their size rapidly. However, when the grains become larger, condensation is less effective and accretion becomes

the dominant process in the growth of these particles. Accretion is the sticking together of solid particles. Gravity is still too small to hold these particles together, however, static electricity generated by the passage of these particles through the solar nebula and sticky coating of carbon based materials would have aided in the growth of these particles until they become large enough to be called planetesimals, about a kilometer in diameter. (pp. 356–358; objectives 4 & 6)

Applications

1. The half-life is the time it takes for the radioactive sample to decrease by half. If the sample contains 200 grams to start, then after the first half life (3.5 minutes) there would be 100 grams remaining. After the second half-life (7.0 minutes), 50 grams. After the third (10.5 minutes), 25 grams; after the fourth (14.0 minutes), 12.5 grams; after the fifth (17.5 minutes), 6.25 grams; and after the sixth (21.0 minutes), 3.125 grams; and after the seventh half-life (24.5 minutes), 1.5625 grams. (pp. 351, 354–355; objective 3)

2. Going backward using each half-life to increase the amount of radioactive carbon 14 until the initial amount is reached. Starting with 100 grams remaining, one half-life earlier (5,370 years) there would have been 200 grams. Two half-lives earlier (10,740 years), there would have been 400 grams. Continuing with this process, three half-lives ago (16,110 years) there would have been 800 grams; four half-lives ago (21,480 years) there would have been 1,600 grams. The sample would be nearly 21,500 years old. (pp. 351, 354–355; objective 3)

3. The velocity of light is 2.99×10^8 m/s and the distance from the sun to Pluto is $5,900 \times 10^6$ km or 5.9×10^{12} m. (pp. 479, 482; objective 3)

   5.900×10^{12} m $\div 2.99 \times 10^8$ m/s $= 19,732$ seconds
   19,732 seconds $\div$ 3,600 seconds/hour $= 7.59$ hours

4. You will need to construct a circle 11.3 inches in diameter.
   $$\frac{\textit{Diameter of Jupiter}}{\textit{Diameter of Earth}} = \frac{142,988 \text{ km}}{12,656 \text{ km}} = 11.29$$
   (pp. 479, 482; objective 3)

5. You will need to cut a piece of adding machine tape 5.2 feet long.
   $$\frac{\textit{Jupiter} - sun - \textit{Distance}}{\textit{Earth} - sun - \textit{Distance}} = \frac{778 \times 10^6 \text{ km}}{149.6 \times 10^6 \text{ km}} = 5.2$$
   (pp. 479, 482; objective 3)

6. You will need to construct a circle approximately 110 inches in diameter.
   $$\frac{\textit{Diameter of sun}}{\textit{Diameter of Earth}} = \frac{1,391,980 \text{ km}}{12,656 \text{ km}} = 109.99$$
   (pp. 479, 482; objective 3)

Lesson Review

Lesson 16: Solar Systems

PLEASE NOTE: Use this matrix to guide your study and achieve the learning objectives of this lesson. It will also help you to view the video, which defines and demonstrates important concepts and principles as they relate to everyday life and actual case studies.

Learning Objective	Textbook	Student Guide
1. Describe the solar nebula theory, noting the observed evidence from our own solar system that supports it.	pp. 348–350, 356–362	Key Terms: 1, 22, 23; Matching I: 1; Completion: 3; Self-Test: 3, 4, 5, 6, 7; Short-Answer: 1.
2. Describe the observations made and conclusions drawn in the search for extrasolar planets.	pp. 362–367	Key Terms: 2, 27; Matching I: 5, 13; Self-Test: 16, 17; Short-Answer: 5.
3. List the characteristic properties of the solar system, and discuss how astronomers have estimated its age.	pp. 351, 354–356	Key Terms: 11; Matching II: 7; Self-Test: 5; Short-Answer: 2; Applications: 1, 2, 3, 4, 5, 6.
4. Describe the characteristics of the terrestrial and Jovian planets and compare the processes that resulted in the formation of each.	pp. 350, 352–353	Key Terms: 5, 6, 7, 12, 13, 14, 16, 17, 19, 20, 21, 24; Matching I: 2, 3, 4, 6, 9; Matching II: 1, 2, 4, 5, 8, 10; Completion: 1, 4; Self-Test: 1, 2, 8, 9, 12, 13, 14, 15; Short-Answer: 4.

Learning Objective	Textbook	Student Guide
5. Identify the minor members of the solar system ("space debris") and describe their characteristics and typical locations.	pp. 350–351	Key Terms: 3, 4, 8, 9, 10, 25, 26; Matching I: 7, 8, 10, 11, 12; Matching II: 2, 12, 13; Completion: 2; Self-Test: 10, 11, 18, 19; Short-Answer: 3.
6. Outline and compare theories on the formation of planetesimals and the growth of protoplanets.	pp. 357–360	Key Terms: 12, 13, 14, 15, 16, 17, 18, 19, 21, 24; Matching I: 2, 9; Matching II: 1, 2, 3, 4, 5, 6, 8; Completion: 5; Self-Test: 12, 13, 14; Short-Answer: 4, 6.

Notes:

LESSON
17

The Terrestrial Planets

Checklist

For the most effective study of this lesson, complete the following activities in this sequence.

Before Viewing the Video

❑ Read the Preview, Learning Objectives, and Viewing Notes below.

❑ Read Chapter 17, "Comparative Planetology of the Terrestrial Planets," pages 370–405, in the *Horizons* textbook.

What to Watch

❑ After reading the textbook chapter, watch the video for Lesson 17, *The Terrestrial Planets*.

After Viewing the Video

❑ Briefly note your answers to questions listed at the end of the Viewing Notes.

❑ Review the Summary below.

❑ Review all reading assignments for this lesson, especially the Chapter 17 summary on page 404 in *Horizons* and the Viewing Notes in this guide.

❑ Write brief answers to the review questions at the end of Chapter 17 in *Horizons*.

❑ Complete the Review Exercises below. Check your answers with the Answer Key and review when necessary.

❑ Use the Lesson Review matrix found at the end of this lesson to review and assess your knowledge of each Learning Objective.

❑ As assigned by your instructor, complete the Applications activities and any additional activities for this lesson.

Preview

In Lesson 16, you surveyed our solar system. You discovered how the terrestrial planets differ from the Jovian planets and you read a preview of how planets are formed. In order to further our study of the formation and evolution of the terrestrial planets, we can compare their characteristics to those of Earth. Earth is currently the most geologically active planet in our solar system, but you'll soon discover that there is interior heat, surface erosion, and recent volcanic activity on other terrestrial planets.

You'll begin to appreciate why Earth is unique as you compare and contrast it to the other terrestrial planets. Earth is the only planet currently with liquid water on its surface, but there is strong evidence that water once flowed on Mars and may still exist as ice under the surface and polar ice caps. While Earth is massive enough to hold on to the thick atmosphere that protects us from the sun's intense rays, we can compare it to Mercury to see how the lack of an atmosphere can affect a planet. You'll also discover that Earth is not the only planet that is impacted by global warming—Venus' atmosphere contains so much carbon dioxide that heat cannot escape into space, making its surface hot enough to melt lead.

In the future lessons, you'll continue your study of the planets and space debris in our solar system. In Lesson 18, you'll learn about the Jovian planets in greater detail, and in Lesson 19 you'll study asteroids, comets, meteorites, and the planet Pluto.

CONCEPTS TO REMEMBER

- Recall from Lesson 16 that *differentiation* is the separation of planetary material according to density. As a planet cools, the dense materials sink to form a core and the lighter materials rise to form the crust (p. 314 in this guide).

- In Lesson 16, you also learned that *heavy bombardment* is the period of intense cratering that occurred during the first half-billion years in the history of the solar system (p. 314 in this guide).

- In Lesson 7, you learned that the *dynamo effect* is the process by which a rotating, convecting body of conducting matter can generate a magnetic field. You also learned that the dynamo effect inside the sun or within the sun's interior causes its magnetic field. A similar process happens inside Earth's outer core (p. 127 in this guide).

Learning Objectives

After you complete this lesson, you should be able to:

1. Describe the four stages of planetary development. HORIZONS TEXTBOOK PAGES 373–374.

2. Identify the observable evidence that supports our model of Earth's interior, its dynamo, and plate tectonics. HORIZONS TEXTBOOK PAGES 374–377, 392–393.

3. Describe and explain the processes that have changed Earth's atmosphere from its earliest formation to the present day. HORIZONS TEXTBOOK PAGES 375, 378.

4. Discuss the physical and geological features of Earth's moon and describe the evidence that explains their origin. HORIZONS TEXTBOOK PAGES 379–385.

5. Describe the evolution of the theory for the origin of Earth's moon. HORIZONS TESTBOOK PAGES 385–386.

6. Discuss the physical and geological features of Mercury and describe the evidence that explains their origin. HORIZONS TEXTBOOK PAGES 386–388.

7. Discuss the physical and geological features of Venus and describe the evidence that explains their origin. HORIZONS TEXTBOOK PAGES 388–395.

8. Describe the characteristics of the atmosphere and clouds of Venus and explain the role of the "greenhouse effect" in the environment on the surface of the planet. HORIZONS TEXTBOOK PAGES 388–389.

9. Discuss the physical and geological features of Mars and describe the evidence that explains their origin. HORIZONS TEXTBOOK PAGES 396–402.

10. Describe the atmosphere of Mars, the loss of gases, and the atmosphere's interaction with the planet's polar regions. HORIZONS TEXTBOOK PAGES 395–396.

11. Identify and describe the two moons of Mars and their possible origin. HORIZONS TEXTBOOK PAGES 402–403.

At this point, read Chapter 17, "Comparative Planetology of the Terrestrial Planets," pages 370–405.

Viewing Notes

The video program addresses the quest to understand how Earth and the other terrestrial worlds formed.

The video program contains the following segments:

- ❂ The Earth & Moon
- ❂ Mercury

The true delight is in the finding out rather than in the knowing.

—Isaac Asimov
American author
and biochemist
(1920–1992)

⚙ Venus

⚙ Mars

The following information will help you better understand the video program:

The differences among the inner planets are quite interesting—to study them we can compare them to the largest and most studied and well known of the terrestrial planets—Earth. Astronomers call this process comparative planetology.

The video program reveals the four stages of planetary development through which all of the terrestrial planets went: differentiation, cratering, flooding, and slow surface evolution. You will see how Earth evolved through these stages and discover the evidence to support this model of planetary development.

After studying Earth in detail, you will discover how the other terrestrial planets evolved through these stages. You will compare the features of Mercury, Venus, Earth's moon, and Mars to the features of Earth.

QUESTIONS TO CONSIDER

- What are the four stages of planet development?

- How do the interiors of the terrestrial planets differ? How are they the same?

- How do the crusts of the terrestrial planets differ? How are they the same?

- How do the atmospheres of the terrestrial planets differ? How are they the same?

- Why is Earth able to hold on to a thick atmosphere, but the moon is not?

Watch the video for Lesson 17, *The Terrestrial Planets*.

Key Terms and Concepts

Page references are keyed to the *Horizons* textbook.

1. **comparative planetology:** The study of planets in relation to one another. (p. 370; video lesson; objective 1)

2. **mantle:** The layer of dense rock and metal oxides that lies between the molten core and Earth's surface. Also, similar layers in other planets. (p. 373; video lesson; objective 2)

3. ***P* waves:** A pressure wave; a type of seismic wave produced in Earth by the compression of the material. (p. 374; video lesson; objective 2)

4. *S* **waves:** A shear wave; a type of seismic wave produced in Earth by the lateral motion of the material. (p. 374; video lesson; objective 2)

5. **plastic:** A material with the properties of a solid but capable of flowing under pressure. (p. 374; objective 2)

6. **plate tectonics:** The constant destruction and renewal of Earth's surface by the motion of sections of crust. (p. 376; video lesson; objective 2)

7. **midocean rise:** One of the undersea mountain ranges that pushes up from the seafloor in the center of the oceans. (p. 376; video lesson; objective 2)

8. **basalt:** Dark igneous rock that is characteristic of solidified lava. (p. 376; video lesson; objective 2)

9. **subduction zone:** A region of planetary crust where a tectonic plate slides downward. (p. 376; video lesson; objective 2)

10. **folded mountain range:** A long range of mountains formed by the compression of a planet's crust. (p. 376; objective 2)

11. **rift valley:** A long straight, deep valley produced by the separation of crustal plates. (p. 376; objective 2)

12. **primeval atmosphere:** Earth's first air. (p. 375; video lesson; objective 3)

13. **secondary atmosphere:** The gases outgassed from a planet's interior; rich in carbon dioxide. (p. 378; video lesson; objective 3)

14. **greenhouse effect:** The process by which a carbon dioxide atmosphere traps heat and raises the temperature of a planetary surface. (p. 378; video lesson; objectives 3 & 8)

15. **global warming:** The gradual increase in the surface temperature of Earth caused by human modifications to Earth's atmosphere. (p. 378; objective 3)

16. **mare (plural: maria):** One of the lunar lowlands filled by successive flows of dark lava. From the Latin for "sea." (p. 379; objective 4)

17. **terminator:** The dividing line between daylight and darkness on a planet or moon. (p. 379; objective 4)

18. **ejecta:** Pulverized rock scattered by meteorite impacts on a planetary surface. (p. 380; video lesson; objective 4)

19. **rays:** Ejecta from meteorite impacts forming white streamers radiating from some lunar craters. (p. 380; video lesson; objective 4)

20. **secondary crater:** An impact crater formed by debris ejected from a larger impact. (p. 380; objective 4)

21. **micrometeorite:** Meteorite of microscopic size. (p. 381; objective 4)

22. **multiringed basin:** Large impact feature (crater) containing two or more concentric rims formed by fracturing of the planetary crust. (p. 381; objective 4)

23. **vesicular basalt:** A porous rock formed by solidified lava with trapped bubbles. (p. 379; objective 4)

24. **anorthosite:** Rock of aluminum and calcium silicates found in the lunar highlands. (p. 379; objective 4)

25. **breccia:** Rock composed of fragments of earlier rocks bonded together. (p. 379; objective 4)

26. **albedo:** The ratio of the light reflected from an object divided by the light that hits the object. Albedo equals zero (0) for perfectly black and 1 for perfectly white. (p. 383; objective 4)

27. **fission hypothesis:** The theory that the moon and Earth formed when a rapidly rotating protoplanet split into two pieces. (p. 385; objective 5)

28. **condensation hypothesis:** The theory that Earth and the moon condensed from the same cloud of material in roughly their present orbital relationship. (p. 385; objective 5)

29. **capture hypothesis:** The theory that Earth's moon formed elsewhere in the solar nebula and was later captured by Earth. (p. 385; objective 5)

30. **large-impact hypothesis:** The theory that the moon formed from debris ejected during a collision between Earth and a large planetesimal. (p. 385; objective 5)

31. **lobate scarp:** A curved cliff such as those found on Mercury. (p. 386; objective 6)

32. **composite volcano:** A volcano formed by successive lava and ash flows. Such volcanoes have steep sides and, on Earth, are found along subduction zones. (p. 392; objective 2)

33. **shield volcano:** Wide, low-profile volcanic cone produced by highly liquid lava. (p. 392; objective 7)

34. **coronae:** On Venus, large, round geological faults in the crust caused by the intrusion of magma below the crust. (p. 391; objective 7)

35. **outflow channel:** Geological feature produced by the rapid motion of floodwaters; usually applied to features on Mars. (p. 398; video lesson; objective 9)

36. **valley network:** A system of dry drainage channels on Mars that resembles the beds of rivers and tributary streams on Earth. (p. 398; objective 9)

Summary

Earth is a distinctive planet in our solar system. Its name is derived from the Hebrew word, *erez*, meaning the ground. With its unique atmosphere, it is the only planet with an environment that is hospitable for human life. No other planet has water in liquid form or has oceans covering much of its surface. Earth is a dynamic planet with volcanic eruptions and it is the only planet with active movements in its crust.

A Travel Guide to the Terrestrial Planets

One of the best ways to analyze the planets in our solar system is to compare them to one another in a process known as **comparative planetology**. We can use Earth as the basis for comparison when studying the other terrestrial worlds: Mercury, Venus, Mars, and Earth's moon. Earth's moon is not considered a planet—it is a satellite of Earth—but the way it was formed, as well as its properties, makes it similar to a terrestrial planet.

The terrestrial worlds differ in size, with Earth being the largest. Next comes Venus, which is about 95 percent of Earth's diameter. Mars is about half of Earth's diameter, Mercury is about one-third, and the moon is about one-quarter the size of Earth. Some of the terrestrial worlds have atmospheres and some do not; whether or not a planet has an atmosphere depends mainly on the mass and temperature of the world.

Continuing your study of each of the terrestrial worlds, you will learn how each of them went through the four stages of planetary development and how the stages created the worlds that we study today.

Earth: Planet of Extremes

By studying the four stages of Earth's development, we can better understand how the other terrestrial worlds formed. All of the terrestrial worlds went through the four stages of planetary development—differentiation, cratering, flooding, and slow surface evolution. The differences in composition, mass, and temperature of each world emphasized some stages over the others and produced very different worlds.

In the first stage of planetary development, called differentiation, the materials in Earth were separated by density to form its core, mantle, and crust. By studying the way seismic waves travel through Earth, scientists have concluded that Earth's inner core is solid and the outer core is liquid. Earth's core is roughly as hot as the sun's photosphere and it is denser than lead; this extreme pressure keeps the metal as a solid in the center of the core. The fact that Earth has a solid iron-nickel core that is surrounded by a liquid outer core is consistent with this first stage of planetary development.

Earth's first atmosphere—the **primeval atmosphere**—seems to have been formed as the planet formed. As the young Earth grew, it accreted planetesimals that were rich in water, ammonia, and carbon dioxide that contributed to the atmosphere. Scientists have concluded that the primeval atmosphere must have been made of water vapor, nitrogen, and carbon dioxide—very different than the air we breathe today. Because the planet was hot, the atmospheric gases were originally dissolved in the magma inside the planet and were later released as the magma reached the surface and cooled.

The period of heavy bombardment allowed the planet to grow. During this period, meteorites crashed into Earth. Once Earth's surface began to cool, the impacts of these objects left their mark and the surface became heavily cratered. As debris from the solar nebula dissipated, the rate of cratering decreased.

In the third stage of planetary development, Earth was flooded first by lava and later by water. As the decay of radioactive materials heated Earth's interior, rock melted in the upper mantle and welled up through fissures in Earth's crust. This flooded the deepest crevices on Earth's surface. When the lava cooled, the gases trapped inside were released and added to the atmosphere.

Ultraviolet radiation broke down the primitive atmosphere and once the hot Earth began to cool, the warm atmosphere was in contact with its cooler surface and the water vapor began to condense. It began to rain, oceans began to form, and the planet was flooded with water. Carbon dioxide in the air dissolved into the water and reacted with the compounds in the ocean water to form mineral sediments. With the oceans transferring the carbon dioxide from the atmosphere to the ocean floor, Earth's atmosphere became cooler.

The process of outgassing, where gases were baked out of Earth's rock, contributed to today's **secondary atmosphere**. But, it wasn't until plant life evolved that oxygen appeared in the young atmosphere. Additionally, free oxygen released by the plants combined to form the ozone layer, which protects the surface from the sun's ultraviolet radiation. The reduction of carbon dioxide in the atmosphere allowed the infrared radiation to escape making Earth cool enough for humans to reside.

The fourth stage of planetary development, that of slow surface evolution, continues today. Earthquakes, volcanoes, and mountain building are all part of the continual evolution of the planet. Earth's surface is also eroded by air and water that wear away the planet's geological features.

Earth is the only world with a highly active crust; we can see evidence of this activity all around us. Earth's crust is divided into sections called plates, and there is constant destruction and renewal of Earth's surface as a result of the motion of these plates. This process is called **plate tectonics**. The driving force of plate tectonics is most likely convection in the mantle.

In addition to plate tectonics and volcanism, Earth's atmosphere and liquid water also contribute to slow surface evolution by wearing away the surface features. Earth's atmosphere continues to evolve—humans have altered it by releasing carbon dioxide into the air. An atmosphere rich in carbon dioxide can trap heat by a process called the **greenhouse effect**. This process is essential in warming the planet; without the greenhouse effect, Earth would be about 40 K colder or about 72°F colder than at present. But, we can look to Venus to see the results of a runaway greenhouse effect—if enough carbon dioxide were not removed from Earth's atmosphere by green plant life, the planet would be uninhabitable by humans.

In addition to increasing the greenhouse gases in the atmosphere, humans have also decreased the amount of protective ozone that shields us from the sun's high-energy photons. Today, plant life is needed to keep a steady supply of free oxygen that combines to form ozone. The deforestation of the planet, in conjunction with the release of chlorofluorocarbons (CFCs) into Earth's atmosphere, has begun to dissipate the ozone layer. This causes the intensity of harmful ultraviolet rays that reach Earth's surface to increase on a yearly basis.

The Moon

The moon is a gray, waterless world that looks quite different than Earth. Because of the moon's small size, its four stages of evolution—differentiation, cratering, flooding, and slow surface evolution—are quite different than those of Earth.

During the stage of differentiation, the dense materials sank to the bottom and the lighter material floated to the top, forming a core, a mantle, and a crust. As soon as the crust solidified, the second stage of cratering began. One of the most interesting features of the moon is its heavily cratered surface, which reveals that the moon was heavily bombarded by meteorites. Traveling at speeds between 10–60 km/s, a meteorite striking the lunar surface could have produced an impact crater 10 or more times larger than the diameter of the meteorite. The heated rock produced an explosion to form a crater.

The impacts from the meteorites cracked the crust as deep as 10 km. The lunar surface is still covered with craters and we can see dark and light areas on its disk. The dark areas are lunar lowlands called **maria** (plural for **mare**). The bright areas are highlands—rugged mountainous areas. Evidence suggests that the majority of the cratering on the moon's surface happened in the first half a billion years after the solar system was formed. As long as humans have studied the moon, no crater has been observed to have been formed.

During the third stage of flooding, molten rock flowed up from the moon's subsurface through the fractures caused by the impacts. The lava flooded these large craters, thus forming the smooth maria. The crust is thicker on the far side of the moon—perhaps because of tidal effects—and much of the lava was unable to flow out from under this thick crust. The crust on the side of the moon that faces Earth is thinner, and the lava easily welled up, flooding the basins on this side of the moon.

Many of the rocks on the moon are igneous—they solidified from molten rock—providing evidence that the surface was flooded with lava. Most of the rocks are typical of hardened lava; some are **vesicular basalt**—a porous rock formed by solidified lava and trapped air bubbles. The presence of this vesicular basalt provides evidence that much of the moon's surface has been covered by successive lava flows, particularly on the lowlands.

Because it is less massive than Earth, the moon has a low escape velocity and its gravity isn't great enough to hold on to an atmosphere. Because the moon has no atmosphere, no water, and it cooled rapidly, the fourth stage of slow surface evolution is minimal. Erosion is limited to that caused by the bombardment of micrometeorites. The moon lost its internal heat rapidly, so volcanism quickly died down and the crust never divided into moving plates, rendering the moon geologically dead.

Studying the rocks on the moon can also give us a clue as to how the moon has evolved. Until the mid-1980s, astronomers had no acceptable theory about how the moon was formed until they developed the **large-impact hypothesis**. This theory proposes that the moon formed when a planetesimal crashed into the proto-Earth in an off-center collision. Both the planetesimal and Earth were already differentiated. Most of the iron from the planetesimal's core became trapped inside the proto-Earth, and iron-poor debris was ejected to form a ring surrounding the protoplanet. The particles in the ring began to accrete into larger bodies and volatiles were lost into space. The moon eventually formed from the iron-poor and volatile-poor matter from the disk. Evidence supports this hypothesis—moon rocks differ in chemical composition from Earth; they are poor in iron and volatiles.

The original impact of the proto-Earth and the planetesimal would have generated lots of heat. The material falling together to eventually form the moon would have been heated—hot enough to form as a sea of magma that formed the abundance of basalt moon rocks. The basalt rocks found on the lunar surface provide additional evidence to support this theory. You will learn more about the moon's formation in the video program in Lesson 19.

Mercury

Mercury is named after the fleet-footed messenger of the gods—its orbital period is just under 88 days. It is the planet closest to the sun and is similar to Earth's moon in many ways. It is a small, hot world that is heavily cratered; it is much more dense than the moon—nearly as dense as Earth. But, because Mercury lies so close to the sun, it is difficult to observe.

In Mercury's first stage of planet formation, it differentiated to form a large metal core, a rocky mantle, and a thin crust. Mercury contains a vast amount of iron, but has a weak magnetic field compared to Earth. It's possible that the fluid outer core contains sulfur, which lowers the melting point and keeps it fluid at a lower temperature than if it were pure iron. It seems as if most of the rocky mantle was shattered and driven away by a catastrophic event. The remaining iron and rock formed a small planet with an unusually large core.

In the second and third stages of planet formation—cratering and flooding— meteorites battered the crust and lava welled up to fill in the lowlands. The metal core shrunk as it cooled and the crust contracted along with it. This created the feature unique to Mercury—long curving ridges called **lobate scarps**. The scarps are up to 3 km high and 500 km long. Cutting through craters, they must have been formed after the period of heavy bombardment.

Like the moon, Mercury has no atmosphere, no water, and it cooled rapidly; the fourth stage of slow surface evolution is therefore limited. The surface never separated into plates, so there are no plate tectonics. Because there is no atmosphere, surface erosion is minimal.

Venus

Venus is named for the Roman goddess of love, but it is a deadly hot desert with flowing volcanoes—its surface temperature is hot enough to melt lead.

Like the other planets, Venus differentiated into a core, mantle, and crust. But Venus' thick atmosphere has altered its geology. You might wonder how a planet roughly Earth's size could contain such a hostile environment. The main differences between Earth and Venus are the lack of water on Venus and the fact that its thick atmosphere creates an extreme greenhouse effect.

During the period of heavy bombardment, Venus' crust was cratered. The damaged crust allowed lava to well up and smooth its surface. There are only 10 percent the number of craters on Venus as there are on the moon—evidence that lava flows renewed its surface during the stage of flooding. There may have been small oceans on Venus when it was young, but because it's closer to the sun, the intense heat may have vaporized them.

Venus and Earth would have outgassed about the same amount of carbon dioxide, but on Earth, most of it has been converted to limestone and sediments by the oceans. The diameter of Venus is 95 percent that of Earth, making it large enough to hold onto an atmosphere. But the air is unbreathable—it is almost 100 times denser than Earth's atmosphere, very hot, and contains roughly 96 percent carbon dioxide. It also contains nitrogen, argon, sulfur dioxide, and traces of sulfuric acid, hydrochloric acid, and water vapor, making it extremely inhospitable.

Slow surface evolution continues on Venus; the entire crust appears as if it has been replaced within the last half-billion years. The old crust may have broken up and sunk while lava flows created the new crust. Or, perhaps an intense volcano raised the surface temperature even higher, softening the crust, increasing the volcanism, and driving the planet into a period of resurfacing.

There are great lava plains and mountains on the surface of Venus; it contains long, narrow lava channels that are thousands of kilometers long. Its surface gets it character by rising convection currents of magma pushing up the crust to form **coronae**—circular bulges up to 2,100 km in diameter bordered by fractures, volcanoes, and lava flows. When the lava recedes, the crust sinks back down, but the circular fractures remain to mark the edge of the coronae. The volcanoes on Venus are shield volcanoes—they are formed by highly fluid lava made of basalt that flows easily and creates low-profile volcanic peaks. The lava can flow easily because the crust of Venus is made flexible by the intense heat; it did not break into rigid plates like Earth's crust did.

Today, Venus seems to be trapped in a runaway greenhouse effect. The lack of water prevents the carbon dioxide in the atmosphere from being dissolved, thereby intensifying the greenhouse effect and causing the planet to grow hotter and hotter. Sunlight is able to filter through the atmosphere and warm the surface of Venus, but the overwhelming amount of carbon dioxide, along with the sulfur dioxide and water vapor, makes the gas opaque to infrared and the heat cannot escape. Comparing the atmosphere of Venus to that of Earth, you can see why Earth is hospitable to life.

Mars

Mars is a red desert planet named for the Roman god of war. Mars developed differently than the other terrestrial planets—its growth was stunted because of its small size. Mars differentiated into a dense core, mantle, and crust, as did the other terrestrial worlds. The young Mars apparently had a molten core, but it must have cooled and solidified quickly because only remnants of a very weak magnetic field can be detected in parts of its thick crust.

During the period of heavy bombardment, the surface of Mars was struck by meteorites, which may have broken or weakened the crust. The southern hemisphere of Mars is a highly cratered highland—its surface is 2–3 billion years old. By contrast, the northern hemisphere is a younger lowland plane with few craters—evidence that lava flows smoothed the surface during the stage of flooding. Outgassing released carbon dioxide into the atmosphere.

Evidence suggests that water once flowed on the surface of Mars, so its atmosphere must have been thicker than it is now, preventing the water from being evaporated by the sun's intense heat. The liquid water on the surface created **outflow channels**—features that are produced by the rapid motion of floodwater. They appear to have swept away the existing geological features and scarred the land. **Valley networks**—systems of dry drainage channels—once carried streams. Analysis of the craters on these features indicates that they formed about 1–3 billion years ago.

Today, liquid water is not evident because of the low-pressure atmosphere. But a significant part of the crust from latitude 60° north and from latitude –60° south is made of water ice. The polar ice caps are made of frozen carbon dioxide (dry ice), and there may be frozen water beneath them as well.

The fourth stage—slow surface evolution—has been slowly declining. The planet lost most of its internal heat and most of the volcanoes erupted billions of years ago. There are no folded mountain ranges on Mars, and it seems like crustal plates were never formed—evidence of shield volcanoes supports this. As the small planet continued to cool, its crust became thick and immobile.

Because its axis is tipped, Mars has seasons. During summer in one hemisphere, the frozen carbon dioxide in the polar cap vaporizes and adds carbon dioxide to the atmosphere. In the opposite hemisphere, winter is freezing the carbon dioxide out of the air and adding a layer to the ice cap. To have liquid water billions of years ago, the climate on Mars must have been drastically different. Analysis of the layers of terrain at the ice caps suggests that the climate on Mars has changed repeatedly. These changes may have been caused by the cyclical changes in the rotation and orbital revolution of the planet. In Lesson 3, you learned that Earth undergoes cyclical changes in rotation and revolution as well.

Aside from Earth, Mars is the only terrestrial planet to have moons. However, the two Martian moons, Phobos and Deimos, are tiny compared to Earth's moon and they are both very dusty.

By using comparative planetology, we have been able to compare and contrast the features of the terrestrial planets. In the next lesson, you'll use comparative planetology again to study the outer planets: the four Jovian worlds and Pluto.

Review Exercises

Matching I

Match each term with the appropriate definition or description.

1. _____ comparative planetology		7. _____ rift valley	
2. _____ mantle		8. _____ plate tectonics	
3. _____ plastic		9. _____ greenhouse effect	
4. _____ dynamo effect		10. _____ primeval atmosphere	
5. _____ basalt		11. _____ secondary atmosphere	
6. _____ midocean rise			

a. A material with the properties of a solid but capable of flowing under pressure.
b. Layer of dense rock and metal oxides between the molten core and Earth's surface.
c. Process by which an atmosphere rich in carbon dioxide entraps heat and raises the temperature of a planetary surface.
d. Constant destruction and renewal of Earth's surface resulting from movement of sections of crust.
e. Earth's first air.
f. Dark igneous rock characteristic of solidified lava.
g. Study of planets in relation to one another.
h. Process by which Earth's magnetic field is generated in the conducting material of its molten core.
i. Long, straight, deep valley formed when continental plates drift apart.
j. Undersea mountains that push up from the seafloor in an ocean's center.
k. The gases outgassed from a planet's interior; rich in carbon dioxide.

Matching II

Match each term with the appropriate definition or description.

1. _____ ejecta		8. _____ valley network	
2. _____ rays		9. _____ fission hypothesis	
3. _____ vesicular basalt		10. _____ condensation hypothesis	
4. _____ mare		11. _____ capture hypothesis	
5. _____ anorthosite		12. _____ large-impact hypothesis	
6. _____ breccia		13. _____ lobate scarp	
7. _____ albedo			

a. The theory that the moon formed from debris ejected during a collision between Earth and a large planetesimal.
b. A lunar lowland filled by successive flows of dark lava.

c. The theory that the moon and Earth formed when a rapidly rotating protoplanet split into two pieces.
d. Pulverized rock scattered by meteorite impacts on a planet.
e. Rock with trapped bubbles formed by solidified lava.
f. A system of dry drainage channels on Mars.
g. Rock composed of fragments of earlier rocks bonded together.
h. White streamers radiating from some lunar craters.
i. The theory that Earth and the moon condensed simultaneously from the same cloud of material.
j. Rock of aluminum and calcium silicates found in the lunar highlands.
k. The theory that Earth's moon formed elsewhere in the solar nebula and was later captured by Earth.
l. A curved cliff, such as those found on Mercury.
m. The ratio of the light reflected from an object divided by the light that hits the object.

Completion

Fill each blank in the sentences below with the most appropriate term from the list of completion answers that follow. A term may be used once, more than once, or not at all. Check your answers with the Answer Key and review when necessary.

differentiation	Mercury	primeval
global warming	moons	solar nebula
greenhouse	ozone layer	Venus
lobate scarps	Pangaea	water vapor
maria		

1. Earth's first atmosphere, sometimes called the _____ atmosphere, was thought to consist of gases originally belonging to the _____.

2. _____, or lunar lowlands, are enormous lava flows that cover 17 percent of the moon's surface.

3. Distinctive curved cliffs, known as _____, formed on Mercury as it cooled and shrank.

4. The atmosphere of _____ contains acid compounds, including sulfuric and hydrochloric acid, as well as large quantities of carbon dioxide, which is responsible for trapping heat through the _____ effect.

5. Mars is the terrestrial planet with the most _____, which are thought to be captured asteroids.

6. About 200 million years ago, Earth's continents were connected to form one land mass called _____.

Self-Test

Select the best answer.

1. Iron and nickel drained to the center of Earth, while at the same time, the lighter silicates rose to the surface, solidified, and formed a thin crust. This process is known as
   a. differentiation.
   b. sedimentation.
   c. liquefaction.
   d. distillation.

2. Which of the following sequences best describes the four-stage history of Earth's formation and evolution?
   a. differentiation, surface evolution, cratering, basin flooding
   b. condensation, cratering, basin flooding, differentiation
   c. differentiation, cratering, basin flooding, surface evolution
   d. cratering, condensation, accretion, surface evolution

3. Earth's core has a solid center and liquid outer parts. Evidence for the liquid component of the core comes from
   a. gravity studies by spacecraft.
   b. seismic wave studies.
   c. physical measurement by subterranean probes.
   d. oil well drillings.

4. Between the crust and liquid core lies a solid material that behaves as a plastic, flowing under high pressure. This material makes up Earth's
   a. mantle.
   b. crust.
   c. inner core.
   d. continental shelf.

5. Earth's magnetic field is thought to result from the dynamo effect, which is
   a. motion in the liquid core that generates seismic waves.
   b. motion in the liquid core that generates electric currents.
   c. thermal cooling in the liquid core that generates temperature gradients.
   d. seismic convection in the liquid core that generates liquefaction.

6. One of the best ways to locate the crustal plate boundaries is to identify
   a. large mountain ranges on the continents.
   b. lakes, major rivers, and oceans.
   c. the outline of the continental shelf.
   d. the locations of earthquakes.

7. The carbon dioxide in Earth's early atmosphere was produced when large amounts of gases were released during the accretion process when Earth's surface temperature was very hot. Most of the carbon dioxide remained in Earth's atmosphere until it was removed when
   a. ultraviolet radiation caused the gas to disassociate into carbon and oxygen.
   b. volcanic activity produced chemicals that neutralized the gas.
   c. liquid water formed on Earth's surface and dissolved the carbon dioxide causing it to react with other compounds to form limestone and other sediments.
   d. infrared radiation heated Earth's atmosphere and evaporated the carbon dioxide.

8. The form of electromagnetic radiation that the ozone layer prevents from reaching Earth's surface is
   a. X rays.
   b. visible light.
   c. ultraviolet radiation.
   d. infrared rays.

9. The temperature of Earth is much warmer and more uniform than it would be if Earth did not have an atmosphere. The gas most responsible for trapping heat in Earth's atmosphere is
   a. carbon dioxide.
   b. oxygen.
   c. helium.
   d. nitrogen.

10. The greenhouse effect best describes the process by which
    a. ultraviolet radiation is filtered from sunlight.
    b. infrared radiation is reflected off the top of Earth's atmosphere.
    c. carbon dioxide is dissolved by liquid water.
    d. heat is trapped by the atmosphere, thus increasing the temperature at a planet's surface.

11. Free oxygen in Earth's atmosphere was produced by
    a. chemical reactions taking place in the oceans.
    b. living organisms during photosynthesis.
    c. chemical reactions taking place on land.
    d. the decay of early plant life.

12. Large, reasonably flat and dark areas that cover a good portion of the side of the moon facing Earth are referred to as lunar
    a. maria.
    b. craters.
    c. escarpments.
    d. mass-concentrations.

13. The craters on the lunar surface were produced by
    a. volcanic activity.
    b. subsurface erosion due to running water.
    c. impacts by asteroids and meteorites.
    d. excavation by aliens.

14. Lunar rocks found in the highlands are
    a. older than the rocks in the maria and composed mostly of basalt.
    b. younger than the rocks in the maria and composed of aluminum and calcium.
    c. rich in anorthocites, a light-colored, low density rock.
    d. younger than the rocks in the maria and composed mostly of basalt.

15. The core of the moon
    a. is large and composed largely of iron.
    b. is small and contains little iron.
    c. mostly pulverized dust, like that found on the surface.
    d. was destroyed billions of years ago by meteorites in the early bombardment epoch.

16. The large-impact hypothesis proposes that the moon formed as a result of
    a. a large amount of material breaking off Earth as a result of Earth's rapid rotation.
    b. two large asteroids colliding, shattering, and condensing to form the moon.
    c. a young moon colliding with Earth, forming a ring system that later condensed to form the moon we see today.
    d. a Mars-sized object colliding with Earth and ejecting material from Earth's mantle, some of which fell back to Earth and the rest of which condensed to form the moon.

17. A problem with the fission hypothesis for the formation of the moon is that
    a. the Earth-moon system has less angular momentum than would be expected from such a formation scenario.
    b. Earth and the moon have similar densities.
    c. it requires too many coincidental events.
    d. the moon is rich in volatiles.

18. The craters on Mercury are the result of
    a. impacts by asteroids and meteorites.
    b. volcanic activity.
    c. subsurface erosion.
    d. solar wind erosion.

19. Lobate scarps on Mercury were probably caused by
    a. meteorite impacts.
    b. shrinkage of the surface as the interior of the planet cooled.
    c. surface erosion by flooding.
    d. wind erosion.

20. Evidence for Mercury's iron core comes from
    a. large impact craters.
    b. unusually low overall density for a planet of its size.
    c. unusually high overall density for a planet of its size.
    d. large lava-filled maria.

21. The weak (although stronger than expected) magnetic field around Mercury detected by the Mariner 10 spacecraft indicates
    a. a solid core.
    b. solar flare activity.
    c. the interior of the planet is similar to that of the moon.
    d. the interior of the planet is similar to that of Earth.

22. Radar mapping of Venus by spacecraft discovered that roughly 85 percent of the surface of Venus consists of
    a. volcanoes.
    b. fault zones.
    c. craters.
    d. basaltic lowlands.

23. Spacecraft images (radar) of Venus offer little evidence for
    a. volcanic activity.
    b. major plate motion.
    c. the existence of mountains on the surface.
    d. continent-sized land masses.

24. Analysis of the surface by spacecraft that have landed on Venus indicates that the surface material consists of
    a. basalt.
    b. water.
    c. liquid magma.
    d. silicate carbonates.

25. The Magellan and Venera spacecraft located similar features, called coronae, on the surface of Venus. These features are
    a. large craters produced by asteroid impacts.
    b. volcanic cones.
    c. circular features probably caused by rising columns of magma under the crust.
    d. eroded shield volcanoes.

26. The major component of the atmosphere on Venus is
    a. nitrogen.
    b. oxygen.
    c. sulfuric acid.
    d. carbon dioxide.

27. The atmosphere of Venus contains much more carbon dioxide than the atmosphere of Earth. This difference is probably the result of
    a. the presence of more carbon dioxide in the primeval atmosphere of Venus.
    b. the slightly warmer temperatures on Venus that prevented the accumulation of great amounts of liquid water, thus preventing carbon dioxide from dissolving into the liquid water.
    c. most of the carbon dioxide in Earth's atmosphere escaping into space.
    d. carbon dioxide still being produced by volcanoes on Venus.

28. Evidence of liquid water existing in the Martian past was found by the discovery of
    a. meandering drainage valleys and elongated flow features exposing ancient craters.
    b. oceans of liquid water currently on the surface.
    c. a dense atmosphere.
    d. volcanic activity.

29. Evidence of extensive volcanic activity on Mars is found in the
    a. highly cratered southern hemisphere of Mars.
    b. Tharsis region in the northern hemisphere where there are numerous faults and lava flows.
    c. polar ice caps, where spectrum analysis shows evidence of basalt rocks.
    d. southern hemisphere where volcanoes such as Olympus Mons are located.

30. The polar caps of Mars consist of
    a. water ice.
    b. solid methane and carbon dioxide.
    c. solid nitrogen ice and liquid water.
    d. carbon dioxide and frozen water.

31. The atmosphere of Mars consists primarily of
    a. carbon dioxide.
    b. oxygen.
    c. nitrogen.
    d. water vapor.

32. The density of the atmosphere on Mars is about
    a. equal to Earth's atmosphere.
    b. 50 percent of Earth's atmosphere.
    c. 10 percent of Earth's atmosphere.
    d. 1 percent of Earth's atmosphere.

33. The moons of Mars are
    a. Phobos and Deimos.
    b. Io and Ganymede.
    c. Hellas and Ishtar Terra.
    d. Maxwell Montes and Elysium.

34. The moons of Mars can be briefly described as
    a. smooth and nearly spherical in shape.
    b. being lightly cratered.
    c. irregular in shape with numerous surface craters.
    d. nearly identical in surface features.

35. Rover Opportunity photographed small spheres of hematite in the
    rocks on Mars. These spheres indicate
    a. the presence of liquid water.
    b. the presence of carbon dioxide in the soil.
    c. that liquid water was once present in standing pools at this location.
    d. that water flowed as a flood at one time in Martian history.

36. Mars' atmosphere was much more dense in its past. Mars lost most of
    its atmosphere because
    a. it is too close to the sun.
    b. its escape velocity is very low.
    c. it had no liquid water on its surface.
    d. the greenhouse effect was insufficient to maintain the atmosphere.

37. The heavy bombardment era had the greatest influence on which stage
    of Earth's development?
    a. differentiation
    b. cratering
    c. flooding
    d. slow surface evolution

Short-Answer Questions

1. Briefly describe the four stages of planetary development.

2. Why is there more cratering visible on the other terrestrial planets and on the moon than on Earth?

3. What are the similarities and differences between the surfaces of Mercury and the moon.

4. What is the "greenhouse effect" and why does it cause much higher temperatures on Venus than it does on Earth?

5. Identify the moons of Mars and briefly describe their surface characteristics.

6. Briefly describe the characteristics of the atmosphere of Mars and explain why it is so thin.

Applications

1. What characteristics make Earth's surface unique among the terrestrial planets?

2. How is plate tectonics related to continental drift and seafloor spreading?

3. How does one explain the differences in the appearance of the crust between the side of the moon facing Earth and the moon's far side?

4. Why are the volcanoes on Mars so much larger and higher than the volcanoes on Earth?

5. What is the evidence that suggests that Mars had large amounts of liquid water flowing on its surface some time in its past?

6. Where is the water that used to flow on the Martian surface now?

Answer Key

Matching I

1. g (p. 370; video lesson; objective 1)

2. b (p. 373; video lesson; objective 2)

3. a (p. 374; objective 2)

4. h (pp. 134, 374; objective 2)

5. f (p. 376; video lesson; objective 2)

6. j (p. 376; video lesson; objective 2)

7. i (p. 376; objective 2)

8. d (p. 376; video lesson; objective 2)

9. c (p. 378; video lesson; objectives 3 & 8)

10. e (p. 375; video lesson; objective 3)

11. k (p. 378; video lesson; objective 3)

Matching II

1. d (p. 380; video lesson; objective 4)

2. h (p. 380; video lesson; objective 4)

3. e (p. 379; objective 4)

4. b (p. 379; objective 4)

5. j (p. 379; objective 4)

6. g (p. 379; objective 4)

7. m (p. 383; objective 4)

8. f (p. 398; objective 9)

9. c (p. 385; objective 5)

10. i (p. 385; objective 5)

11. k (p. 385; objective 5)

12. a (p. 385; objective 5)

13. l (p. 386; objective 6)

Completion

1. primeval, solar nebula (p. 375; objective 3)

2. maria (p. 379; objective 4)

3. lobate scarps (p. 386; objective 6)

4. Venus, greenhouse (pp. 388–389; objective 8)

5. moons (p. 402; objective 11)

6. Pangaea (p. 377; objective 2)

Self-Test

1. a (p. 374; objective 1)

2. c (p. 374; objective 1)

3. b (p. 374; objective 2)

4. a (p. 373; objective 2)

5. b (pp. 134, 374; objective 2)

6. d (pp. 376–377; objective 2)

7. c (p. 378; objective 3)

8. c (p. 379; objective 3)

9. a (p. 378; objective 3)

10. d (p. 378; objective 3)

11. b (p. 379; objective 3)

12. a (p. 379; objective 4)

13. c (pp. 380–381; objective 4)

14. c (p. 379; objective 4)

15. b (p. 383; objective 4)

16. d (p. 385; objective 5)

17. a (p. 385; objective 5)

18. a (p. 386; objective 6)

19. b (p. 386; objective 6)

20. c (p. 387; objective 6)

21. d (p. 387; video lesson; objective 6)

22. d (p. 390; objective 7)

23. b (pp. 390–391; objective 7)

24. a (p. 390; objective 7)

25. c (p. 391; objective 7)

26. d (p. 388; objective 8)

27. b (pp. 388–389; objective 8)

28. a (pp. 397–400; objective 9)

29. b (p. 397; objective 9)

30. d (p. 395; objective 9)

31. a (p. 395; objective 10)

32. d (p. 395; objective 10)

33. a (p. 402; objective 11)

34. c (p. 402; objective 11)

35. c (pp. 399, 401; objective 9)

36. b (pp. 395–396; objective 10)

37. c (p. 374; video lesson; objective 3)

Short-Answer Questions

1. After the gas and dust materials within the solar nebula have condensed and formed planetesimals and the planetesimals have collided to form larger worlds, the process of planetary formation begins. The first stage is differentiation. Heat generation from infalling planetesimals and radioactive decay melts the interior. The more dense matter, iron and nickel, sinks to the core, while the less dense silicates rise to the surface to form a crust.

   The second stage is cratering. After the crust solidifies, debris that has yet to collide with the early planetesimals is now attracted by the larger gravity of the forming planets. These objects now impact the surface and leave craters as evidence of this period in the history of the solar system.

   The third stage is flooding. The larger impacts left deep craters that allowed the liquid interior of the planets to swell up through the cracks and fill these basins. This flooding of lava is evidenced by the "seas" on the surface of the moon and Mercury. On Earth, as the atmosphere cooled, water vapor turned to liquid water and began the period of flooding that eventually formed Earth's oceans.

The final stage of planetary development is slow surface evolution. This evolution is most evident on Earth and Mars and less evident on Venus. Water, wind, volcanic activity, and plate tectonics have destroyed the original surface of Earth. The surface of Venus has suffered a similar fate, however, its evolution has been limited to volcanic activity. No water and very little wind exist on its surface. The surface of Mars has evolved as the result of water erosion and wind activity, in addition to lava flows. Mercury and the moon have surfaces that date back to the period of heavy bombardment and flooding of the impact basins. No erosion has occurred on these surfaces for the last 3.5 billion years. (p. 374; objective 1)

2. Earth is unique among the terrestrial planets because most of its surface is water (75 percent) and although many impacts have occurred in the past, 75 percent of them have occurred in the oceans. Furthermore, Earth's atmosphere and the plate tectonics of its crust produce a significant amount of surface erosion and evolution that has destroyed most evidence of early cratering. (pp. 373–374, 376–377; objective 1)

3. Mercury and the moon have roughly the same number of craters and maria. The main difference is that the material within the maria of Mercury is lighter in color than the material in the maria of the moon. Furthermore, Mercury has cliffs called "lobate scarps" resulting from a shrinking of the crust during the planet's cooling period. (pp. 379–382, 386–387; objectives 4 & 6)

4. The greenhouse effect occurs when sufficient carbon dioxide is in the atmosphere of a planet and prevents infrared radiation from escaping into space. As a result, the surface temperature of the planet rises significantly. The greenhouse effect on Venus is much more significant than on Earth because Venus has much more carbon dioxide in its atmosphere. Liquid water on Earth dissolved much of the carbon dioxide that formed in Earth's early atmosphere. This is not the case on Venus, probably due to the fact that Venus is closer to the sun and warmer. (pp. 378–379, 388–389; objectives 3 & 8)

5. The two moons of Mars are Phobos and Deimos. These names are translated as Fear and Panic respectively. Phobos is observed to have craters and grooves on its surface, probably the result of impacts forming the craters. Deimos is much smoother with apparently less cratering. However, evidence shows that Deimos has a thicker dust layer than Phobos which probably fills in the craters and grooves making them less visible. (pp. 402–403; objective 11)

6. The atmosphere of Mars is mostly carbon dioxide and very thin. During the summer months, the carbon dioxide in the polar caps turns from solid to vapor and adds to the atmosphere. Evidence shows that Mars must have had a much denser atmosphere in its past. It lost

most of its atmosphere because the temperature at the orbit of Mars is high enough that the molecules in its atmosphere had a velocity that exceeded the escape velocity of the planet. Furthermore, since Mars has no ozone layer, ultraviolet radiation can break up atmospheric molecules allowing them to escape more easily. (pp. 395–396; objective 10)

Applications

1. The two most salient characteristics that make Earth's surface unique among the terrestrial planets is the presence of liquid water and crustal plates. No other planet in the solar system has liquid water on its surface or undergoes plate tectonics. In addition to these characteristics, Earth has the least number of craters on its surface. Of course, that is primarily the result of erosion by wind, water, and plate movements. (pp. 373–375; objective 2)

2. New crust forms at the midocean rises, and the plates, floating on the plastic mantle, move farther apart. The continents sitting on these plates must also move farther apart. The separation of continents is called continental drift. Of course, as continents separate in one region, they move closer in another. The subcontinent of India eventually moved away from the continent of Antarctica and eventually collided with the continent of Asia, forming the Himalayan Mountains. (pp. 376–377; objective 2)

3. Studies show that the side of the moon facing Earth is thinner than the far side. This is probably the result of tidal (gravitational) effects. The moon keeps one side pointed toward Earth at all times. The thinner crust allowed for molten lava to reach the surface and flood the large impact craters, whereas on the far side, no such flooding occurred. (p. 384; objective 4)

4. The crust of Mars is much stronger than the crust of Earth. This is probably the result of the more rapid cooling of Mars than of Earth. There is no evidence of plate tectonics on Mars, so with a thicker crust, the volcanoes grew much larger than those on Earth. (p. 397; objective 9)

5. The two most obvious pieces of evidence to demonstrate that Mars had liquid water on its surface are the valley networks and outflow channels. Valley networks are the result of liquid water flowing for extended periods of time. They are similar to river beds and streams on Earth. The outflow channels are the result of massive floods carrying water many times the volume of the Mississippi River. (pp. 397–399; objective 9)

6. Most of the water now present on the Martian surface is located at the north and south Martian poles frozen solid under the carbon dioxide layer and under the Martian soil as permafrost. (p. 395; objective 9)

Notes:

Lesson Review

Lesson 17: The Terrestrial Planets

PLEASE NOTE: Use this matrix to guide your study and achieve the learning objectives of this lesson. It will also help you to view the video, which defines and demonstrates important concepts and principles as they relate to everyday life and actual case studies.

Learning Objective	Textbook	Student Guide
1. Describe the four stages of planetary development.	p. 374	Key Terms: 1; Matching I: 1; Self-Test: 1, 2; Short-Answer: 1, 2.
2. Identify the observable evidence that supports our model of Earth's interior, its dynamo, and plate tectonics.	pp. 374–377	Key Terms: 2, 3, 4, 5, 6, 7, 8, 9, 10, 11, 32; Matching I: 2, 3, 4, 5, 6, 7, 8, 11, 12; Completion: 6; Self-Test: 3, 4, 5, 6; Applications: 1, 2.
3. Describe and explain the processes that have changed Earth's atmosphere from its earliest formation to the present day.	pp. 375, 378	Key Terms: 12, 13, 14, 15; Matching I: 9, 10; Completion: 1; Self-Test: 7, 8, 9, 10, 11, 37; Short-Answer: 4.
4. Discuss the physical and geological features of Earth's moon and describe the evidence that explains their origin.	pp. 379–385	Key Terms: 16, 17, 18, 19, 20, 21, 22, 23, 24, 25, 26; Matching II: 1, 2, 3, 4, 5, 6, 7; Completion: 2; Self-Test: 12, 13, 14, 15; Short-Answer: 3; Applications: 3.
5. Describe the evolution of the theory for the origin of Earth's moon.	pp. 385–386	Key Terms: 27, 28, 29, 30; Matching II: 9, 10, 11, 12; Self-Test: 16, 17.

Learning Objective	Textbook	Student Guide
6. Discuss the physical and geological features of Mercury and describe the evidence that explains their origin.	pp. 386–388	Key Terms: 31; Matching II: 13; Completion: 3; Self-Test: 18, 19, 20, 21; Short-Answer: 3.
7. Discuss the physical and geological features of Venus and describe the evidence that explains their origin.	pp. 388–395	Key Terms: 33, 34; Self-Test: 22, 23, 24, 25.
8. Describe the characteristics of the atmosphere and clouds of Venus and explain the role of the "greenhouse effect" in the environment on the surface of the planet.	pp. 388–389	Key Terms: 14; Completion: 4; Self-Test: 26, 27; Short-Answer: 4.
9. Discuss the physical and geological features of Mars and describe the evidence that explains their origin.	pp. 396–402	Key Terms: 35, 36; Matching II: 9; Self-Test: 28, 29, 30, 35; Applications: 4, 5, 6.
10. Describe the atmosphere of Mars, the loss of gases, and the atmosphere's interaction with the planet's polar regions.	pp. 395–396	Self-Test: 31, 32, 36; Short-Answer: 6.
11. Identify and describe the two moons of Mars and their possible origin.	pp. 402–403	Completion: 11; Self-Test: 33, 34; Short-Answer: 5.

LESSON
18

The Jovian Worlds

Checklist

For the most effective study of this lesson, complete the following activities in this sequence.

Before Viewing the Video

- ❑ Read the Preview, Learning Objectives, and Viewing Notes below.
- ❑ Read Chapter 18, "Comparative Planetology of the Outer Planets," pages 406–437, in the *Horizons* textbook.

What to Watch

- ❑ After reading the textbook chapter, watch the video for Lesson 18, *The Jovian Worlds*.

After Viewing the Video

- ❑ Briefly note your answers to questions listed at the end of the Viewing Notes.
- ❑ Review the Summary below.
- ❑ Review all reading assignments for this lesson, especially the Chapter 18 summary on pages 435–436 in *Horizons* and the Viewing Notes in this guide.
- ❑ Write brief answers to the review questions at the end of Chapter 18 in *Horizons*.
- ❑ Complete the Review Exercises below. Check your answers with the Answer Key and review when necessary.
- ❑ Use the Lesson Review matrix found at the end of this lesson to review and assess your knowledge of each Learning Objective.
- ❑ As assigned by your instructor, complete the Applications activities and any additional activities for this lesson.

Preview

In Lesson 17, we explored the terrestrial worlds—planets that might seem more familiar because they include the Earth and planets similar to it. In this lesson, you'll explore the four outer Jovian worlds—enormous planets composed primarily of gas.

These Jovian worlds are as interesting as they are beautiful. These giants formed in the solar nebula at the same time as the inner terrestrial planets. But, colder temperatures in the outer nebula—beyond the snow line—produced very different results. Jupiter and Saturn, being closer to the sun, are gas giants. Their small cores are surrounded by liquid that dissipates into gas. Uranus and Neptune, at the far reaches of our solar system, are sometimes referred to as ice giants—their rock-ice interiors are surrounded by a slushy layer that is then surrounded by a thick layer of gas.

You'll soon discover that these four giants—not just Saturn—have ring systems. Each of the planets has many moons, some of which guide the rings and prevent their particles from escaping into space. In the next lesson, you'll learn about the objects in far reaches of the solar system—Pluto and the Kuiper belt objects as well as asteroids, and comets.

CONCEPTS TO REMEMBER

- Recall from Lesson 7 that *differential rotation* is the rotation of a body in which different parts of the body have different periods of rotation. This is true of the sun and the Jovian planets (p. 127 in this guide).

- In Lesson 7, you also learned that the *dynamo effect* is the process by which a rotating, convecting body of conducting matter can generate a magnetic field (p. 127 in this guide).

Learning Objectives

After you complete this lesson, you should be able to:

1. Discuss the physical and atmospheric characteristics of Jupiter and describe the evidence that explains their origin. *HORIZONS* TEXTBOOK PAGES 409–413, 417–418.

2. Discuss the physical and atmospheric characteristics of Saturn and describe the evidence that explains their origin. *HORIZONS* TEXTBOOK PAGES 418–419, 422, 424.

3. Discuss the physical and atmospheric characteristics of Uranus and describe the evidence that explains their origin. *HORIZONS* TEXTBOOK PAGES 424–425, 427.

4. Discuss the physical and atmospheric characteristics of Neptune and describe the evidence that explains their origin. *HORIZONS* TEXTBOOK PAGES 427, 430–432.

5. Describe the characteristics and origins of the ring and satellite systems of the Jovian planets. *HORIZONS* TEXTBOOK PAGES 414–417, 419–421, 423, 425–431.

At this point, read Chapter 18, "Comparative Planetology of the Outer Planets," pages 406–437.

Viewing Notes

The outer worlds formed in the solar nebula at the same time as the planets in the inner solar system, but with contrasting results. These beautiful gas and ice giants, with their elaborate satellite and ring systems, are very different from the terrestrial worlds.

The video program contains the following segments:

- ⊛ Building a Gas Giant
- ⊛ Jupiter
- ⊛ Jupiter's Galilean Moons
- ⊛ Saturn
- ⊛ The Ice Giants

The following information will help you better understand the video program:

The planets in the outer region of the solar system are far enough away from the sun that "volatile" gases are able to condense. Volatile gases are those that evaporate readily at normal temperatures and pressures. It is cold enough in the outer region of the solar system for gases like water vapor to condense into ice.

The video program briefly discusses the term *snowline* (first introduced in Lesson 16). This is the demarcation of the inner solar system where the terrestrial planets reside from the outer solar system where the gaseous Jovian planets reside. Beyond this boundary, water is stable and could condense and form icy particles. In the outer reaches of our solar system, beyond Neptune, astronomers have discovered thousands of icy bodies that orbit the sun in a region known as the **Kuiper belt**. Pluto is one of the largest objects in the belt. You will learn more about the Kuiper belt in Lesson 18.

In order to learn more about the formation of our solar system, astronomers can study other solar systems and **extrasolar planets** that orbit stars other than our own sun. These are sometimes called **exosolar planets**. Astronomers can detect them by observing the Doppler shift of the star's spectrum—as the planet

Space isn't remote at all. It's only an hour's drive away if your car could go straight upwards.
—Sir Fred Hoyle
British astronomer and author
(1915–2001)

orbits the star, it tugs on the star and causes it to move slightly. Astronomers can also use the method of photometry to detect extrasolar planets—as a planet orbits in front of a star, the amount of light from the star that makes it to Earth is diminished. Because they are the easiest to detect, only massive, Jovian-like worlds unexpectedly close to their stars have been discovered so far.

Questions to Consider

- How does the solar nebula theory explain how our solar system was formed?

- What are the primary differences between the terrestrial planets and the Jovian planets?

- How is the solar nebula cleared away?

- How can studying extrasolar planets help us to understand our own solar system?

Watch the video for Lesson 18, *The Jovian Worlds*.

Key Terms and Concepts

Page references are keyed to the *Horizons* textbook.

1. **belt-zone circulation:** The atmospheric circulation typical of Jovian planets. Dark belts and bright zones encircle the planet parallel to its equator. (p. 408; video lesson; objectives 1, 2, 3, & 4)

2. **liquid metallic hydrogen:** A form of liquid hydrogen that is a good electrical conductor, found in the interiors of Jupiter and Saturn. (p. 410; video lesson; objectives 1 & 2)

3. **magnetosphere:** The volume of space around a planet within which the motion of charged particles is dominated by the planetary magnetic field rather than the solar wind. (p. 410; video lesson; objectives 1, 2, 3, & 4)

4. **forward scattering:** The optical property of finely divided particles to preferentially direct light in the original direction of the light's travel. (p. 414; video lesson; objective 5)

5. **Roche limit:** The minimum distance between a planet and a satellite that holds itself together by its own gravity. If a satellite's orbit brings it within its planet's Roche limit, tidal forces will pull the satellite apart. (p. 414; video lesson; objective 5)

6. **gossamer rings:** Jupiter's largest and most tenuous rings of dust. (p. 415; video lesson; objective 5)

7. **grooved terrain:** Regions of the surface of Ganymede consisting of parallel grooves. Scientists believe such terrain was formed by repeated fracturing of the icy crust. (p. 416; objective 5)

8. **tidal heating:** The heating of a planet or satellite because of friction caused by tides. (p. 417; video lesson; objective 5)

9. **oblateness:** The flattening of a spherical body usually caused by rotation. (p. 419; video lesson; objective 5)

10. **shepherd satellite:** A satellite that, by its gravitational field, confines particles to a planetary ring. (p. 421; objective 5)

11. **occultation:** The passage of a larger appearing body in front of a smaller appearing body. (p. 425; objective 5)

12. **ovoid:** The oval features found on Miranda, a satellite of Uranus. (p. 427; objective 5)

Summary

In Lesson 17, you used the method of comparative planetology to study the terrestrial planets by comparing them to Earth. In your study of the Jovian worlds, you'll use Jupiter as the basis of comparison.

A Travel Guide to the Outer Planets

The four Jovian worlds have a lot in common—they are large, gaseous planets with rings. Although Pluto is a part of the outer solar system, it has more differences than similarities with the planets, and therefore will be studied in Lesson 19.

These worlds formed at the same time as the terrestrial planets, but they are vastly different. Jupiter and Saturn are considered gas or liquid giants; they are composed of a solid core that is surrounded by liquid hydrogen. Uranus and Neptune are ice giants; they too have a solid core, but the planet is composed primarily of a slushy mixture of water, ices, and other minerals.

Jupiter

The most massive of the Jovian planets, Jupiter, is named after the Roman king of the gods. It contains 71 percent of all the planetary mass in our solar system, but it is not very dense. It formed far enough from the sun to capture large numbers of icy planetesimals. Because of its mass, it was able to grow even larger by gravitational collapse—it captured gas directly from the solar nebula and grew rich in hydrogen and helium. Its tremendous mass allows it to hold on to all of its gases.

Jupiter's center contains heavier elements, such as iron, nickel, and silicon. The core is under extreme pressure and its temperature is four times that of the surface of the sun. Because Jupiter emits twice as much energy than it absorbs from the sun, it appears that it still has heat left over from when it was formed. Surrounding the planet's core is a liquid interior composed mostly of hydrogen and helium. Under extreme pressure, the hydrogen in Jupiter's interior has become **liquid metallic hydrogen**, which is an excellent conductor of electricity.

Jupiter doesn't have a solid surface. The pressure at the base of the atmosphere is so great and the temperature so hot, the liquid interior gradually dissipates into the atmosphere. Farther away from the core, where the pressure and the temperature are reduced, the atmosphere becomes less dense. Jupiter's upper atmosphere is a very thin, turbulent layer of gases and clouds with high- and low-pressure areas. The planet's rapid rotation stretches the high- and low-pressure areas into beautiful belts and zones that we can detect from Earth. The form of atmospheric circulation that creates these cloud formations is called **belt-zone circulation**. Jupiter's high-pressure weather systems can remain stable for decades or centuries and create visible spots—such as the Great Red Spot.

Because the planet is composed primarily of gas and liquid, it rotates differentially. Jupiter's rotation and convection currents stir the conductive liquid in its interior and drive a *dynamo effect*, which produces a powerful magnetic field. This magnetic field creates Jupiter's **magnetosphere**—the volume of space around the planet within which the motion of charged particles is dominated by the planet's magnetic field. The magnetosphere traps high-energy particles in the solar wind to form the doughnut-shaped radiation belts that surround the planet. The interaction between the solar wind and Jupiter's magnetic field produces auroras around the north and south magnetic poles. They are larger than the diameter of Earth and are visible only in ultraviolet wavelengths.

Although the icy rings around Saturn are the best known, Jupiter also has a ring system. The dark, reddish ring around Jupiter is bright when it is illuminated from behind. This indicates that the diameters of particles within it are about the same size as the wavelength of visible light. This effect is called **forward scattering** and tells us that the ring is made of particles about the size of those found in cigarette smoke. The ring is within the planet's **Roche limit**—the orbital distance from the planet within which a moon cannot hold itself together by its own gravity. If a moon fell inside the Roche limit, it would be ripped apart by the planet's tidal forces. Because the billions of particles in the ring around Jupiter lie within its Roche limit, they cannot pull themselves together to form a moon.

The pressure of sunlight and the powerful magnetic field cause the dust specks in the ring to eventually spiral downward toward the planet. They also are forced to rise above and below the plane of the ring into a low-density halo by electromagnetic effects. In about one hundred years, the intense radiation around the planet grinds the dust specks down to nothing. Therefore, we can conclude that the particles that are currently within the ring are young. As particles spiral down into Jupiter's atmosphere, the ring is replenished with particles of dust that are knocked loose from Jupiter's moons. The ring is most dense near the outer edge where the small moon Adrastea orbits, providing evidence that the ring is replenished by the dust from the moon. Jupiter's faint gossamer rings extend twice as far from the planet as the main ring and are densest at the orbits of two of its small moons.

In addition to its rings, Jupiter also has an elaborate satellite system—it has four large moons and dozens of smaller moons. The four largest—the Galilean moons, named for their discoverer—are clearly related to one another and probably formed when Jupiter was formed. These large moons are Io, Europa, Ganymede, and Callisto. They orbit closer to Jupiter than all but four of the smaller moons. Jupiter's small outer moons are icy and less dense than the inner moons, because they presumably contain a greater percentage of ice.

The large moon closest to Jupiter—Io—seems to suffer from **tidal heating**. As the moon's elliptical orbit carries it nearer to and farther from the planet, the gravitational pull from Jupiter also changes. The variation in the gravitational pull squeezes and flexes the moon, which causes friction that heats its interior and makes it almost entirely molten. Because of this internal heating, it has more than 100 active volcanoes on its surface that spew sulfur-rich gas and ash. As the ash falls back to the surface, it fills in any impact craters and gives Io a relatively smooth appearance.

Europa also shows signs of geological activity—there are mountain-like folds across the relatively smooth surface. The pattern of folds and cracks in the ice suggests that the moon flexes, in a way similar to Io although not to the same extreme. Evidence suggests that a liquid-water ocean lies beneath Europa's icy crust.

Ganymede is the largest moon in the solar system. The bright regions of **grooved terrain** and older, cratered areas are believed to be systems of faults in the brittle crust. Some of the grooves overlap, which provides evidence that there were extended periods of geological activity in Ganymede's past.

Callisto is Jupiter's outermost large moon; astronomers suspect that it, too, has an ocean of liquid water beneath its icy crust. Many impact craters remain on the surface from the period of heavy bombardment, indicating the moon is geologically dead.

Saturn

Saturn—named for the protector of the sowing of seed—is perhaps the most distinct planet because of its beautiful icy rings. The sixth planet from the sun,

Saturn formed in the cold, outer reaches of the solar nebula where ice particles were more stable and may have contained more trapped gases than in the inner nebula.

Saturn is nearly the size of Jupiter in diameter, but it is much less massive. Saturn is less dense than water, which suggests that heavier elements sank to form a very small, solid metallic core while hydrogen and helium were captured by the nebula, allowing the planet to grow by gravitational collapse. Convection in the liquid mantle produces a magnetic field, which is 20 times weaker than that of Jupiter.

From Earth, it appears as if there are three large rings around Saturn. However, there are actually more than a thousand ringlets within the rings. The three rings that we can see from Earth have been labeled A, B, and C and have several large divisions in between. The A and B rings contain ice particles that are mostly the size of golf balls, while the C ring—the one closest to the planet—has larger, dark particles that contain more minerals and less ice than in the outer rings. Upon further inspection, astronomers have discovered a ring outside of A that they have labeled F, which is clumpy and braided.

Because they are made of billions of particles of ice, the rings can't have been formed when the planet was formed—the heat from the forming planet would have vaporized them. Astronomers suspect that they are made of debris from passing comets that have collided with Saturn's icy moons. As a result of the collision, ice would scatter and settle on the equatorial plane with some of the ice becoming trapped in the rings. Like Jupiter's rings, Saturn's rings can't pull themselves together to form a moon because they lie within the planet's Roche limit.

Saturn has nearly 50 moons that are composed of mixtures of rock and ice. Some of these moons are probably captured asteroids; others formed along with the planet. Some of them act as **shepherd satellites**—they gravitationally marshal the outer edges of the rings and keep them from spreading outward. These shepherd moons are locked in their orbits by the gravitational force of the larger moons. The planet's moons control the rings, like they do in the other Jovian planets.

Saturn's largest moon, Titan, is slightly bigger than Mercury. Saturn's other moons are small, icy worlds that are heavily cratered. Most of their surfaces are ancient indicating that any geological activity ceased long ago. Enceladus is an exception—some regions on its surface contain 1000 times fewer craters than others indicating activity since the period of heavy bombardment. Like small moons around other planets, tidal heating likely caused this geological activity.

Uranus

Uranus—named after the oldest of the Greek gods—is a giant icy world far from the sun. The most interesting feature of Uranus is that it rotates nearly on its side—

its equator is inclined to its orbit nearly 98°. This inclination may have been caused by a catastrophic event—a large planetesimal may have collided with Uranus.

The planet's orbital period is 84 Earth years. Because it is tipped on its side, the 21-year-long seasons are extreme. When the Voyager 2 passed it in 1986, Uranus' south pole was pointed almost directly at the sun.

Uranus is only one-third the diameter of Jupiter and has only one-twentieth Jupiter's mass. Models suggest that it has a small core of heavy elements. In the outer reaches of the solar nebula where the material was less dense, Uranus must have grown slowly. Unlike Jupiter and Saturn, it was not massive enough to grow by gravitational collapse. Where Jupiter and Saturn are rich in helium and hydrogen, Uranus is rich in water and ice. The deep mantle contains rocky materials and ammonia and methane dissolved in partly frozen water. Its magnetic field—highly inclined to its axis of rotation—may be generated by circulation in the electrically conductive mantle.

Uranus' atmosphere is a deep layer of hydrogen and helium with traces of methane. This methane absorbs red light and makes the atmosphere a beautiful shade of greenish blue. There is no visible belt-zone circulation except for a few changing cloud bands in both hemispheres that are presumably related to the seasons.

Another fascinating feature of this icy planet is that it, as well as Neptune, may have continuous hailstorms of diamonds. Models suggest that methane can decompose in the atmosphere due to its unique pressure and temperature. The carbon that is released can form crystals of diamonds the size of pebbles that can fall to the core.

Like the other Jovian planets, Uranus has rings; small satellites shepherd the particles and keep them from drifting off into space. The rings are not easily visible from Earth—they are very narrow, like hoops of wire. Astronomers were able to detect them during an **occultation**—when the planet passed in front of a star. Decreases in the amount of light from the star that reached the Earth revealed that ring material must be present. Because the rings were not bright when illuminated from behind, astronomers concluded that they must not contain much dust—the nine main rings contain meter-sized boulders of rock and ice. The rings may be dark because radiation breaks down the methane-rich ices to release carbon and darken the ices. The dust ejected by impacts on the planet's icy moons replenishes them.

The small moons that shepherd the rings are held in place by the larger outer moons. Two small moons—Ophelia and Cordelia—orbit just inside and outside the outermost ring. The two outer moons—Oberon and Titania—are less than half the diameter of Earth's moon. They have old, cratered surfaces and long faults. Liquid water may have once flowed over these moons and resurfaced them by filling in the craters. Moving inward, Umbrial has an ancient, cratered surface that shows no sign of geological activity. By contrast, Ariel has a cratered crust and flowing ice may have formed its smooth valleys. The innermost of the large moons, Miranda, is marked by **ovoids**—oval features believed to be caused by convection in the icy mantle

Neptune

The bright blue planet of Neptune was named for the god of the sea. It's slightly smaller in diameter than Uranus and it has a small core of heavy elements and a slushy mantle of water, ice, and rock. Because of its size, it grew slowly and was not large enough to capture great amounts of hydrogen and helium from the solar nebula.

Neptune is bluer than Uranus because it contains more methane in its hydrogen-rich atmosphere. Winds on Neptune blow up to 2,000 kilometers (1,200 miles) an hour. Heat flowing from the interior causes convection in the atmosphere and the rapid rotation of the planet produces high-speed winds, white clouds made of methane ice crystals, and spots caused by rotating storms.

Like the rings around Uranus, Neptune's are faint and difficult to detect from Earth and several small satellites marshal them. Some of Neptune's rings do not completely encircle the planet; rather, they consist of a series of arcs—stable clumps of dust separated by space. The gravitational effects of Galatea—a moon just inward from the ring—are thought to cause the ring arcs visible in the outer ring. The particles in the rings are dark, which provides evidence that they contain methane-rich ice that has been darkened by radiation. When illuminated from behind, the forward-scattered light indicates that Neptune's rings contain a lot of dust.

Neptune's two largest moons have unique orbits, which are probably a result of an interaction with a massive planetesimal. The largest moon, Triton, orbits Neptune backwards—clockwise as viewed from above. Triton is about 80 percent the diameter of Earth's moon and has a thin atmosphere composed of nitrogen and methane. You might wonder how a moon smaller than Earth's moon can hold onto an atmosphere. Because it is so far from the sun, its atmospheric temperature is only 37 K (−393°F) making the gases less likely to dissipate into space.

Triton's surface has been renewed by volcanism within the last million years and might still be active today. Radioactive decay may be the energy source for this volcanism. Instead of spurting molten rock, the cold planet's volcanoes may spew water and ammonia—this mixture can melt at low temperatures and erupt through its icy crust. The dark smudges visible in the southern polar ice cap, which are the effect of active geysers, also provide evidence that Triton is geologically active.

As you've discovered, the striking and beautiful Jovian planets are quite different from the terrestrial planets. Each of them has a ring system that is shepherded by a few of the many satellites that surround each planet. These giants are spheres of gas and liquid or ice, which hide the evidence of ever being stricken by wayward celestial objects.

Review Exercises

Matching

Match each term with the appropriate definition or description.

1. _____ magnetosphere	8. _____ forward scattering
2. _____ belt-zone circulation	9. _____ methane
3. _____ Roche limit	10. _____ occultation
4. _____ grooved terrain	11. _____ ovoids
5. _____ tidal heating	12. _____ ring arcs
6. _____ oblateness	13. _____ gossamer rings
7. _____ shepherd satellite	14. _____ liquid metallic hydrogen

a. The heating of a planet or satellite because of friction caused by tidal forces.

b. The minimum distance between a planet and a satellite that holds itself together by its own gravity.

c. A volume of space around a planet within which the motion of charged particles is dominated by the planetary magnetic field rather than by the solar wind.

d. Atmospheric behavior typical of Jovian planets.

e. The flattening of a spherical body.

f. Regions of Ganymede's surface produced by repeated fracture and refreezing of the icy crust.

g. An object that, by its gravitational field, confines particles to a planetary ring.

h. An optical property of finely divided particles to preferentially direct light in the original direction of the light's travel.

i. Oval features found on Miranda.

j. The passage of one celestial body in front of another.

k. The gas that makes the atmospheres of Uranus and Neptune appear blue.

l. Sections of incomplete rings, unique to the planet Neptune.

m. Jupiter's largest and most tenuous rings of dust.

n. This particular material is a good electrical conductor and is found in the interiors of Jupiter and Saturn.

Completion

Fill each blank in the sentences below with the most appropriate term from the list of completion answers that follow. A term may be used once, more than once, or not at all. Check your answers with the Answer Key and review when necessary.

belts	grooves	shepherd satellites
Callisto	Io	tidal heating
Europa	liquid metallic hydrogen	Titan
Galilean	Miranda	Voyager
Galileo	ovoids	zones

1. The four largest moons of Jupiter were discovered by
   _____ and are referred to as the
   _____ satellites.

2. Saturn's _____ and _____ are
   obscured by high haze in its atmosphere.

3. Although Uranus and Neptune have no _____
   in their interiors, they do have magnetic fields.

4. Evidence of active volcanism can currently be found on Jupiter's moon
   _____, and is believed to be caused by

   _____.

5. The smallest of the large moons of Uranus is _____,
   whose surface is marked by oval features known as

   _____.

Self-Test

Select the best answer.

1. The largest planet in the solar system is
   a. Jupiter.
   b. Saturn.
   c. Uranus.
   d. Neptune.

2. The oblateness of Jupiter and Saturn is the result of their
   a. ring system.
   b. rapid rotation.
   c. numerous satellites.
   d. gravitational resonances.

3. As viewed through a telescope, Jupiter appears as a
   a. spherical and heavily cratered object.
   b. bland and colorless disk.
   c. disk composed of various colored belts and bands.
   d. highly elliptical and bland disk.

4. Saturn, as viewed through a telescope, appears as a
   a. spherical and heavily cratered object.
   b. bland and nearly featureless disk.
   c. disk composed of brightly colored belts and bands.
   d. highly elliptical and bright blue disk.

5. Spacecraft photos revealed active volcanoes on the surface of Jupiter's moon
   a. Callisto.
   b. Titan.
   c. Ganymede.
   d. Io.

6. The grooved terrain on Ganymede is believed to be
   a. volcanic activity on the surface.
   b. systems of faults in the brittle crust.
   c. asteroid impacts.
   d. comet impacts.

7. The ring particles around Jupiter are very small, as evidenced by the observation that light
   a. produces a rainbow of colors.
   b. reflects off the particles in the backward direction.
   c. scatters in the forward direction.
   d. is absorbed by the ring particles.

8. The icy rings of Saturn are most likely not composed of particles left over from the formation of the planet but rather
   a. remnants of a moon that came within the planet's Roche limit.
   b. debris from collisions between the heads of comets and icy moons.
   c. captured micrometeorites.
   d. material captured as comets pass within the planet's Roche limit.

9. The core of Jupiter is composed of
   a. rock and ice.
   b. liquid and gaseous hydrogen.
   c. liquid helium.
   d. heavy elements such as iron, nickel, and silicon.

10. Jupiter's belts and bands are the result of atmospheric features that have been
    a. brought up by convection currents from deep within Jupiter's interior.
    b. refracted through the hydrogen and helium atmosphere.
    c. distorted by the impacts of comets.
    d. stretched and elongated as a result of Jupiter's rapid rotation.

11. The atmosphere of Titan is composed primarily of
    a. oxygen.
    b. nitrogen.
    c. methane.
    d. nothing; there is no atmosphere on Titan.

12. The Huygens probe to the surface of Titan revealed
    a. pebble-sized chunks of ice.
    b. liquid water lakes.
    c. liquid methane oceans.
    d. erupting volcanoes of sulfur.

13. The Cassini division is a large gap in the rings of Saturn. This gap is most likely produced by
    a. the reflection of sunlight off dark material.
    b. resonances between ring particles and the moon Mimas.
    c. shepherding satellites.
    d. the absorption of sunlight by dark material.

14. Two small _____ interact gravitationally with wandering ring particles to keep Saturn's irregularly twisted F ring confined.
    a. gaps
    b. shepherd satellites
    c. artificial satellites
    d. comets

15. Jupiter's Great Red Spot probably has an extended life because it
    a. originates from deep within the interior of Jupiter.
    b. is related to the position of the moon Io.
    c. lies continually over a liquid-water "surface."
    d. generates its own thermal energy.

16. The belts and zones in the atmosphere of Saturn are less visible than those on Jupiter because
    a. of thermal radiation.
    b. of extreme temperatures, both high and low.
    c. of the gravitational pull of Titan.
    d. they occur deeper in the cold atmosphere below a layer of methane.

17. Examination of photographs of Europa taken by the Galileo spacecraft in orbit around Jupiter support the conclusion that
    a. liquid water exists beneath the icy crust.
    b. the core of Europa is rocky.
    c. volcanoes are currently erupting.
    d. life exists beneath the surface.

18. Which of the following planets has its axis of rotation tipped the most?
    a. Uranus
    b. Neptune
    c. Saturn
    d. Jupiter

19. The discovery of Neptune was made as the result of
    a. a study of planetary nebulae.
    b. a search for comets.
    c. predictions following mathematical calculations.
    d. a search for near-Earth asteroids.

20. The very bland appearance of the atmosphere of Uranus is probably caused by
    a. the absence of hydrogen and helium.
    b. extremely cold temperatures and limited heat flow from its interior.
    c. the absence of methane in the atmosphere.
    d. the presence of methane in the atmosphere.

21. Neptune's blue appearance is the result of the presence of
    a. hydrogen.
    b. helium.
    c. ammonia.
    d. methane.

22. The rings of Neptune were discovered as the result of an occultation of a star by Neptune. Astronomers concluded that the rings were not complete because
    a. only one side of the planet yielded the expected dimming of the starlight.
    b. only half the rings were visible from Earth.
    c. theoretical calculations predicted these results.
    d. only half the rings were observed by the Voyager spacecraft.

23. The ring systems of Uranus and Neptune consist of very narrow rings. This narrowness is probably the result of
    a. random collisions that have driven particles out of the ring system.
    b. magnetic constrictions resulting from the interaction of the particles with the magnetic fields of the planets.
    c. shepherding satellites controlling the positions of the ring particles.
    d. solar radiation pressure.

24. The oval patterns of grooves (ovoids) on the surface of Uranus's moon Miranda is explained by
    a. gravitational influences produced by Uranus.
    b. tidal influences produced by the other nearby moons Umbriel and Ariel.
    c. collisions between Miranda and the ring particles.
    d. internal heat that produced convection in the icy mantle.

25. Dark smudges on the surface of Neptune's moon Triton appear to be the result of
    a. organic polymers resulting from ammonia interacting with Neptune's magnetic field.
    b. impact debris from asteroid collisions releasing organic matter under the icy crust falling back to the surface as smudges behind the craters.
    c. methane, converted by sunlight into dark deposits, ejected into space during the eruption of geysers of liquid nitrogen.
    d. disintegration of ring particles falling on the planet and confined to thin smudges by the planet's magnetic field.

Short-Answer Questions

1. What are the primary components of Jupiter and Saturn? What are the gases?

2. How is the physical appearance of Saturn's clouds different from that of Jupiter's clouds?

3. What is the Great Red Spot?

4. What constituent of the atmosphere of Neptune makes it appear blue? Why does this blue color appear?

Applications

1. How do scientists explain the volcanic activity on Io?

2. How does the rapid rotation of Jupiter affect its cloud patterns?

3. Saturn is the most oblate planet in the solar system. Why is Saturn so oblate and what does this tell you about Saturn's interior?

4. An occultation of a star revealed that both Uranus and Neptune had rings. However, the rings of these two planets are not identical. How are these rings different and what explanation can account for these differences?

Answer Key

Matching I

1. c (p. 410; video lesson; objectives 1, 2, 3, & 4)

2. d (p. 408; video lesson; objectives 1, 2, 3, & 4)

3. b (p. 414; video lesson; objective 5)

4. f (p. 416; objective 5)

5. a (p. 417; video lesson; objective 5)

6. e (p. 419; video lesson; objective 5)

7. g (p. 421; objective 5)

8. h (p. 414; video lesson; objective 5)

9. k (pp. 425, 430; objectives 3 & 4)

10. j (p. 425; objective 5)

11. i (p. 427; objective 5)

12. l (p. 429; objective 5)

13. m (p. 415; video lesson; objective 5)

14. n (p. 410; video lesson; objectives 1 & 2)

Completion

1. Galileo, Galilean (pp. 415–416; objective 5)

2. belts, zones [in any order] (pp. 412–413; video lesson; objective 2)

3. liquid metallic hydrogen (pp. 410, 425, 427, 430–431; video lesson; objectives 3 & 4)

4. Io, tidal heating (p. 417; video lesson; objective 5)

5. Miranda, ovoids (p. 427; objective 5)

Self-Test

1. a (p. 409; video lesson; objective 1)

2. b (p. 419; video lesson; objectives 1 & 2)

3. c (pp. 412–413; video lesson; objective 1)

4. b (p. 418; objective 2)

5. d (p. 417; video lesson; objective 5)

6. b (p. 416; objective 5)

7. c (p. 414; video lesson; objective 5)

8. b (p. 420; video lesson; objective 5)

9. d (p. 410; objective 1)

10. d (p. 413; objective 1)

11. b (p. 423; video lesson; objective 5)

12. a (p. 423; video lesson; objective 5)

13. b (p. 421; objective 5)

14. b (p. 421; objective 5)

15. a (p. 413; video lesson; objective 1)

16. d (p. 418; objective 2)

17. a (p. 417; video lesson; objective 5)

18. a (p. 425; objective 3)

19. c (pp. 429, 431; objective 4)

20. b (p. 425; objective 3)

21. d (p. 430; objective 4)

22. a (pp. 428–429; objective 5)

23. c (pp. 428–429; objective 5)

24. d (p. 427; objective 5)

25. c (p. 431; objective 5)

Short-Answer Questions

1. The interiors of both Jupiter and Saturn consist of a rocky core made of iron, nickel, and other heavy elements. Because of the lower density of Saturn, its core is much smaller than Jupiter's. The cores of both planets are covered by liquid metallic hydrogen. Again, the amount of the liquid metallic hydrogen in Saturn is much less than that in Jupiter, as evidenced by the weaker magnetic field of Saturn. Beyond the liquid metallic hydrogen is liquid hydrogen, the predominant component of both planets. The atmospheres of the two planets consist of hydrogen and helium gas. One thing should be remembered—there is no surface to either planet. The atmosphere transitions from a gaseous to a liquid state without any sharp boundaries. (pp. 410–413, 418–419, 422; video lesson; objectives 1 & 2)

2. Because Saturn is much farther from the sun than Jupiter (nearly twice the distance), its belts and zones are only faintly visible from Earth. They are present but are deeper in the atmosphere and covered with a layer of methane because of the much lower temperatures in the atmosphere. (pp. 418–419; objectives 1 & 2)

3. The Great Red Spot is a giant circulating storm, similar to a hurricane, within the atmosphere of Jupiter. It has been visible for more than 300 years. Its stability (permanence) is probably the result of its origination deep within the planet's interior. (pp. 412–413; video lesson; objective 1)

4. The constituent of the atmosphere that makes Neptune appear blue is methane. This gas removes red photons from the white light from the sun leaving only the blue light to be reflected. A similar explanation gives Uranus its blue color as well. (p. 430 objective 4)

Applications

1. Because Io has an elliptical orbit, the gravitational force on this moon varies from place to place, flexing the moon with tides. This flexing of Io causes friction, and friction causes heat. There is enough heat to melt the interior and produce the volcanoes observed by the Voyager and Galileo spacecraft. (p. 417; video lesson; objective 5)

2. The rapid rotation of Jupiter causes the cloud patterns that are observed to stretch into belts and bands. Only those clouds that originate

deep within the interior retain their oval and hurricane like shape for extended periods of time. (pp. 412–413; objective 1)

3. The rapid rotation of Saturn, once every 10.5 hours causes the planet to become flattened or oblate. The oblateness of Saturn reveals that the planet is mostly liquid. Because the composition is mostly hydrogen, the liquid is hydrogen in liquid form with some metallic hydrogen, a form of hydrogen that conducts electricity. By measuring the density of the planet, a small rocky core must be added to complete the model of Saturn's interior. (pp. 418–419; objective 2)

4. The rings of Uranus and Neptune are very thin. The ring particles are confined to narrow bands as the result of the gravitational influence of shepherding satellites or moons. However, there is a larger satellite in orbit around Neptune called Galatea, that seems to add an additional influence. This moon, along with other small ones, causes some of the particles around Neptune to concentrate in certain regions and avoid others. This produces what are referred to as ring arcs. In other words, the rings around Neptune are not as continuous as the rings around Uranus. (pp. 425–426, 428–429, 430–431; objective 5)

Lesson Review

PLEASE NOTE: Use this matrix to guide your study and achieve the learning objectives of this lesson. It will also help you to view the video, which defines and demonstrates important concepts and principles as they relate to everyday life and actual case studies.

Learning Objective	Textbook	Student Guide
1. Discuss the physical and atmospheric characteristics of Jupiter and describe the evidence that explains their origin.	pp. 409–413, 417–418	Key Terms: 1, 2, 3; Matching: 1, 2, 14; Self-Test: 1, 2, 3, 9, 10, 15; Short-Answer: 1, 2, 3; Applications: 2.
2. Discuss the physical and atmospheric characteristics of Saturn and describe the evidence that explains their origin.	pp. 418–419, 422, 424	Key Terms: 1, 2, 3; Matching: 1, 2, 14; Completion: 2; Self-Test: 2, 4, 16; Short-Answer: 1, 2; Applications: 3.
3. Discuss the physical and atmospheric characteristics of Uranus and describe the evidence that explains their origin.	pp. 424–425, 427	Key Terms: 1, 3; Matching: 9; Completion: 3; Self-Test: 18, 20.
4. Discuss the physical and atmospheric characteristics of Neptune and describe the evidence that explains their origin.	pp. 427, 430–432	Key Terms: 1, 3; Matching: 9; Completion: 3; Self-Test: 19, 21; Short-Answer: 4.
5. Describe the characteristics and origins of the ring and satellite systems of the Jovian planets.	pp. 414–417, 419–421, 423, 425–431	Key Terms: 4, 5, 6, 7, 8, 9, 10, 11, 12; Matching: 3, 4, 5, 6, 7, 8, 10, 11, 12, 13; Completion: 1, 4, 5; Self-Test: 5, 6, 7, 8, 11, 12, 13, 14, 17, 22, 23, 24, 25; Applications: 1, 5.

Notes:

LESSON
19

Solar System Debris

Checklist

For the most effective study of this lesson, complete the following activities in this sequence.

Before Viewing the Video

❑ Read the Preview, Learning Objectives, and Viewing Notes below.

❑ Read Chapter 19, "Meteorites, Asteroids, and Comets," pages 438–455, and Section 18-6, "Pluto and the Dwarf Planets," pages 432–434, in the *Horizons* textbook.

What to Watch

❑ After reading the textbook chapter, watch the video for Lesson 19, *Solar System Debris*.

After Viewing the Video

❑ Briefly note your answers to questions listed at the end of the Viewing Notes.

❑ Review the Summary below.

❑ Review all reading assignments for this lesson, especially the Chapter 19 summary on page 454 in *Horizons* and the Viewing Notes in this guide.

❑ Write brief answers to the review questions at the end of Chapter 19 in *Horizons*.

❑ Complete the Review Exercises below. Check your answers with the Answer Key and review when necessary.

❑ Use the Lesson Review matrix found at the end of this lesson to review and assess your knowledge of each Learning Objective.

❑ As assigned by your instructor, complete the Applications activities and any additional activities for this lesson.

Preview

In previous lessons, you discovered how the planets and the sun in our solar system were formed. The exploration of our solar system concludes with this lesson as we look at some of the objects left over from the time of its formation. Objects such as asteroids, meteoroids, and comets might be considered "space debris," but they can be responsible for affecting life on Earth.

The ghost-like appearances of comets in the night sky have influenced kings and civilizations throughout history. Comets acquired a reputation over time as being portents of doom; bad things seemed to happen on Earth when comets appeared. Once believed to be the omen of death and destruction, modern science reveals that the fragile, icy comet could very well be one of the sources of Earth's first oceans. Although asteroids and meteoroids don't hold the same amount of intrigue as the comet, they still hold valuable information about when and how the solar system was formed.

On occasion, a piece of space debris strikes a celestial body. Unlike the moon, where impact craters are easily visible, impact craters on Earth are more difficult to find. This is the result of surface erosion by wind, water, and plate tectonics that you learned about in Lesson 17.

CONCEPTS TO REMEMBER

- Recall from Lesson 16 that the *Kuiper belt* is the collection of icy planetesimals believed to orbit in a region from just beyond Neptune out to 100 AU or more from the sun. Some comets that we see today may have originated in this area (p. 314 in this guide).

- In Lesson 16, you also learned that the *Oort cloud* is a swarm of icy bodies believed to lie in a spherical shell extending to 100,000 AU from the sun. Scientists have theorized that this region is the source of comets (p. 314 in this guide).

Learning Objectives

After you complete this lesson, you should be able to:

1. Describe the broad categories of meteorites, their compositions, and possible origins. *HORIZONS* TEXTBOOK PAGES 440–441.

2. Describe the characteristics of meteors and meteor showers, and explain how they relate to cometary orbits. *HORIZONS* TEXTBOOK PAGES 441–443.

3. Describe the physical properties, orbital characteristics, and possible origins of asteroids. *HORIZONS* TEXTBOOK PAGES 444–447.

4. Describe the parts and observed properties of comets and explain the effect of solar wind on the tails of a comet. *HORIZONS* TEXTBOOK PAGES 445, 448–450.

5. Explain the significance of the Oort cloud and Kuiper belt. *HORIZONS* TEXTBOOK PAGES 433–434, 450–451.

6. Explain the possible effects of an asteroid or comet collision with Earth and describe efforts to find near-Earth objects. *HORIZONS* TEXTBOOK PAGES 451–453.

7. Describe the physical and geological characteristics of Pluto and summarize the debate over its classification as a planet versus a Kuiper belt object. *HORIZONS* TEXTBOOK PAGES 432–434.

At this point, read Chapter 19, "Meteorites, Asteroids, and Comets," pages 438–455, and Section 18-6, "Pluto and the Dwarf Planets," pages 432–434.

I came in with Halley's Comet in 1835. It is coming again next year, and I expect to go out with it...
—Mark Twain
American author and humorist
(1835–1910)

Viewing Notes

Although the period of heavy bombardment ended 4 billion years ago, space debris continues to fall on the planets and moons in our solar system. Astronomers believe these leftovers of the solar nebula played a major role in the establishment of life on Earth and later, in the mass extinction of various species.

The video program contains the following segments:

- ❂ Meteors
- ❂ Asteroids
- ❂ Comets
- ❂ Pluto & Kuiper-Belt Objects
- ❂ Near-Earth Objects

The following information will help you better understand the video program:

Throughout history, objects from space have collided with Earth, sometimes with devastating results. However, not all objects that enter Earth's atmosphere do harm on Earth. In fact, the "shooting stars" that we sometimes see in the night sky are actually little grains of dust so small that they burn up in the atmosphere and produce flashes of light called meteors.

The asteroid belt is a geological treasure trove containing evidence that supports the solar nebula theory of the formation of the solar system. The belt that lies between Mars and Jupiter is believed to contain the remnants of material that never accreted into a small planet because they were unable to resist the pull of Jupiter's gravity.

Comets were thought to be harbingers of bad luck by ancient civilizations. Made primarily of water-ice mixed with dust and other organic materials, they may be precursors of life on Earth and possibly other planets. The beautiful tail that you see is actually dust and gas from the nucleus of the comet.

QUESTIONS TO CONSIDER

- What is the most accepted explanation for the origin of Earth's moon?

- How do meteors, meteoroids, and meteorites differ?

- What is an asteroid?

- Where is the asteroid belt? What is believed to be the origin of this belt?

- What is a comet? Where do comets originate?

- Why is there a debate as to whether Pluto is a planet or one of the largest objects in the Kuiper belt?

Watch the video for Lesson 19, *Solar System Debris*.

Key Terms and Concepts

Page references are keyed to the *Horizons* textbook.

1. **Widmanstätten pattern:** The bands in iron meteorites that appear because of the presence of large crystals of nickel-iron alloys. (p. 441; objective 1)

2. **chondrite:** A stony meteorite that contains chondrules. (p. 441; objective 1)

3. **chondrule:** A round, glassy body found in some stony meteorites. Chondrules are believed to have solidified very quickly from molten drops of silicate material. (p. 441; objective 1)

4. **carbonaceous chondrite:** A stony meteorite that contains both chondrules and volatiles. These chondrites may be the least-altered remains of the solar nebula still present in the solar system. (p. 441; video lesson; objective 1)

5. **achondrite:** A stony meteorite containing no chondrules or volatiles. (p. 441; objective 1)

6. **stony-iron meteorite:** A meteorite composed of stone and iron mixed together. (p. 441; video lesson; objective 1)

7. **meteor shower:** A multitude of meteors that appear to come from the same region of the sky and are believed to be caused by comet debris. (p. 441; video lesson; objective 2)

8. **Oort cloud:** A swarm of icy bodies believed to lie in a spherical shell extending to 100,000 AU from the sun. Scientists have advanced the hypothesis that this region is the source of long-period comets (p. 450; video lesson; objective 5)

9. **gas tail (Type I):** A comet tail composed of ionized gas atoms released from the nucleus and carried outward by the solar wind. (p. 448; video lesson; objective 4)

10. **dust tail (Type II):** A comet tail composed of dust released from the nucleus and pushed away by the pressure of sunlight. (p. 448; video lesson; objective 4)

11. **sublimation:** The process of a solid going directly to a gas without becoming a liquid first. (video lesson; objective 4)

12. **coma:** The glowing head of a comet. (p. 448; video lesson; objective 2)

13. **Kuiper belt:** The collection of icy planetesimals believed to orbit in a region from just beyond Neptune out to 100 AU or more. (p. 443; video lesson; objective 3)

14. **dwarf planets:** The family of icy worlds that orbit beyond Neptune in our solar system. (p. 433; objectives 5 & 7)

15. **Plutinos:** Icy Kuiper belt objects that, like Pluto, are caught in a 3:2 orbital resonance with Neptune. (p. 443; video lesson; objective 3)

Summary

Our study of the solar system concludes with comets, asteroids, meteorites, and Kuiper belt objects. This "space debris" can provide us with clues as to how the solar system was formed.

Pluto

Pluto is a unique world that orbits Earth in the far reaches of the solar system. It is a small icy world that is difficult to detect from Earth and has not yet been visited by a spacecraft. It was previously categorized as one of the 9 planets of the solar system, but in 2006, the IAU (International Astronomical Union) voted to demote Pluto to **dwarf planet** status. Some astronomers considered it a planet and others considered it one of the largest known objects in the **Kuiper belt**—a region beyond Neptune where thousands of icy bodies orbit the sun.

Pluto is 65 percent the diameter of Earth's moon and has areas of light and dark terrain. It has a thin atmosphere composed of nitrogen and carbon monoxide with traces of methane. It's cold enough to freeze most compounds that are gases on Earth.

On average, Pluto is usually the farthest planet from the sun, but its orbit is highly elliptical. Pluto's orbit crosses inside of Neptune's bringing the small planet closer to the sun for 20 years out of its 247.7 Earth-year orbit. Pluto is tidally locked to its moon Charon so they always keep the same side toward one another; Charon's highly inclined orbit causes Pluto to rotate on a highly inclined axis.

These characteristics don't sound that unusual, so you might wonder why some people don't consider Pluto to be a planet. To understand this, we must consider Pluto's neighbors—the objects in the Kuiper belt. Pluto is the one of the largest known bodies in the belt that lies just beyond Neptune, but there probably are other objects in this belt that are as big or larger. A few of them appear to have their own moons or they may be two bodies orbiting in a binary system.

More than a dozen of the objects in the Kuiper belt, including Pluto, are locked in a 3:2 orbital resonance with Neptune. These bodies have become known as **Plutinos** and they may have become caught in resonance with Neptune by gravitational interactions. It's possible that Uranus and Neptune may have formed closer to the sun than they are now and may have shifted outwards due to gravitational interactions with Jupiter and Saturn. Pluto may have started in the Kuiper belt and was pulled out of its original orbit as Neptune migrated outwards toward its current orbit making it seem more like a planet.

Meteorites

In Lesson 16, you learned that the solar system is filled with *meteoroids*— most of which are specks of dust, grains of sand, or tiny pebbles. When meteoroids fall into Earth's atmosphere, the friction with the atmosphere vaporizes them and they create streaks of light in the sky called *meteors*. If the object is large and strong enough to survive passage through Earth's atmosphere and make it to the surface, it is called a *meteorite*.

Astronomers can study meteorites to learn about the history of our solar system and the universe—their composition can help us understand their origins. **Stony-iron meteorites** are mixtures of iron and stone and appear to have formed when a mixture of molten iron and rock cooled and solidified. There are two other categories of meteorites: iron and stony.

As the name implies, an *iron meteorite* is a solid chunk of iron and nickel that makes it dense and heavy. The passage through Earth's atmosphere may form it into a unique shape with a dark and rusted surface. When an iron meteorite is sliced open, polished, and etched with nitric acid, it will reveal **Widmanstätten patterns**—regular bands formed by slow cooling of nickel-iron alloys. These patterns indicate that the rocks may have cooled from a molten state no faster than a few degrees every million years.

Stony meteorites are made of silicates and resemble Earth rocks. A stony meteorite will tend to have a crust that has been fused by melting in Earth's atmosphere. There are two types of stony meteorites: chondrite and achondrite. A **chondrite** has a chemical composition that resembles a cooled lump of matter originating from the sun but that has all of its volatile gases removed. Different properties found in various chondrites may suggest how they were formed.

Most chondrites contain **chondrules**—rounded bits of glassy rock. These chondrules appear to have originated when the solar system was young as droplets of molten rock cooled and hardened rapidly. Some chondrules show evidence that a chondrite was slightly heated—hot enough to release the volatile gases in the meteorite, but not hot enough to melt the chondrules. A **carbonaceous chondrite** generally contains both chondrules and volatile compounds, including carbon. These volatile-rich chondrites were probably never heated enough to vaporize their gases and are perhaps the most pristine objects in our solar system.

Another type of stony meteorite is called an **achondrite**, which contains no chondrules and has a composition similar to Earth's lavas. This type of meteorite may have been subjected to intense heat that vaporized the chondrules and volatile gases.

One way we tried to understand the origin of meteorites was by studying **meteor showers**—a multitude of meteors that appear to come from the same region of the sky. Studies of meteors reveal that most originate from tiny bits of debris from comets. Meteoroids from comets are small, not structurally strong, and don't reach Earth. But, astronomers can study the meteors and deduce that their motion matches the orbits of comets.

As you've discovered in the last few lessons, the planets and the moons in our solar system all show signs of being struck by space debris. In order for a meteoroid to reach Earth, it must be made of much stronger material than comet debris; evidence reveals that they come from asteroids. The iron and stone meteorites appear to be fragments of planetesimals that were large enough to evolve and differentiate to form iron and nickel cores with rocky mantles. The types of meteorites can reveal what part of the planetesimal they were from. Iron meteorites appear to be remains of the iron core, stony meteorites appear to come from the mantle, and stony-iron meteorites appear to come from the boundary where the stony mantle meets the iron core. Smaller bodies that never differentiated probably formed the chondrites; carbonaceous chondrites may have been formed in small, cold bodies far from the sun.

There are more than 150 meteorite craters on Earth—evidence that large impacts have occurred in the past and are likely to occur in the future.

Scientists have studied sediments on Earth that settled about 65 million years ago—at the time that the dinosaurs became extinct. Analysis of the samples indicates that the sediment contains a large amount of iridium, which is rare on Earth, but common in meteorites. Many scientists agree that the impact of a large meteorite on Earth may have caused the dinosaurs to become extinct and destroyed 75 percent of all life on Earth. Another impact that occurred 250 million years ago may have extinguished 95 percent of life in the oceans and 80 percent of life on land. Many creatures would have died from the initial impact of the meteorite, but many more would have died from the effects that the collision had on global climate and on the atmosphere.

Asteroids

As you learned in Lesson 16, some asteroids can be considered minor planets. Roughly 20,000 asteroids have been identified. Most asteroids are small, but tens of thousands are larger than 10 km (6 mi), and about 200 are larger than 100 km (60 mi) in diameter. Many orbit around the sun in the asteroid belt between Mars and Jupiter and provide evidence that they are the remains of a planet that failed to form there.

There are three main types of asteroids: S-type, M-type, and C-type. S-type asteroids are the most common and appear to be the source of the most common chondrites. They are bright and appear reddish. M-type asteroids may be mostly iron-nickel alloys and are not very dark or red. C-type asteroids appear to be carbonaceous and are as dark as a lump of coal. They are more common in the outer asteroid belt where it is cooler.

Many asteroids are irregular in shape and heavily cratered, which is evidence of past collisions. Some are composed of floating piles of debris with large spaces between the fragments. A few asteroids may have developed molten cores, differentiated, and became geologically active. Some may have incorporated short-lived radioactive elements from a supernova explosion, which provided their interior heat.

Comets

You discovered in Lesson 16 that comets are objects that orbit the sun and produce tails of gas and dust when they approach it. The ices in the comet are made of water and other volatile compounds such as carbon dioxide, carbon monoxide, methane, and ammonia. These gases would have been typical in the outer solar nebula, which seems to indicate that comets were formed along with the planets.

The nucleus of the comet is irregular in shape and is composed of a fluffy mixture of ices and dust. The nucleus is darker than a lump of coal, indicating that the dust is composed of carbonaceous chondrites. The cloud of gas and dust that surrounds the nucleus of the comet is called the **coma**. Although the nucleus is typically quite small, the coma can be up to 1,000,000 km in diameter.

Comets typically have two tails: **type I**, also known as a **gas tail**, and **type II**, also known as a **dust tail**. The tails generally face away from the sun because the pressure of the sunlight and the solar wind pushes them away. The precise direction of the tails depends on the orbital motion of the nucleus and the direction of the solar wind. Gas carried away from the nucleus is ionized by ultraviolet sunlight. This ionized gas is swept away from the nucleus by the solar wind to produce the gas tail. Dust from the vaporizing nucleus is pushed gently outward by the pressure of sunlight to create the dust tail. The dust tail is typically curved—the dust particles enter into their own orbit around the sun as they are blown away from the nucleus.

About 600 comets have been well studied and most of them have orbital periods of more than 200 years; these are deemed long-period comets. Their orbits do not follow the disk shape of the solar system and only about half revolve in the counterclockwise direction like the other objects in our solar system. Long-period comets are believed to come from the **Oort cloud**—a spherical cloud that extends from 10,000 to 100,000 AU from the sun and contains trillions of icy bodies.

About 100 comets have orbital periods of less than 200 years; these are known as short-period comets. Most orbit counterclockwise around the sun and their orbital planes lie within 30° of the plane of the solar system. Some of these comets may have originated in the Oort cloud, but it is believed that most of these short-period comets originated in the Kuiper belt.

As the comet passes the sun, it loses a great amount of material. The nucleus slowly loses its ice until there is nothing left of the comet. Because of its delicate nucleus of dust, ice, and empty space, a comet may only last between 100 and 1,000 orbits around the sun. This tells us that the short-period comets that we see today can't have survived in their current orbits since the solar system was formed—there must be a continuous supply of new comets coming from the Kuiper belt and the Oort cloud.

These comets, asteroids, and meteoroids may hold the final clues to the formation of our solar system and may alter the course of our planet's future. These objects may have contributed to life on Earth by delivering the organic material that the young planet lacked. On the other hand, they have altered Earth's geology and may have caused mass-destruction of life on Earth.

Review Exercises

Matching

Match each term with the appropriate definition or description.

1. _____ achondrite		11. _____ meteor	
2. _____ carbonaceous chondrite		12. _____ meteorite	
3. _____ chondrite		13. _____ meteoroid	
4. _____ chondrule		14. _____ Type I comet tail	
5. _____ coma		15. _____ Plutino	
6. _____ meteor shower		16. _____ Type II comet tail	
7. _____ Oort cloud		17. _____ stony-iron meteorite	
8. _____ Widmanstätten pattern		18. _____ asteroid	
9. _____ Kuiper belt		19. _____ sublimation	
10. _____ radiant point		20. _____ dwarf planets	

a. Bands in iron meteorites due to large crystals of nickel-iron alloys.

b. A stony meteorite that contains chondrules.

c. A round, glassy body found in some stony meteorites.

d. A stony meteorite that contains both chondrules and volatiles.

e. An event that occurs when Earth passes near the orbit of a comet, probably caused by dust and debris released by a comet.

f. The glowing head of a comet.

g. The hypothetical source of long-period comets.

h. A small bit of matter heated by friction to incandescent (glowing) vapor as it falls into Earth's atmosphere.

i. The region beyond Neptune that is a reservoir for short-period comets.

j. A small object or particle in space before it enters Earth's atmosphere.

k. A small region in the sky from which meteor showers appear to originate.

l. A small object or particle that survives passage through Earth's atmosphere and reaches the ground.

m. One of the icy objects that is caught in a 3:2 orbital resonance with Neptune.

n. A comet tail composed of dust released from the nucleus and pushed away by the pressure of sunlight.

o. A comet tail composed of ionized gas atoms released from the nucleus and carried outward by the solar wind.

p. A meteorite composed of stone and iron mixed together.

q. A stony meteorite containing no chondrules or volatiles.

r. A small, rocky world that is primarily found between the orbits of Mars and Jupiter.

s. Small icy worlds that orbit beyond Neptune.

t. The transition of a substance from the solid phase directly to the gas phase.

Completion

Fill each blank in the sentences below with the most appropriate term from the list of completion answers that follow. A term may be used once, more than once, or not at all. Check your answers with the Answer Key and review when necessary.

closer	farther	Oort cloud
comets	gas	stone
dust	Kuiper belt	sun
elliptical	meteors	Widmanstätten

1. _____ patterns are found in iron meteorites.

2. The Oort cloud is believed to be the origination point for long period _____.

3. Scientists hypothesize that the _____ is the location where most short-period comets formed.

4. The type II, or _____, tail of a comet always faces away from the _____.

5. Because Pluto's orbit is highly _____, that planet is sometimes _____ to the sun than is Neptune.

Self-Test

Select the best answer.

1. Pluto has characteristics that astronomers feel may link it and its moon to the
    a. asteroids.
    b. ovoids.
    c. Kuiper belt objects.
    d. protoplanets.

2. The mass of Pluto was determined
    a. by direct observation from Voyager.
    b. from the size of the orbit and period of its moon, Charon.
    c. by direct observation from Earth-based telescopes.
    d. by direct observation from the Hubble Space Telescope.

3. The Kuiper belt objects called Plutinos are in a 3:2 resonance with the planet Neptune. This means that
   a. these objects orbit the sun three times as Neptune orbits twice.
   b. these objects orbit the sun twice as Neptune orbits three times.
   c. these objects orbit the sun once as Neptune orbits six times.
   d. these objects orbit the sun three times as Neptune orbits six times.

4. Which of the following are classifications of meteorites?
   a. basalt and igneous
   b. irons and magnetite
   c. iron and stony
   d. shower and crystal

5. Because of the delicate nature of the carbonaceous chondrites, it is believed that these meteorites came from planetesimals that
   a. orbit the sun near the planet Mercury.
   b. collided with Earth billions of years ago.
   c. orbited far from the sun in regions of the solar system that are very cold and are probably the least altered remains of the solar nebula.
   d. collided with Mars, causing material to be ejected from the planet and eventually land on Earth.

6. The iron meteorites probably formed from
   a. large planetesimals that melted and differentiated to form an iron core.
   b. condensing iron atoms during the cooling period of the early solar nebula.
   c. collisions with Mars-sized planetesimals.
   d. cometary impacts with asteroids.

7. Most often, a meteor is
   a. a particle from outer space.
   b. a particle from outer space that lands on Earth.
   c. produced by tiny bits of debris from comets.
   d. a small asteroid.

8. A meteor shower occurs when
   a. an asteroid collides with Earth.
   b. Earth passes through the dust and bits of material left in the orbit of a passing comet.
   c. meteorites are found in a short period of time.
   d. Earth reaches its nearest position to the sun (perihelion).

9. M-type asteroids appear to
   a. orbit in a fixed position between Jupiter and Mars.
   b. be made of stone and silicates and are red in color.
   c. consist of carbon and look very dark.
   d. be made of iron-nickel alloy and are very bright.

10. The formation of asteroids is the result of
    a. a collision between two planets.
    b. planetary material that never formed into large planets.
    c. an exploded planet.
    d. the escape of Jovian-type satellites.

11. Many collisions caused planetesimals to fragment into smaller pieces.
    These pieces probably make up the material known as
    a. meteor showers.
    b. meteorites.
    c. asteroids.
    d. planetary moons.

12. The solar wind directs the gas tail of a comet
    a. outward, away from the sun at all times.
    b. away from the sun when the comet enters the solar system and
       toward the sun when it leaves the solar system.
    c. toward the sun when the comet enters the solar system and away
       from the sun when it leaves.
    d. toward the sun at all times.

13. The nucleus of the comet contains mostly the ices of water, carbon
    dioxide, and ammonia. The coma is produced when the
    a. nucleus explodes.
    b. comet reaches its most distant point from the sun (aphelion) and the
       ices vaporize.
    c. tail of the comet compresses back toward the nucleus.
    d. sun causes sufficient heating and the ices of the nucleus vaporize to
       produce a vast cloud of gas and dust.

14. Studies show that the short-period comets cannot originate at a distance
    thousands of astronomical units from the sun. A closer source for
    short-period comets is the
    a. Kuiper belt.
    b. Oort cloud.
    c. Lagrangian points.
    d. asteroid belt.

15. The spacecraft that intercepted Comet Halley in 1985 and 1986 discovered that the nucleus of Comet Halley is
    a. perfectly round and smooth.
    b. bright and irregularly shaped.
    c. very dark and irregularly shaped.
    d. very dark and perfectly smooth, with no obvious relief.

16. Jets of high speed gas are seen to be emitted from the nucleus of Comet Halley only
    a. from the side of the comet point away from the sun.
    b. when the comet is at aphelion.
    c. during the period of time when the tail is visible.
    d. from the sunlit side of the comet.

17. Evidence that Earth has collided with an asteroid-sized object is suggested by a layer of the rare element
    a. ytterbium.
    b. plutonium.
    c. iridium.
    d. planetarium.

18. The "smoking gun" crater that fits the description of the asteroid impact that may have caused the extinction of the dinosaurs was found
    a. off the coast of Hawaii.
    b. off the coast of the Yucatán peninsula.
    c. outside of Flagstaff, Arizona.
    d. in Crater Lake, Oregon.

Short-Answer Questions

1. What is a Plutino and how are these objects connected to Pluto?

2. Explain why, at times, Pluto is not the most distant planet in the solar system.

3. Of the two types of comet tails, which one emits light and which one reflects sunlight?

4. Describe the characteristics of the most common type of asteroid.

5. What is a NEO?

Applications

1. What evidence suggests that the meteorites appear to be fragments of planetesimals that have differentiated?
2. What are the differences between the Kuiper belt of objects and the Oort cloud?
3. What is the evidence that suggests that comets or asteroids have struck Earth?
4. How did the discovery of its orbiting moon Charon help to develop a better model of Pluto's interior?

Answer Key

Matching

1. q (p. 441; objective 1)
2. d (p. 441: video lesson; objective 1)
3. b (p. 441; objective 1)
4. c (p. 441; objective 1)
5. f (p. 448; video lesson; objective 4)
6. e (p. 441; video lesson; objective 2)
7. g (p. 450; video lesson; objective 5)

8. a (p. 441; objective 1)

9. i (pp. 433–434, 451; video lesson; objectives 5 & 7)

10. k (pp. 432–433; objective 2)

11. h (p. 440; objective 2)

12. l (p. 440; objective 1)

13. j (p. 440; objective 1)

14. o (p. 448; video lesson; objective 4)

15. m (p. 433; video lesson; objective 7)

16. n (p. 448; video lesson; objective 4)

17. p (p. 441; video lesson; objective 1)

18. r (p. 444; objective 1)

19. t (p. 433; objectives 5 & 7)

20. s (p. 444; video lesson; objective 4)

Completion

1. Widmanstätten (p. 441; objective 1)

2. comets (p. 450; video lesson; objective 5)

3. Kuiper belt (p. 451; video lesson; objectives 3 & 5)

4. dust, sun (p. 448; video lesson; objective 4)

5. elliptical, closer (p. 450; objective 7)

Self-Test

1. c (pp. 432–433; objective 7)

2. b (p. 433; objective 7)

3. b (p. 433; objective 7)

4. c (p. 441; video lesson; objective 1)

5. c (p. 441; objective 1)

6. a (pp. 441–443; objective 1)

7. c (p. 442: video lesson; objective 2)

8. b (p. 442; video lesson; objective 2)

9. d (p. 447; objective 3)

10. b (p. 444; objective 3)

11. b (p. 443, Figure 19-4; objective 4)

12. a (p. 448; objective 4)

13. d (pp. 445, 448, 450: video lesson; objective 4)

14. a (pp. 450–451; video lesson; objective 5)

15. c (p. 445; objective 4)

16. d (pp. 448–450; video lesson; objective 4)

17. c (p. 451; objective 6)

18. b (p. 452; video lesson; objective 6)

Short-Answer Questions

1. A Plutino is a Kuiper belt object that is in a gravitational resonance with the planet Neptune. That means that the period of the Plutino's orbit is related to the period of Neptune's orbit. The Plutino orbits the sun twice as Neptune orbits the sun three times. Pluto is also locked in this resonance with the other Plutinos. (p. 443; video lesson; objective 7)

2. Pluto's orbit is very elliptical. In fact, its orbit is more elliptical than that of Neptune and as a result, there is a period of time, during the orbit of Pluto, that Pluto comes closer to the sun than Neptune. That happened between January 21, 1979, and March 14, 1999. (p. 432; objective 7)

3. The gas, or Type I, tail is made up of ionized gases that produce an emission spectrum of the gases of which it is comprised. This tail "emits" light. The dust, or Type II, tail is made up of dust particles and reflects sunlight and the observed spectrum is that of sunlight. (p. 448; video lesson; objective 4)

4. The most common type of asteroid is the S-type. They are brighter and have a red color. They are probably the source of the most common meteorites, the chondrites. (p. 447; objective 3)

5. A NEO is a near-Earth object. These are asteroids that cross Earth's orbit and are at present the most likely threat of an impact with Earth. It is estimated that a few thousand of these objects may be larger than 1 km in size. (p. 444; objectives 3 & 6)

Applications

1. The various types of meteorites—stony, stony-iron, and iron—have characteristics and properties that would originate from a planetesimal that has differentiated. Iron meteorites would have originated from the core and stony meteorites would have originated from the crust. (pp. 441–444; video lesson; objective 1)

2. The Kuiper belt is located between 30 to 50 AU from the sun. The Oort cloud is located more than 100,000 AU away. The Kuiper belt is believed to be the source of the short-period comets, whereas the Oort

cloud is believed to be the source of long period-comets. Kuiper belt objects most likely formed at the outer edges of the young solar system near the forming Jovian planets with icy planetesimals solidifying in the cold environment. The Oort cloud probably formed of icy planetesimals that were ejected out of the immediate solar system by the Jovian planets as these planets grew in size. (pp. 450–451; video lesson; objective 5)

3. There are 150 meteorite craters on the surface of Earth that demonstrate that Earth has been struck by asteroids and comets. Further evidence is provided by the numerous craters observed on the surface of the moon. It is close to Earth and smaller. That would mean that the Earth, being larger, has been struck more often. In 1994, observers watched as Comet Shoemaker-Levy 9 struck Jupiter. You may wish to look up the history of what is believed to be an impact of a comet on Earth in 1908 in the Tunguska region of Siberia. There were eyewitnesses to the 1908 event. (pp. 451–452; video lesson; objective 6)

4. The discovery of the moon Charon about Pluto allowed astronomers to treat the system as they would a binary star system. Using Kepler's third law, they could determine the mass of Pluto. Since the moon eclipses Pluto as viewed from Earth, they applied the techniques used to study an eclipsing binary star system to determine the moon's and Pluto's diameter. With the size of Pluto and its mass now known, the density of the planet could be calculated and a model of the interior developed. (p. 433; objective 7)

Notes:

Lesson Review

Lesson 19: Solar System Debris

PLEASE NOTE: Use this matrix to guide your study and achieve the learning objectives of this lesson. It will also help you to view the video, which defines and demonstrates important concepts and principles as they relate to everyday life and actual case studies.

Learning Objective	Textbook	Student Guide
1. Describe the broad categories of meteorites, their compositions, and possible origins.	pp. 440–441	Key Terms: 1, 2, 3, 4, 5, 6; Matching: 1, 2, 3, 4, 8, 12, 13, 17, 18; Completion: 1; Self-Test: 4, 5, 6; Applications: 1.
2. Describe the characteristics of meteors and meteor showers, and explain how they relate to cometary orbits.	pp. 441–443	Key Terms: 7, 12; Matching: 6, 10, 11; Self-Test: 7, 8.
3. Describe the physical properties, orbital characteristics, and possible origins of asteroids.	pp. 444–447	Key Terms: 13, 15; Completion: 3; Self-Test: 9, 10; Short-Answer: 3, 4.
4. Describe the parts and observed properties of comets and explain the effect of solar wind on the tails of a comet.	pp. 445, 448–450	Key Terms: 9, 10, 11; Matching: 5, 14, 16, 19; Completion: 4; Self-Test: 11, 12, 13, 15, 16; Short-Answer: 3.
5. Explain the significance of the Oort cloud and Kuiper belt.	pp. 433–434, 450–451	Key Terms: 8, 14; Matching: 7, 9, 20; Completion: 2, 3; Self-Test: 14; Applications: 2.

Learning Objective	Textbook	Student Guide
6. Explain the possible effects of an asteroid or comet collision with Earth and describe efforts to find near-Earth objects.	pp. 451–453	Self-Test: 17, 18; Short-Answer: 5; Applications: 3.
7. Describe the physical and geological characteristics of Pluto and summarize the debate over its classification as a planet versus a Kuiper belt object.	pp. 432–434	Key Terms: 14; Matching: 9, 15, 20; Completion: 5; Self-Test: 1, 2, 3; Short-Answer: 1, 2; Applications: 4.

LESSON
20

The Search for Life Beyond Earth

Checklist

For the most effective study of this lesson, complete the following activities in this sequence.

Before Viewing the Video

❑ Read the Preview, Learning Objectives, and Viewing Notes below.

❑ Read Chapter 20, "Life on Other Worlds," pages 456–473, and the "Afterword," pages 475–476, in the *Horizons* textbook.

What to Watch

❑ After reading the textbook chapter, watch the video for Lesson 20, *The Search for Life Beyond Earth.*

After Viewing the Video

❑ Briefly note your answers to questions listed at the end of the Viewing Notes.

❑ Review the Summary below.

❑ Review all reading assignments for this lesson, especially the Chapter 20 summary on page 472 in *Horizons* and the Viewing Notes in this guide.

❑ Write brief answers to the review questions at the end of Chapter 20 in *Horizons.*

❑ Complete the Review Exercises below. Check your answers with the Answer Key and review when necessary.

❑ Use the Lesson Review matrix found at the end of this lesson to review and assess your knowledge of each Learning Objective.

❑ As assigned by your instructor, complete the Applications activities and any additional activities for this lesson.

Preview

Is life here on Earth just a cosmic accident, or is the universe filled with life? There are more than 100 billion stars in the Milky Way Galaxy alone, and perhaps 100 billion times more in the universe. With that many stars, surely some of them must have planets orbiting them and perhaps there is life on some of these planets. If we could estimate the number of planets in the universe, perhaps we could calculate the number of those that could sustain life.

The search for life begins with the search for water. Water is necessary for any life form that we are aware of—where there is water in the universe, there might be life. Astronomers search for not only intelligent life with whom we could communicate, but also the simplest forms of life. Some organisms can survive in the harshest of conditions—from the bottom of the icy lakes in Antarctica to boiling hot springs that spew acidic water. Life appears to be tenacious.

The search for intelligent life involves radio telescopes that scan the sky for signals from extraterrestrial intelligence. Astronomers on Earth occasionally send out signals, hoping that an intelligent life form will receive them. But, when we take into consideration how far away even the nearest star is, it might be tens of thousands of years before any life form receives our signals. In our lifetimes, we may never find out if life exists beyond Earth—that might be a discovery made by future generations.

Concepts to Remember

- Recall from Lesson 16 that *heavy bombardment* is the intense cratering that occurred during the first 0.5 billion years in the history of our solar system. During this period, some theories suggest that comets or asteroids brought organic matter to Earth (p. 314 in this guide).

- In Lesson 17, you learned that the *greenhouse effect* is the process by which a carbon dioxide atmosphere traps heat and raises the temperature of a planetary surface. Greenhouse gases in the atmosphere of a planet can regulate the temperature, making it possible to sustain life (p. 341 in this guide).

- In Lesson 6, you learned that a *spectral type* identifies a star's position in the temperature classification system—O, B, A, F, G, K, M—ranging from hot to cool, based on the appearance of the star's spectrum. Only stars of the late F or G and early K spectral type have a lifespan long enough and temperature ranges that would most likely have planets with the conditions necessary to have life evolve (p. 104 in this guide).

Learning Objectives

After you complete this lesson, you should be able to:

1. Explain the criteria scientists use to define life. *HORIZONS* TEXTBOOK PAGE 458.

2. Explain the nature of DNA and RNA and the role they serve in the reproduction of life. *HORIZONS* TEXTBOOK PAGES 459–461.

3. Describe the current understanding of how life originated on Earth and the significance of the Miller experiment. *HORIZONS* TEXTBOOK PAGES 462–465.

4. Discuss the possibility that the conditions for life may exist (or may have existed) elsewhere in our solar system. *HORIZONS* TEXTBOOK PAGES 465–466.

5. Describe the characteristics of a star and its planetary system that would provide the physical conditions necessary for life to begin and evolve. *HORIZONS* TEXTBOOK PAGES 466–468.

6. Describe attempts to communicate with extraterrestrial civilizations and to detect radio transmissions from them. *HORIZONS* TEXTBOOK PAGES 468–470.

7. Explain how the Drake equation is used to estimate the number of possible technological civilizations in the galaxy. *HORIZONS* TEXTBOOK PAGES 470–471.

At this point, read Chapter 20, "Life on Other Worlds," pages 456–473, and the "Afterword," pages 475–476.

Viewing Notes

The video program explores one of the most fascinating questions in all of the sciences: Are we alone in the universe?

The video program contains the following segments:

- ✪ The Nature of Life
- ✪ The Origins of Life
- ✪ Life in Our Solar System
- ✪ Life Among the Stars
- ✪ Communicating with Distant Civilizations

The following information will help you better understand the video program:

We shall not cease from exploration
And the end of all our exploring
Will be to arrive where we started
And know the place for the first time.
—T. S. Eliot
"Little Gidding"
(1942)

Earth's first oceans were filled with a rich mixture of organic compounds that is referred to as a **primordial soup**. The compounds dissolved in this primordial soup then linked together to form larger molecules. Amino acids linked together to form proteins to form the building blocks of life. Although these molecules were not alive, the more stable molecules bonded together and this **chemical evolution** led to smaller molecules joining together to form larger ones. Eventually, somewhere in the oceans a molecule formed that could reproduce itself. At that point, the chemical evolution of molecules became the biological evolution of living things.

Life on Earth depends on water. The **life zone**, also known as the habitable zone, is the region around a star within which the existence of liquid water on planets can be maintained.

If intelligent life does exist on other planets, they might be trying to communicate with others in the universe. The **water hole** is a small range of frequencies through which communication with intelligent life on other planets seems most likely. The water hole is between the 21-cm line of neutral hydrogen (H) and the 18-cm line of OH. If you combine H with OH, you get water (H_2O).

QUESTIONS TO CONSIDER

- What defines that something is alive?

- Why do astronomers search for water to determine if life might exist somewhere else in the solar system?

- Aside from Earth, where might life exist in our solar system?

- Does a habitable zone exist around every star?

- How do astronomers estimate the number of planets in the universe that might be able to sustain life?

- What is SETI and how does it use radio waves to listen for extraterrestrial communications?

Watch the video for Lesson 20, *The Search for Life Beyond Earth.*

Key Terms and Concepts

Page references are keyed to the *Horizons* textbook.

1. **natural selection:** The process by which the best traits are passed on, allowing the most able to survive. (p. 459; video lesson; objective 1)

2. **mutant:** Offspring born with altered DNA. (p. 459; video lesson; objective 1)

3. **DNA (deoxyribonucleic acid):** The long carbon-chain molecule that records information to govern the biological activity of the organism. DNA carries the genetic data passed to offspring. (p. 460; video lesson; objective 2)

4. **amino acid:** Carbon-chain molecule that is the building block of protein. (p. 460; video lesson; objective 2)

5. **protein:** Complex molecule composed of amino acid units. (p. 460; video lesson; objective 2)

6. **enzyme:** Special protein that controls processes in an organism. (p. 460; objective 2)

7. **RNA (ribonucleic acid):** Long carbon-chain molecules that use the information stored in DNA to manufacture complex molecules necessary to the organism. (p. 461; objective 2)

8. **chromosome:** A body within a living cell that contains genetic information responsible for the determination and transmission of hereditary traits. (p. 461; objective 2)

9. **gene:** A unit of DNA—or sometimes RNA—information responsible for controlling an inherited physiological trait. (p. 461; objective 2)

10. **Cambrian period:** A geological period 0.6 to 0.5 billion years ago during which life on Earth became diverse and complex. Cambrian rocks contain the oldest easily identifiable fossils. (p. 464; objective 3)

11. **Miller experiment:** An experiment that reproduced the conditions under which life began on Earth and manufactured amino acids and other organic compounds. (p. 462; video lesson; objective 3)

12. **primordial soup:** The rich solution of organic molecules in Earth's first oceans. (p. 462; video lesson; objective 3)

13. **chemical evolution:** The chemical process that led to the growth of complex molecules on primitive Earth. This process did not involve the reproduction of molecules. (p. 463; video lesson; objective 3)

14. **stromatolite:** A layered fossil formation caused by ancient mats of algae or bacteria, which build up mineral deposits season after season. (p. 464; objective 3)

15. **life zone:** A region around a star within which a planet can have temperatures that permit the existence of liquid water; also known as the habitable zone. (p. 467; video lesson; objective 5)

16. **water hole:** The interval of the radio spectrum between the 21-cm hydrogen radiation and the 18-cm OH radiation. Likely wavelengths to use in the search for extraterrestrial life. (p. 469; video lesson; objective 6)

17. **SETI:** The Search for Extra-Terrestrial Intelligence; an exploratory science that seeks evidence of life in the universe by looking for some signature of its technology. (p. 469; video lesson; objective 6)

18. **Drake equation:** The equation that estimates the total number of communicative civilizations in our galaxy. (p. 470; video lesson; objective 7)

Summary

You may have wondered if life exists elsewhere in the universe. As we look into the night sky, we are able to see a few thousand stars. Surely, life must exist in our solar system, elsewhere in our galaxy, or on a distant planet somewhere in the universe.

The Nature of Life

Before we determine if life exists on other planets, we must define what life is. When we think of something that's alive, we might think of an organism that uses energy, has some sort of metabolism and is able to reproduce.

Scientists have determined that life's characteristics depend on two important aspects: a physical basis and a unit that stores and duplicates information. The physical basis for life on Earth is carbon. Carbon can bond with other atoms to form complex chains that can extract, use, and store energy. These carbon-based chains can combine to form **amino acids** that, when combined in a specific pattern, can store and duplicate information. Thus, life as we know it must be carbon-based.

The information that guides all the processes in an organism—how it breathes or digests food, for example—is stored in its DNA. **DNA (deoxyribonucleic acid)** is made of amino acids that combine to form proteins. The DNA molecule looks like a spiral ladder and acts as a template that guides the amino acids to join together in the correct order to build proteins. The rungs on the ladder are made of the four base amino acids: adenine, cytosine, guanine, and thymine. The bases always pair in the same way: adenine always pairs with thymine and cytosine always pairs with guanine. The order in which they combine and repeat creates a code that gives the living being its characteristics.

While these DNA molecules can store information that provides a blueprint of life, life can only exist if these molecules duplicate information, or reproduce. DNA is stored safely in the nucleus of each cell of the organism; the code is copied to create a molecule of **RNA (ribonucleic acid)** that carries the instructions of life to the area of the cell where proteins and enzymes are manufactured. The cells can reproduce by division—the DNA ladder splits and as the cell begins to divide, the DNA duplicates itself. The cell divides to produce two identical cells, each containing a full set of DNA code—enabling the DNA code to be passed down to the organism's offspring.

In order for generations of the organism to survive, the information stored in the DNA must change to adapt to changes in the environment. Those species that survive in a changing world evolve by **natural selection**—the process by which the best traits are passed on allowing the species to survive.

The Origin of Life

We now understand how life is based on the storage and duplication of information from generation to generation of organisms. But, that doesn't explain how life originated on Earth. By studying the oldest fossils on Earth, scientists have concluded that life began in the sea at least 2.5 billion years ago.

A popular hypothesis about how life began on Earth is that organic compounds and water came to Earth on asteroids or comets during the period of heavy bombardment. In the presence of water, the organic compounds were able to combine to form simple organisms, such as bacteria. Simple bacteria evolved into life forms that are more complex, with natural selection preserving the most advantageous traits. Life evolved slowly in the first 2 billion years, until a little over a half-billion years ago life suddenly began to become more complex and diverse. This was perhaps the result of a change in the environment and marks the beginning of the Cambrian period. The Cambrian creatures survived in the sea; life forms eventually evolved and were able to live on land beginning about 400 million years ago.

To understand how life could have begun on the young Earth, we can look to the **Miller experiment**. The experiment sought to simulate the primitive conditions on Earth. The experimenters constructed a closed glass container that contained liquid water, as well as the gases hydrogen, ammonia, and methane to simulate the oceans and the atmosphere on Earth. They ran an electric current through the glass enclosure to simulate lightning bolts that may have been present in the young Earth's atmosphere.

The scientists allowed the experiment to run for a week and they analyzed the material in the glass enclosure. They discovered many compounds including amino acids, fatty acids, and urea—a molecule common to life processes. This experiment shows that complex organic (carbon-based) molecules can form naturally in diverse conditions. Its success was in demonstrating that the gases in Earth's primitive atmosphere, when energized by electricity, could generate some of the basic components of life such as amino acids and urea.

The Miller experiment can help us understand how life began on Earth. The oceans were filled with a **primordial soup**—a rich solution of organic materials. Compounds dissolved in the oceans and linked together to form more complex molecules. The young Earth's surface was changing through extensive volcanism and meteorite impacts. The climate was not stable, and any life forms exposed on the surface were destroyed. Molecules in the oceans were protected from the hostile atmosphere and were able to link together to form increasingly complex molecules. Eventually, molecules formed that were able to reproduce themselves and life on Earth began. The earliest life forms were simple—they were likely single-cell organisms similar to bacteria, but they eventually evolved into the complex life forms that exist today.

The most important requirement for life as we know it is liquid water. In addition to water, biologists believe that life requires an energy source and an assortment of organic elements, such as carbon and nitrogen. Once scientists were able to determine the conditions under which life could exist, they were able to begin searching for life within our solar system.

NASA and the European Space Agency have been searching for life on other planets in our solar system. Since water is critical to life as we know it, their searches are based on locating it. Venus is too hot to maintain liquid water, and the worlds in the outer solar system are too cold. Airless worlds like Earth's moon and Mercury cannot maintain liquid water on their surfaces. The Jovian planets have no surfaces on which oceans can survive, so life on those planets seems unlikely.

Some of the moons around the Jovian planets, however, might sustain conditions hospitable to life. It seems that there may be an ocean of liquid water beneath the icy surface of Jupiter's moon Europa. Minerals dissolved in the ocean might provide the organic materials necessary for life and radiation from Jupiter's magnetic field may provide a source of needed energy. But conditions on Europa may not have been stable for a long enough period of time to allow life to evolve beyond the microscopic stage.

If life ever existed anywhere else in our solar system, the most likely place is on Mars. Liquid water once flowed on the surface of the planet, but conclusive evidence of life on Mars has not been discovered. There is an abundance of methane gas on Mars. This seems to indicate that there are bacteria that are releasing the methane or there's volcanic activity on the planet. Scientists have yet to determine the source of the methane. As far as we currently know, life in our solar system exists only on Earth.

With billions of other stars in our universe, we might imagine that life may exist on a distant planet. For life to exist on another planet, several conditions must be met. First, the planet must have a stable orbit around its sun for it to exist long enough to nurture life. Planets that orbit stars in a binary system do not last long enough for life to develop before the planet is broken apart by one of the stars or ejected from the binary system. Second, conditions suitable for life must remain satisfactory for long periods of time. It took at least 0.5 to 1 billion years for the first cells to evolve on our own planet and it's possible that the circumstances would be similar on other planets.

Although life can exist in extremely harsh environments on Earth, a third condition must be met—liquid water must still be present for life as we know it to exist. A main-sequence star might be able to warm a planet, but the planet must be far enough away from the star that the water doesn't vaporize. The planet would have to be in the habitable zone, or **life zone**—a region within which planets have temperatures that permit the existence of liquid water.

Should all three of these conditions be met on a planet elsewhere in our galaxy, relatively simple chemical processes could breed life on other planets.

Communication with Distant Civilizations

Although intelligent life in our solar system is limited to Earth, it's possible that it exists in other solar systems. If intelligent life does exist on other planets, communication with distant civilizations might be possible. Radio signals might be the most effective form of communication, but it has its limitations. Distant civilizations may be a few light-years away or thousands of light-years away. Since radio waves can travel only as fast as the speed of light, it would be difficult to carry on two-way communications with life forms on other planets.

Since two-way communication with extraterrestrial civilizations could be quite difficult, astronomers on Earth have sent a few messages into the universe with a detailed description of life on Earth. Only a certain range of wavelengths—between 1 cm and 30 cm—is open for communication. Wavelengths longer than 30 cm would be lost in the background radio noise in the galaxy and wavelengths shorter than 1 cm are absorbed in our atmosphere.

Even though the wavelengths available for communication are limited, the radio window is still wide and it would take a long time for any civilization to scan the available wavelengths for a signal. Within the radio window is a portion of wavelengths that could convey a message—between the 21-cm line of neutral hydrogen (H) and the 18-cm line of OH. This is known as the **water hole** because the combination of hydrogen and OH is water (H_2O). Since water is fundamental to our life form, it might be fundamental to other life forms as well and they might search for a message between these wavelengths.

Communication with intelligent life in the universe is challenging: the distances to other planets are vast, the window of available wavelengths is limiting, and the timing of the evolution of the society might not coincide with our society. Other civilizations might not be advanced enough to receive and interpret the signals. Or perhaps their civilization declined hundreds or thousands of years ago and intelligent life no longer exists.

Determining how many civilizations might be able to communicate with us is a daunting task. There are many variables of which we are uncertain, such as the number of years a technological society can exist on a planet or the fraction of life forms that ever evolve to intelligence. The **Drake equation** is a formula that can help scientists estimate the number of civilizations that may be able to communicate in any given galaxy (see p. 470 in the *Horizons* textbook).

Sophisticated searches are underway by **SETI**, the Search for Extra-Terrestrial Intelligence, to detect radio transmissions from other worlds. To communicate with other civilizations, they must be as technologically advanced as we are within a similar timeframe, depending on how far away they are. For example, if a distant civilization is 26,000 light-years away from us, they must be capable of receiving a transmission from Earth 26,000 years from now. The civilization may not have evolved yet in order to receive and interpret the signal or life on the planet may have already been extinguished as a result of societal or environmental collapse.

As far as we know, we are the only intelligent life in the universe. We are able to think and ask questions and communicate with one another. The inhabitants of planet Earth may be the only ones that can study and admire the universe—we have a great responsibility to do it justice.

Review Exercises

Matching I

Match each term with the appropriate definition or description.

1. _____ natural selection	6. _____ enzyme
2. _____ mutant	7. _____ RNA
3. _____ DNA	8. _____ gene
4. _____ amino acid	9. _____ chromosome
5. _____ protein	

a. Long carbon-chain molecules that use the information stored in DNA to manufacture complex molecules necessary to the organism.

b. Long carbon-chain molecule that records information to govern the biological activity of the organism.

c. A special protein that controls processes in an organism.

d. Offspring born with altered DNA.

e. Complex molecule composed of amino acid units.

f. The basic building block of protein.

g. The process by which the most adaptive traits are passed on, allowing the most able to survive.

h. A unit of DNA—or sometimes RNA—information responsible for controlling an inherited physiological trait.

i. A body within a living cell that contains genetic information responsible for the determination and transmission of hereditary traits.

Matching II

Match each term with the appropriate definition or description.

1. _____ Cambrian period		6. _____ life zone	
2. _____ Miller experiment		7. _____ water hole	
3. _____ primordial soup		8. _____ SETI	
4. _____ chemical evolution		9. _____ Drake equation	
5. _____ stromatolite			

a. A shorthand reference for the search for extraterrestrial intelligence.

b. An interval of the radio spectrum believed to include the likely wavelength to use in the search for extraterrestrial life.

c. A chemical process that led to the growth of complex molecules on primitive Earth.

d. Rocks from this geologic time contain the oldest identifiable fossils.

e. The rich solution of organic molecules found in Earth's first oceans.

f. The region around a star within which a planet can have temperatures that permit the existence of liquid water.

g. Structure produced by communities of blue-green algae or bacteria.

h. An experiment that reproduced the conditions under which life may have begun on Earth.

i. A formula that estimates the total number of communicative civilizations in our galaxy.

Completion

Fill each blank in the sentences below with the most appropriate term from the list of completion answers that follow. A term may be used once, more than once, or not at all. Check your answers with the Answer Key and review when necessary.

amino acids   enzymes   ribonucleic acids
carbon   extraterrestrial   SETI
communicative   galaxy   silicon
deoxyribonucleic acids   intelligent   universe
energy zone   Miller   water hole

1. The physical basis of all life on Earth is the _____ atom.

2. The _____ experiment succeeded in creating _____, the building blocks for proteins.

3. The portion of the electromagnetic spectrum between the 21-cm line of neutral hydrogen and the 18-cm line of OH is known as the

   _____.

4. Radio astronomer Frank Drake is credited with creating the Drake equation, which estimates N_c, or the total number of _____ civilizations in any given

   _____.

5. Current efforts to detect a signal from an _____ intelligence are called _____. Someday we may have an answer from life on another world.

Self-Test

Select the best answer.

1. The basic unit of life as we know it is
   a. the nucleus.
   b. the cell.
   c. DNA.
   d. RNA.

2. The element carbon is the physical basis for life on Earth. This atom
   a. is the most abundant atom on the planet.
   b. is the most abundant element in the universe.
   c. is capable of bonding to other atoms and makes long, stable chains that are capable of extracting, storing, and utilizing energy.
   d. dissolves in water.

3. The information that enables the organism to survive and is transferred from parent to offspring is found in
   a. DNA.
   b. the cell membrane.
   c. a spiral ladder.
   d. proteins.

4. The information a cell needs to function is found on the DNA molecule in the form of
   a. multiple combinations of unknown chemicals.
   b. RNA subgroups.
   c. a spiral helix.
   d. four bases and the order in which they appear.

5. The information regarding cell functioning is carried out of the nucleus by
   a. amino acids.
   b. RNA.
   c. proteins.
   d. enzymes.

6. Evidence that life began in the oceans has been found in
   a. surface rocks.
   b. sediments on the ocean floor.
   c. Cambrian period fossils.
   d. the primordial soup.

7. The Miller experiment consisted of various gases, believed to have been present in the early atmosphere of Earth, exposed to an electrical arc. This experiment resulted in
   a. nothing.
   b. carbon dioxide gas and water vapor.
   c. amino acids, fatty acids, and urea.
   d. rain.

8. Hot stars, such as O and A stars, are unlikely to harbor planets suitable for sustaining life because
   a. their stable energy-producing lifespan is too short.
   b. they are too hot.
   c. they are too cold.
   d. they produce an unusually high amount of deadly radiation.

9. The region of space around a star in which water can remain in liquid form is called the
   a. water hole.
   b. temperate zone.
   c. main-sequence stage.
   d. life (or habitability) zone.

10. Scientists think Jupiter's moon Europa might be able to support life because
    a. the Voyager and Galileo spacecraft sent back pictures showing areas of vegetation on its surface.
    b. it gets far more life-giving sunlight than Jupiter's other moons.
    c. there is evidence that below its ice surface there may be liquid water.
    d. its atmosphere contains large quantities of oxygen.

11. Several meteorites found in Antarctica have been determined to come from Mars. The evidence for this conclusion lies in the
    a. analysis of trapped gases in these meteorites with the same abundance as gases found in the Martian atmosphere.
    b. residual magnetism in these meteorites, which is identical to the residual magnetism on Mars itself.
    c. orbital trajectories of these meteorites, which coincide with the position of Mars 16 million years ago.
    d. similarities of organic structures found in rocks on Mars as analyzed by various spacecraft that have explored the Martian surface.

12. The Drake equation is used to
    a. determine the number of stars in the Milky Way Galaxy.
    b. count the number of stars visible in the night sky.
    c. determine the spacing distance of planets in our solar system.
    d. estimate the number of technological civilizations within our galaxy.

13. Of the following factors, the one that is a part of the Drake equation is the
    a. temperature of the star during its main-sequence stage.
    b. number of O and B stars in the galaxy that exhibit the possible existence of orbiting planets.
    c. fraction of the lifetime of a star during which a technological civilization is capable of communicating.
    d. total number of planets in a solar system orbiting about a binary star.

14. One of the most likely wavelengths of radio energy being used to search for extraterrestrial signals is near the
    a. scattered blue light of the interstellar medium.
    b. 21-centimeter line of neutral hydrogen.
    c. 3 degrees Kelvin microwave background energy.
    d. Balmer wavelengths of neutral hydrogen.

15. The water hole refers to
    a. a range of radio wavelengths (frequencies) that fall between the wavelengths of neutral hydrogen and the OH molecule.
    b. the region around a star in which water can remain in liquid form.
    c. planets that have been confirmed to contain traces of liquid or frozen water.
    d. nebulae of interstellar matter that have water or OH molecules.

Short-Answer Questions

1. Why are O, B, and A stars unlikely candidates to harbor planets with intelligent civilizations?

2. Why is an M type star an unlikely candidate to harbor planets with intelligent civilizations?

3. Why are F, G, and K stars likely candidates to harbor planets with intelligent civilizations?

4. What are adenine, cytosine, guanine, and thymine?

5. What characteristics of the carbon atom make it a suitable atom for the physical basis for life on Earth?

6. Why is the Miller experiment important even though it did not create life nor did it duplicate the conditions of Earth's early atmosphere?

7. What are the arguments for and against life existing inside the moon Europa?

Applications

1. What is the strongest evidence that suggests that life may have originated on Mars?

2. What is the Arecibo message?

3. From "The Nearest Stars" in Appendix A, Table A-9, on page 481 in the *Horizons* textbook, select the stars nearest the sun most likely to shelter civilizations. Be sure to consider spectral type, luminosity class, and whether they are single or multiple star systems.

4. The Drake equation attempts to calculate the number of communicative civilizations in the galaxy. What does the factor F_s represent and why is it considered to be the most uncertain of all the factors in this equation?

5. What is an anticoded message?

Answer Key

Matching I

1. g (p. 459; video lesson; objective 1)

2. d (p. 459; video lesson; objective 1)

3. b (p. 460; video lesson; objective 2)

4. f (p. 460; video lesson; objective 2)

5. e (p. 460; video lesson; objective 2)

6. c (p. 460; objective 2)

7. a (p. 461; objective 2)

8. h (p. 461; objective 2)

9. i (p. 461; objective 2)

Matching II

1. d (p. 464; objective 3)

2. h (p. 462; video lesson; objective 3)

3. e (p. 462; video lesson; objective 3)

4. c (p. 463; video lesson; objective 3)

5. g (p. 464; objective 3)

6. f (p. 467; video lesson; objective 5)

7. b (p. 469; video lesson; objective 6)

8. a (p. 469; video lesson; objective 6)

9. i (p. 470; video lesson; objective 7)

Completion

1. carbon (p. 458; objective 1)

2. Miller, amino acids (pp. 462–463; video lesson; objective 3)

3. water hole (p. 469; video lesson; objective 6)

4. communicative, galaxy (p. 470; video lesson; objective 7)

5. extraterrestrial, SETI (p. 469; video lesson; objective 6)

1. b (p. 458; objective 1)

2. c (p. 458; objective 1)

3. a (pp. 458–459; video lesson; objective 2)

4. d (pp. 460–461; objective 2)

5. b (pp. 460–461; objective 2)

6. c (p. 464; objective 3)

7. c (pp. 462–463; objective 3)

8. a (p. 467; objective 5)

9. d (p. 467; video lesson; objective 5)

10. c (p. 466; video lesson; objective 4)

11. a (p. 466; objective 4)

12. d (p. 470; video lesson; objective 7)

13. c (p. 470; objective 7)

14. b (p. 469; objective 6)

15. a (p. 469; video lesson; objective 6)

Short-Answer Questions

1. The O, B, and A stars are very large and very hot. They remain on the main sequence for only a few million years. This means that the period of time during which their energy production rates remain stable is relatively short, too short for an intelligent civilization to evolve. On our planet it took at least 0.5 billion years for the first cell to evolve and another 4 billion for intelligent life to evolve. (p. 467; objective 5)

2. The M type stars are relatively cool, which would require a life-sustaining planet to orbit fairly close to the star. Some astronomers have argued that a planet so close would become tidally locked to the star keeping one side facing towards the star at all times. This would cause water to freeze on the dark side making life very difficult to evolve. Furthermore, M type stars are subject to solar flares. These bursts of energy would certainly make the evolution of life very difficult if not impossible. (p. 467; objective 5)

3. The F, G, and K stars have relatively large life zones and are the likely candidates to harbor planets capable of having life evolve on their surfaces. Furthermore, their energy production rates remain stable for billions of years which would provide enough time for life to evolve to some degree of intelligence. We also have an example of a star that has met this condition: our sun, spectral type G2. (p. 467; objective 5)

4. Adenine, cytosine, guanine, and thymine are the four bases that make up the rungs of the DNA molecule. They are paired and become the steps of the spiral ladder shaped molecule. (p. 460; objective 2)

5. The carbon atom is the physical basis for life on planet Earth. It is the basic element for organic compounds. Carbon has the ability to bond to other atoms in such a way that it allows the formation of long, complex, and stable molecules. What is important about these molecules is that they are capable of extracting, storing, and utilizing energy. (p. 458; objective 1)

6. The Miller experiment is important because it shows that complex organic (carbon-based) molecules can form naturally in a wide variety of conditions. (pp. 462–463; video lesson; objective 3)

7. Jupiter's moon Europa experiences tidal heating and therefore its interior is warmed to the point that scientists believe there is a liquid water ocean beneath the icy crust. How far below the crust is still a question. The minerals dissolved in water may provide the chemicals needed for chemical evolution to begin. Although these are arguments in favor of life beginning "inside" Europa, there are other arguments against that possibility. Since Europa's internal heat comes from tidal heating, it is not known how long this process has been occurring nor how long it will last. Changes in the orbit of Europa as a result of the gravitational influence of the other moons may cause the interior of this moon to freeze. It is unlikely that the liquid ocean under the crust of Europa has lasted for the billions of years necessary for life to evolve beyond the microscopic stage. (p. 466; video lesson; objective 4)

Applications

1. Spacecraft images of the Martian surface have revealed that water once flowed over the surface. However, the rovers on Mars have detected areas that were once lakes or oceans. In other words, the water pooled and remained stationary. Although this does not mean life actually formed on the surface, it provides the physical conditions necessary for the possible evolution of life. (p. 466; video lesson; objective 4)

2. In 1974, astronomers at the Arecibo radio telescope sent a message directed toward the globular cluster M13 that contained information about life on planet Earth. The message was coded in such a way that an intelligent civilization could decode the signal and, hopefully, realize that it came from another intelligent civilization. (p. 468; objective 6)

3. ε Indi, ε Eri, and τ Ceti are three stars that meet the requirements for the possible evolution of life about a planet in orbit around these stars. They are single stars and of spectral type K2, K5, and G8 respectively. These spectral types indicate they are relatively warm stars (similar to our sun) and exist on the main sequence for billions of years. (pp. 151–152, 467, 481; objective 5)

4. F_s is the fraction of a star's life during which the life form is communicative. The big question is how long an intelligent civilization remains communicative. Our present society needs to consider two important factors. Will it destroy itself as a result of violent events and political upheavals or will it destroy itself as a consequence of the neglect of the planet upon which it exists? The answer to the complicated question is unknown since we are in the process of experiencing the question and its answer at the present time. (p. 470; video lesson; objective 2)

5. An anticoded message is one that is designed to be easily decoded. The message sent by the Arecibo Telescope in 1974 was anticoded. The scientists who developed it did so with the hope that an intelligent civilization would detect it, decode its contents, and understand the information it contained. (p. 470; objective 6)

Notes:

Lesson Review

Lesson 20: The Search for Life Beyond Earth

PLEASE NOTE: Use this matrix to guide your study and achieve the learning objectives of this lesson. It will also help you to view the video, which defines and demonstrates important concepts and principles as they relate to everyday life and actual case studies.

Learning Objective	Textbook	Student Guide
1. Explain the criteria scientists use to define life.	p. 458	Key Terms: 1, 2; Matching I: 1, 2; Completion: 1; Self-Test: 1, 2; Short-Answer: 5.
2. Explain the nature of DNA and RNA and the role they serve in the reproduction of life.	pp. 459–461	Key Terms: 3, 4, 5, 6, 7, 8, 9; Matching I: 3, 4, 5, 6, 7, 8, 9; Self-Test: 3, 4, 5; Short-Answer: 4; Applications: 4.
3. Describe the current understanding of how life originated on Earth and the significance of the Miller experiment.	pp. 462–465	Key Terms: 10, 11, 12, 13, 14; Matching II: 1, 2, 3, 4, 5; Completion: 2; Self-Test: 6, 7; Short-Answer: 6.
4. Discuss the possibility that the conditions for life may exist (or may have existed) elsewhere in our solar system.	pp. 465–466	Self-Test: 10, 11; Short-Answer: 7; Applications: 1.
5. Describe the characteristics of a star and its planetary system that would provide the physical conditions necessary for life to begin and evolve.	pp. 466–468	Key Terms: 10, 11, 12, 13, 14; Matching II: 1, 2, 3, 4, 5; Self-Test: 6, 7; Short-Answer: 6.

Learning Objective	Textbook	Student Guide
6. Describe attempts to communicate with extraterrestrial civilizations and to detect radio transmissions from them.	pp. 468–470	Completion: 3, 5; Self-Test: 17, 18; Short-Answer: 5; Applications: 3.
7. Explain how the Drake equation is used to estimate the number of possible technological civilizations in the galaxy.	pp. 470–471	Matching: 9, 15; Completion: 4; Self-Test: 1, 2, 3; Short-Answer: 1, 2; Applications: 4.